Ductile Fracture in Metal Forming

Ductile Fracture in Metal Forming

Modeling and Simulation

KAZUTAKE KOMORI
Department of Mechanical Systems Engineering,
Daido University, Nagoya, Japan

ELSEVIER

ACADEMIC PRESS
An imprint of Elsevier

Academic Press is an imprint of Elsevier
125 London Wall, London EC2Y 5AS, United Kingdom
525 B Street, Suite 1650, San Diego, CA 92101, United States
50 Hampshire Street, 5th Floor, Cambridge, MA 02139, United States
The Boulevard, Langford Lane, Kidlington, Oxford OX5 1GB, United Kingdom

British Library Cataloguing-in-Publication Data
A catalogue record for this book is available from the British Library

Library of Congress Cataloging-in-Publication Data
A catalog record for this book is available from the Library of Congress

ISBN: 978-0-12-814772-6

For Information on all Academic Press publications
visit our website at https://www.elsevier.com/books-and-journals

Publisher: Matthew Deans
Acquisition Editor: Dennis McGonagle
Editorial Project Manager: Michelle W. Fisher
Production Project Manager: Selvaraj Raviraj
Cover Designer: Mark Rogers

Typeset by MPS Limited, Chennai, India

Contents

Preface

From the macroscopic point of view, ductile fracture in metal forming occurs due to fracturing, such as central bursting, surface cracking, and rupture; whereas it occurs through void nucleation, void growth, and void coalescence from the microscopic point of view. Research on ductile fracture in metal forming has been performed mainly from the macroscopic viewpoint; however, it should be performed from a microscopic viewpoint to understand the physical meaning of ductile fracture and improve the predictability of ductile fracture.

A simulation method for predicting ductile fracture in metal forming is proposed, wherein ductile fracture is predicted microscopically using an ellipsoidal void model, and metal forming is simulated macroscopically using a conventional finite-element method. The ellipsoidal void model has been derived with reference to two seminal papers: Thomason (1968) and Melander and Ståhlberg (1980). In these two papers, the following two assumptions were made: the void shape was a rectangle and the direction of maximum principal strain coincided with that of maximum principal stress during plastic deformation. However, a rectangular void does not exist and the direction of maximum principal strain does not generally coincide with that of maximum principal stress during plastic deformation. Hence, the following two assumptions are made in the proposed ellipsoidal void model: the void shape is an ellipsoid and the direction of maximum principal strain does not coincide with that of maximum principal stress during plastic deformation. While it is difficult to predict ductile fracture in metal forming using the void model presented in these two papers, it is easy to predict ductile fracture in metal forming using the proposed ellipsoidal void model. Micromechanical modeling using the ellipsoidal void model is described in Chapter 5, Micromechanical modeling using an ellipsoidal void model (author's model), whereas simulation results using the ellipsoidal void model are described in Chapter 6, Simulation results using an ellipsoidal void model (author's model).

I have been researching ductile fracture in metal forming since 1990, when I obtained a doctor of engineering degree at the Tokyo Institute of Technology, and then I moved to the Daido Institute of Technology. While the two seminal papers mentioned above impressed me in around 1995, they had been published several decades before. Since these two seminal papers have been indispensable for me since then, I also introduce some classical papers on ductile fracture that were published a decade or more ago and could be beneficial to readers. In Chapter 1, Macroscopic ductile fracture phenomena and Chapter 2, Macroscopic ductile fracture criteria, classical papers on ductile fracture are introduced from the macroscopic viewpoint for each of the metal-forming processes, that is, rolling, forging, and sheet forming, among others. Macroscopic ductile fracture phenomena are described in Chapter 1, Macroscopic ductile fracture phenomena, whereas macroscopic ductile fracture criteria are described in Chapter 2, Macroscopic ductile fracture criteria. In Chapter 3, Microscopic ductile fracture phenomena and Chapter 4, Microscopic ductile fracture criteria, classical papers on ductile fracture are introduced from the microscopic viewpoint for each of the following processes: void nucleation, void growth, and void coalescence. Microscopic ductile fracture phenomena are described in Chapter 3, Microscopic ductile fracture phenomena, and microscopic ductile fracture criteria are described in Chapter 4, Microscopic ductile fracture criteria. Since each paper is summarized in terms of ductile fracture using around 200 words with no equations, I would recommend readers who find a paper of interest to also read the original paper.

I would like to express my gratitude to a number of undergraduate students who performed graduate research on ductile fracture in metal forming under my supervision at Daido Institute of Technology or Daido University, as well as the library in Daido University which collected hundreds of papers on ductile fracture. Finally, I would like to thank my wife Hiromi who has been understanding and unwaveringly supportive of my research life since 1990.

Kazutake Komori
May 2019

CHAPTER 1

Macroscopic ductile fracture phenomena

1.1 Introduction

Deep drawing (Dieter, 1988), in which a cylindrical cup is produced from a circular sheet, is considered. Deep drawing is performed by placing the circular sheet over a die and pressing the circular sheet into the die using a punch. A blank holder is usually used to press the circular sheet, which is also called a blank, against the die. When the pressure to hold the circular sheet is appropriate, a sound defect-free cylindrical cup is obtained. However, when the pressure to hold the circular sheet is insufficient, the circular sheet buckles and wrinkles. Furthermore, when the pressure to hold the circular sheet is excessive, the circular sheet fractures and is occasionally broken into two parts.

Fig. 1.1 shows the forming limits for the deep drawing of a cylindrical cup. The vertical axis indicates the blank holder pressure, whereas the horizontal axis indicates the drawing ratio, which is defined as the diameter of the circular sheet divided by the punch diameter. The circular sheet buckles and wrinkles in the region below the wrinkling limit curve, that is, in the regions II and IV. The circular sheet fractures and is occasionally broken into two parts in the region above the fracturing limit curve, that is, in the regions III and IV. Hence, a sound defect-free cylindrical cup is obtained only in the region I. The horizontal coordinate of the point at

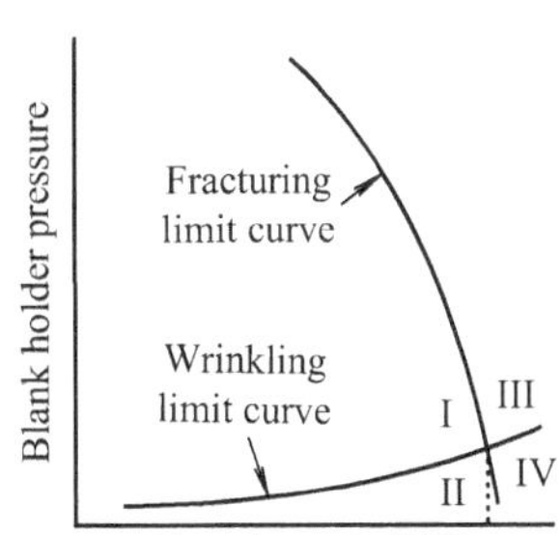

Figure 1.1 Forming limit for deep drawing of a cylindrical cup.

Ductile Fracture in Metal Forming
DOI: https://doi.org/10.1016/B978-0-12-814772-6.00001-1

which the wrinkling limit curve intersects the fracturing limit curve, is called the limiting drawing ratio, which is abbreviated to LDR and indicates the maximum drawing ratio in the case that the pressure to hold the circular sheet is optimized.

Buckle and fracture are the two representative shape defects in metal-forming processes. If productivity is required to increase in metal-forming processes, the possibility of the occurrence of either buckle or fracture increases. Because productivity will be required to increase limitlessly in future metal-forming processes, researches on the prevention of the occurrence of either buckle or fracture will be required endlessly in the future. However, the cause of the occurrence of buckle differs from the cause of the occurrence of fracture, as described in the following.

Representative shape defects due to buckle are center buckle and edge buckle in sheet rolling, wrinkle in deep drawing, and buckle in upsetting of a cylinder having large initial height/diameter ratio. Buckle generally occurs under compressive stress and has no relevance to voids. Hence, increasing the mean normal stress in the region at which buckle occurs is generally required to prevent the occurrence of buckle.

Representative shape defects due to fracture are edge crack in strip rolling, central burst and surface crack in drawing, central burst and surface crack in extrusion, crack in deep drawing, and surface crack in upsetting of a cylinder having small initial height/diameter ratio. Fracture generally occurs under tensile stress and has relation to voids. Hence, decreasing the mean normal stress in the region at which fracture occurs is generally required to prevent the occurrence of fracture. Therefore, the method for preventing and predicting the occurrence of buckle differs from the method for preventing and predicting the occurrence of fracture. Hence, buckle in metal-forming processes is not dealt with in this book.

Fracture is divided into following two types: brittle fracture and ductile fracture. Brittle fracture is a fracture in which the material fractures after little plastic deformation, whereas ductile fracture is a fracture in which the material fractures after large plastic deformation. Because this book deals with the fracture in metal-forming processes, ductile fracture is mainly dealt with in this book.

Working is divided into following two types: hot working and cold working. Hot working is a working in which metal forming is performed above the recrystallization temperature of the material, whereas cold working is a working in which metal forming is performed below the recrystallization temperature of the material. Because the workability of the material in hot working is much higher than the workability of the material in cold working, researches on the fracture of the material in hot working are less required than researches on the fracture of the material in cold working. Hence, fracture of the material in cold working is mainly discussed in this book.

In dynamic plastic deformation, an adiabatic shear band (Zener and Hollomon, 1944) occasionally appears. Although strain rate has only a slight effect upon the isothermal stress–strain relationship, an isothermal deformation is subjected to change to an adiabatic deformation with increasing the strain rate. When the material deforms plastically, the majority of the energy dissipated in the material is converted into heat. Hence, when the material is subjected to deform adiabatically, the heat generated in the material is hardly conducted to surrounding material and the temperature increases drastically in the material. Therefore, with increasing the strain, stress increases due to the strain hardening of the material, whereas stress decreases due to the increase of the temperature. If the magnitude of the stress increase is lower than the magnitude of the stress decrease, stress decreases with increasing the strain, that is, the strain softening of the material occurs and the region where the material deforms plastically is localized.

When the localization of the adiabatic deformation occurs in steels, a white band of martensite appears, which yields when the high-temperature face-centered cubic austenite is rapidly quenched. Hence, the adiabatic shear band is not a slip line, because in the slip-line field theory (Johnson et al., 1982), the material is assumed to be rigid, perfectly plastic. The adiabatic shear band in metal-forming processes is described in a few books (Bai and Dodd, 1992; Dodd and Bai, 1987). Hence, the adiabatic shear band in metal-forming processes is not dealt with in this book.

In Chapter 1, Macroscopic ductile fracture phenomena, macroscopic ductile fracture phenomena are observed experimentally using an optical microscope and are mainly described to utilize observed phenomena in Chapter 2, Macroscopic ductile fracture criteria.

1.2 Physical defects in metal-forming processes

Johnson and Mamalis (1977, 1985), and Mamalis and Johnson (1987) examined the common principal physical defects associated with various metal-forming processes. Table 1.1 shows the physical defects in metal-forming processes (Mamalis and Johnson, 1987). The physical defects

Table 1.1 Physical defects in metal-forming processes.

ROLLING	Impact-extrusion
Flat-and section-rolling	Multiple tensile "necks"
Edge cracking	Thermal break off
Transverse-fire cracking	Drawing of rod, sheet, wire and tube
Alligatoring (crocodiling)	Internal bursts (cup and cone
Fish-tail	chevron)
Folds, laps	Transverse surface cracking
Flash, fins	Edge cracking
Laminations	Chips of metal, bulging
Ridges-spouty material	Poor surface finish
Ribbing	Folding and buckling
Sinusoidal fracture	Fins, laps, spills
Zippering	Chatter (vibration) marks
Cross-, transverse- and helical-rolling	Season cracking
Contral cavity (axial or annular	Island-like welding
fissure)	Pulling out
Overheated ball bearing	Run out
Roll mark	Deep drawing
Folding (or seaming), laps, fringes	Wrinkling
Squaring	Puckering
Necking	Tearing (necking)
Triangulation and triangular fish-tail	Edge cracking
Ring-rolling	Orange peel
Cavities	Stretcher-strains (Lüders lines)
Fish-tail	Earing
Edge cracking	Bending and contour forming
Straight-sided forms	Cracking
FORGING	Wrinkling
Open- and closed-die forging,	Springback
Upsetting, Indentation	Hole flanging
Longitudinal cracking	Lip formation
Hot tears and tears	Petal formation
Edge cracking	Plug formation
Central cavity	Blanking and cropping
Center bursts	Distortion of the part (doming and
Cracks due to t.v.ds and thermal	dishing)
cracks	Cracking
Folds, laps	Martensitic lines
Flash, fins	Eyes, ears, warts, beards, tongues
Laminations	Spinning, flow turning, shear forming
Orange peel	Springback
Shearing fracture	Wall fracture (shear and
Piping	circumferential splitting)

(*Continued*)

Table 1.1 (Continued)

Rotary-forging	Wrinkling
Mushrooming	Buckling
Central fracture	Back extrusion (over reduction)
Flaking	Under reduction
High energy rate forging	Peen forming, ball-drop forming
Piping	Overlapping dimples
Dead metal region	Orange peel
Laps	Surface tearing
Turbulent metal flow	Break-up of surface grains
Extrusion, piercing	Intergranular cracking
Christmas tree (fir tree)	Wrinkling
Hot and cold shortness	Microfissures
Radial and circumferential cracking	Folds
Internal cracking	
Central burst (chevrons)	
Piping (cavity formation)	
Sucking-in	
Corner lifting	
Skin inclusions (side and bottom of the billet)	
Longitudinal streaks	
Laps	
Laminated fractures	
Mottled appearance	
Extrusion defect	

From Mamalis, A.G., Johnson, W., 1987. Defects in the processing of metals and composites. In: Predeleanu, M. (Ed.), Computational Methods for Predicting Material Processing Defects. Elsevier, Amsterdam, Netherlands, pp. 231–250.

include not only fracture but also other physical defects such as buckle and adiabatic shear band, which are not dealt with in this book. Hence, the physical defects on fracture in Table 1.1 are reshown in the following:

Edge cracking, transverse-fire cracking, alligatoring (crocodiling), sinusoidal fracture, zippering, central cavity (axial or annular fissure), and overheated ball bearing in rolling; longitudinal cracking, hot tears and tears, edge cracking, central cavity, central bursts, cracks due to tangential velocity discontinuities, thermal cracks, shearing fracture, and central fracture in forging; Christmas tree (fir tree), hot shortness, cold shortness, radial cracking, circumferential cracking, internal cracking, central burst (chevrons), laminated fracture, and thermal break-off in extrusion and piercing; internal bursts (cup and cone chevron), transverse surface cracking, edge cracking, and season cracking in drawing; tearing (necking), and edge cracking in deep drawing; cracking in bending and contour forming; petal formation in hole flanging; cracking, eyes, ears, warts, beards, and tongues in blanking and cropping; wall fracture (shear and circumferential splitting) in spinning, flow turning, shear forming; and surface tearing, intergranular cracking, and microfissures in peen forming, ball-drop forming.

Ghosh (1981) described some physical defects encountered in sheet metal-forming and provided appropriate literatures, whereas Al-Mousawi et al. (1992) described some physical defects in rolling and forging to determine the reasons for the occurrences of physical defects and to suggest the remedies for physical defects. Johnson and Kudo (1962) performed a comprehensive survey of defects observed during extrusion on the surface of a bar or inside a bar. Schey (1980) reviewed fracture in rolling processes and suggested means of avoiding or minimizing fracture.

In the following sections, classical papers on macroscopic ductile fracture phenomena are introduced for each of the metal-forming processes, that is, rolling, forging, and sheet forming, among others. Because dealing with all the physical defects on fracture reshown is unrealistic, some of the physical defects on fracture are dealt with. Researches on a specific physical defect in a metal-forming process had been performed until the 1980s, whereas researches on material fracture in a specific stress state of a material have been performed since the 1990s. Hence, some of the researches on material fracture in a specific stress state of a material are also dealt with.

1.3 Edge cracking in strip rolling

Edge cracking in strip rolling, in which the edge of the strip in the width direction of the strip fractures during rolling, is one of the fractures in rolling. Dodd and Boddington (1980) performed a review on the causes of edge cracking in cold rolling. Several researches on edge cracking in strip rolling are summarized in the following.

Schey (1966) assumed from published data that the following three causes contributed to the occurrence of edge cracking: limited ductility of the strip, uneven deformation at the edges (bulging or concave edges), and variations in stresses along the width of the strip particularly near the edges. Schey performed the edge-restraint rolling, in which restraining bars, guided into and supported by the grooves machined in the rolls, moved together with the strip and prevented the spread of the strip. Consequently, no edge cracking occurred in the edge-restraint rolling. Hence, Schey proved that

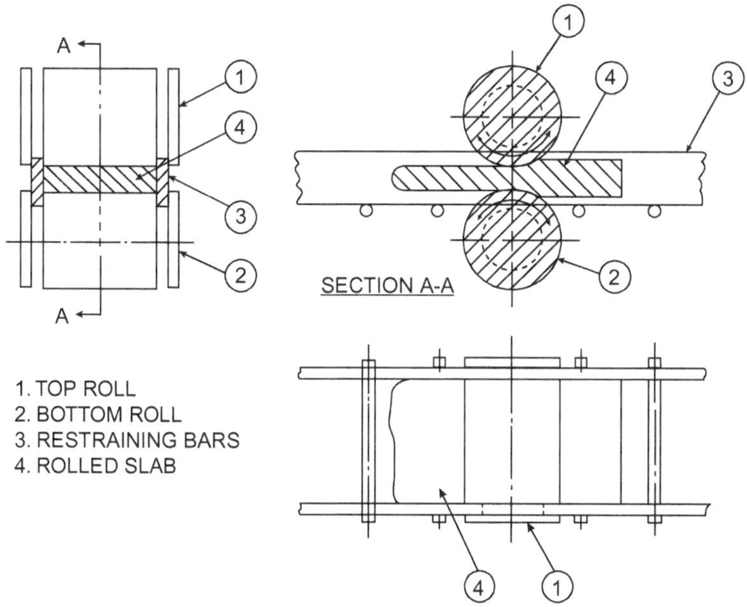

Figure 1.2 Illustrating principle of edge-restraint rolling. *From Schey, J.A., 1966. Prevention of edge cracking in rolling by means of edge restraint. J. Inst. Met. 94, 193–200.*

the concept of eliminating edge cracking by means of the edge-restraint rolling was feasible. Fig. 1.2 shows the illustrating principle of edge-restraint rolling (Schey, 1966). When strip rolling is performed by the edge-restraint rolling, plane-strain deformation occurs; stress distribution is uniform in the width direction of the strip. Hence, no edge cracking by means of the edge-restraint rolling implies that the edge cracking does not occur when the stress distribution at the edge in the width direction of the strip coincides with the stress distribution at the center in the width direction of the strip.

Cusminsky and Ellis (1967) indicated the influence of the edge shape on edge cracking during strip rolling. Table 1.2 shows the longitudinal strain at edge cracking (Cusminsky and Ellis, 1967). With increasing the chamfer angle at the edge of the strip in the width direction of the strip, the longitudinal strain at edge cracking increased. The edge of the chamfer whose angle is equal to 180 degrees implies the squared edge. Cusminsky and Ellis measured the strain distribution in the central plane in the thickness direction of the strip by riveting two strips together to form a specimen, and calculated the stress distribution in the central plane. Cusminsky and Ellis insisted that the incidence of edge cracking was controlled by the longitudinal stress; edge cracking occurred when the longitudinal stress at the edge of the strip was in accord with the stress obtained from the stress—strain relationship of the strip. This accordance implies that edge cracking occurs when the stress distribution at edge cracking is almost the same as the stress distribution in the tensile test.

Oh and Kobayashi (1976) indicated the influence of the width/height ratio of the strip on the reduction in thickness at edge cracking during strip rolling. Fig. 1.3 shows the reduction in thickness at edge cracking

Table 1.2 Longitudinal strain at edge cracking.

Chamfer angle (α)	Copper	Aluminum
60 degrees	0.705	0.820
90 degrees	0.918	1.110
120 degrees	1.240	1.470
180 degrees	No fracture	

From Cusminsky, G., Ellis, F., 1967. An investigation into the influence of edge shape on cracking during rolling. J. Inst. Met. 95, 33–37.

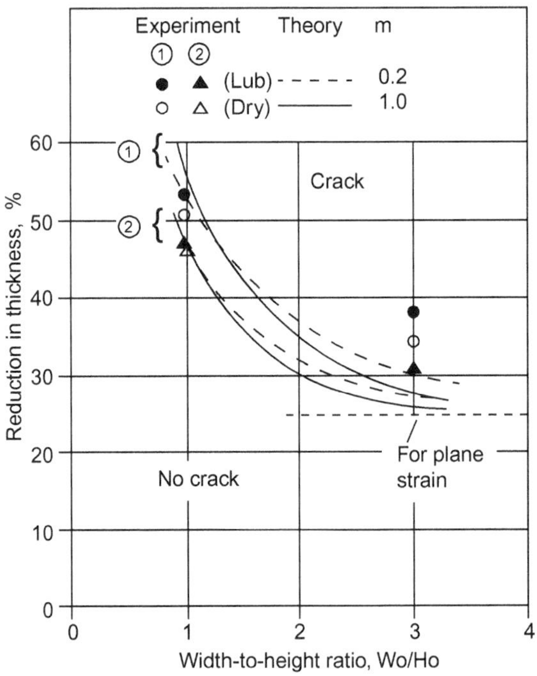

Figure 1.3 Reduction in thickness at edge cracking. *From Oh, S.I., Kobayashi, S., 1976. Workability of aluminum alloy 7075-T6 in upsetting and rolling. Trans. ASME J. Eng. Ind. 98(3), 800–806.*

(Oh and Kobayashi, 1976). With increasing the width/height ratio of the strip, the reduction in thickness at edge cracking decreased. The curve indicated by "theory" in Fig. 1.3 was calculated as follows. First, with reference to Lee and Kuhn (1973), the principal strains at fracture in the plane of a free surface were assumed to satisfy the following equation: the summation of the tensile principal strain at fracture and half of the compressive principal strain at fracture was equal to a certain value, which was a material constant. This was called the equation on the principal strains at fracture hereafter. Next, the material constant was determined by the plane-strain tensile test of a sheet (Clausing, 1970) and the upsetting of a cylinder, and was applied to edge cracking during strip rolling. Furthermore, the strain distribution during strip rolling was calculated by the three-dimensional analysis of rolling of a bar (Oh and Kobayashi, 1975). The strain in the rolling direction was assumed to be the tensile principal strain, whereas the strain in the thickness direction was assumed to be the compressive principal strain. When the tensile principal strain in the plane of a free surface and the compressive principal strain in the plane of a free surface, which were calculated from the strain distribution during strip rolling, satisfied the equation on the principal strains at fracture, edge cracking was assumed to occur. The reduction in thickness at which the tensile principal strain and the compressive principal strain satisfied the equation on the principal strains at fracture, agreed with the reduction in thickness at edge cracking obtained experimentally. Oh and Kobayashi indicated that the effect of the friction between the strip and the roll on the reduction in thickness at edge cracking during strip rolling was relatively small.

Thomson and Burman (1980) obtained edge cracking in Al−Mg alloys both in industrial hot rolling and in laboratory hot rolling. In the introduction, although edge cracking was controlled by optimizing both the ductility of the material and the edge shape of the material, the mechanism of edge cracking was shown to be disputed after an extensive survey of preceding studies. Most edge cracking that occurred during industrial hot rolling of Al−Mg alloys was attributed to the presence of a segregation band at the edge of the ingot. The segregation band contained many inclusions and precipitation particles which initiated and assisted the propagation of the edge crack. A large number of Al−Mg alloy ingots in industrial hot rolling showed the following three types of edge crack: small numerous cracks initiated and propagated within the segregation band, large cracks initiated in the segregation band and propagated into the bulk of the ingot, and massive cracks initiated internally although first observed on the edge of the ingot. In laboratory hot rolling of Al–Mg alloys, both sodium and hydrogen were found to have a deleterious effect on the incidence of edge cracking in Al–Mg alloys, although the mechanism of the incidence of edge cracking was not clarified.

1.4 Alligatoring in strip rolling

Alligatoring in strip rolling, in which the central plane of the strip in the thickness direction of the strip fractures during rolling and the front end of the strip splits into two parts along the central plane of the strip like the jaws of an alligator, is one of the fractures in rolling. Alligatoring is also called crocodiling. Alligatoring in strip rolling is known to occur in hot rolling and in materials of limited ductility such as aluminum–magnesium alloys. Although alligatoring in strip rolling was often observed in industrial rolling mills (Kasz and Varley, 1949), few researches on alligatoring in strip rolling were performed in laboratory rolling mills.

Schey (1966) obtained not only the experimental result on edge cracking in strip rolling but also the experimental result on alligatoring in strip rolling. An aluminum–magnesium alloy containing 8% Mg was used, and alligatoring in strip rolling was observed when the (strip thickness)/(length of contact) ratio was equal to 0.66. A photograph of alligatoring in strip rolling was shown in Schey (1980). Although alligatoring in strip rolling was mentioned to be related to the (strip thickness)/(length of contact) ratio, further researches on alligatoring in strip rolling are required because of limited experimental results.

1.5 Central burst in wire drawing

Central burst in wire drawing, in which the center in the radial direction of the wire fractures during drawing, usually periodically in the length direction of the wire, is one of the fractures in drawing. Because the shape of a cavity, which yields due to fracture, resembles a chevron or a cup, central burst is also called a chevron crack or a cuppy crack. The external shape of the drawn wire in which central burst occurs scarcely differs from the external shape of the drawn wire in which central burst does not occur. However, when central burst in wire drawing occurs, the load to draw the wire fluctuates; the load to draw the wire decreases when a chevron enters the die, whereas the load to draw the wire increases when a chevron emerges from the die. Fig. 1.4 shows the longitudinal section of the wire containing central burst. Several researches on central burst in wire drawing are summarized in the following.

Remmers (1930) performed the multipass wire drawing of copper in various die angles and in various oxygen contents of the copper. With increasing the die angle, the oxygen content of the copper, above which a cuppy wire was obtained, decreased, whereas with increasing the oxygen content of the copper, the die angle, above which a cuppy wire was obtained, decreased. Finally, the relationship between the die angle and the oxygen content of the copper, which determined whether a sound wire or a cuppy wire was obtained, was demonstrated.

Jennison (1930) dealt with two types of fractures in wire drawing of copper: central burst and surface fracture, which is called check mark or crowfeet; and is a V-shaped cavity, which appears intermittently in the length direction of the wire. Fig. 1.5 shows the die and the name of the die's each part. To prevent the occurrence of central burst and check

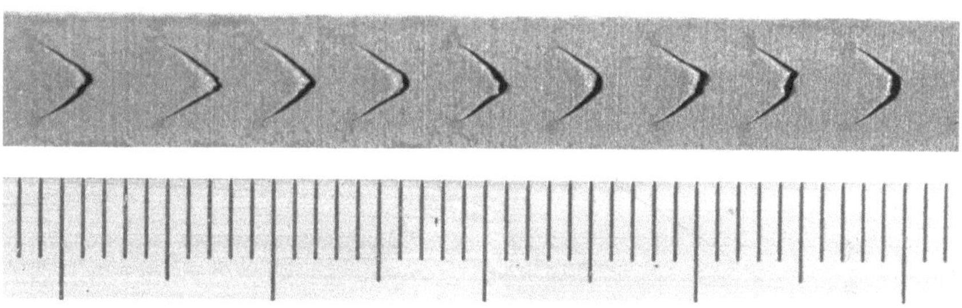

Figure 1.4 Longitudinal section of wire containing central burst. *From* JIS S35C

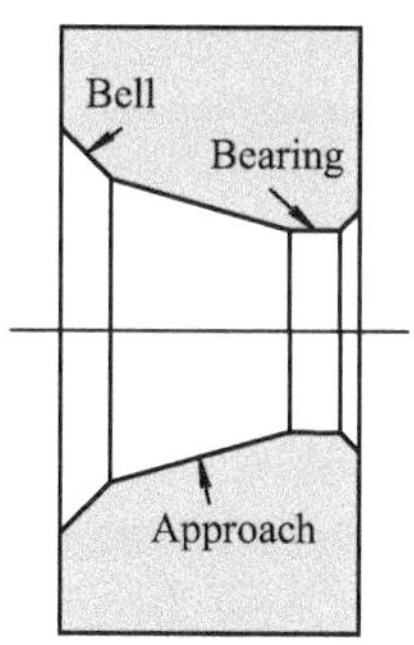

Figure 1.5 Die and name of die's each part.

mark, an appropriate die shape was proposed; a bearing in the die was proved to be indispensable to prevent the occurrence of not only central burst but also check mark.

Tanaka (1952) performed the multipass bar drawing of copper alloy using the following two kinds of composite bars: a bar that consisted of the core made of Al–10Si–8Mg alloy and the mantle made of Al–4Cu alloy, and a bar that consisted of the core made of Al–4Cu alloy and the mantle made of Al–10Si–8Mg alloy. Al–10Si–8Mg alloy was known to be more brittle than Al–4Cu alloy. When a drawing pass schedule was specified, central burst appeared when the bar in which the core was made of Al–10Si–8Mg alloy was drawn, whereas no central burst appeared when the bar in which the core was made of Al–4Cu alloy was drawn. Hence, central burst was shown to be caused by the defect that existed originally in the material.

Orbegozo (1968) performed the multipass wire drawing of an aluminum alloy. The back tension was applied to the wire by drawing through two dies in tandem or in sequence. Crack nucleation was caused by the mean normal stress at the center in the radial direction of the wire inside the die, whereas crack propagation appeared to depend on the drawing stress outside the die. A dimensionless parameter Δ, which is the diameter of the wire halfway through the die divided by the slant length of the cone of the die, was defined. With increasing the die angle or with decreasing the reduction in area in each die, the dimensionless parameter Δ increases. Hence, the dimensionless parameter Δ was shown to be effective to predict the occurrence of central burst.

1.6 Central burst in bar extrusion

Central burst in bar extrusion, in which the center in the radial direction of the bar fractures during extrusion, usually periodically in the length direction of the bar, is one of the fractures in extrusion. The mechanism of the occurrence of central burst in bar extrusion is considered to be the same as the mechanism of the occurrence of central burst in wire drawing. However, when the die angle and the reduction in area in each die are specified, the mean normal stress at the center in the radial direction of the bar in bar extrusion is smaller than the mean normal stress at the center in the radial direction of the wire in wire drawing. Fig. 1.6 shows the longitudinal section of the bar containing central burst.

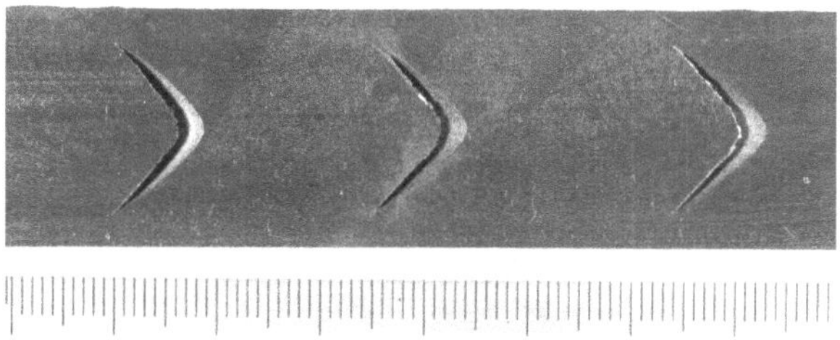

Figure 1.6 Longitudinal section of bar containing central burst. *From JIS SMn438H, by courtesy Tsuda Industries Co. Ltd.*

Pepe (1976) performed the multipass bar hydrostatic extrusion of AISI 1080 steel. The relationship between the bar displacement, the extrusion force, and the central burst shape was examined to clarify the mechanism of the formation of the central burst. Fig. 1.7 shows the relationship between the bar displacement, the extrusion force, and the central burst shape (Pepe, 1976). The cycle of the fluctuation of the central burst shape was consistent with the cycle of the fluctuation of the extrusion force. Position 5 was defined as the end of the cycle of the central burst formation, whereas Position 1 was defined as the beginning of the cycle of the central burst formation. Position 5 was identical to Position 1. The condition of Position 1 was assumed. First, with increasing the bar displacement, the extrusion force increased because no central burst was formed inside the die, although microcracks were formed inside the die. Next, with increasing the bar displacement, the extrusion force decreased because a central burst was formed inside the die. Finally, the condition of Position 5 was reached. The formation of the central burst was proved to occur not only inside the die but also outside the die. The formation of the central burst was also discussed from a metallographic examination.

1.7 Christmas-tree cracking in bar extrusion

Christmas-tree cracking in bar extrusion, in which a surface crack appears in the circumferential direction of the bar during extrusion, usually periodically in the length direction of the bar, is one of the fractures in extrusion. Because the external shape of the extruded bar in which the surface crack appears, resembles the external shape of a fir tree, which is usually used as a Christmas tree, Christmas-tree cracking is also called fir-tree cracking. Christmas-tree cracking in bar extrusion is known to occur in various materials of limited ductility. Although Christmas-tree cracking in

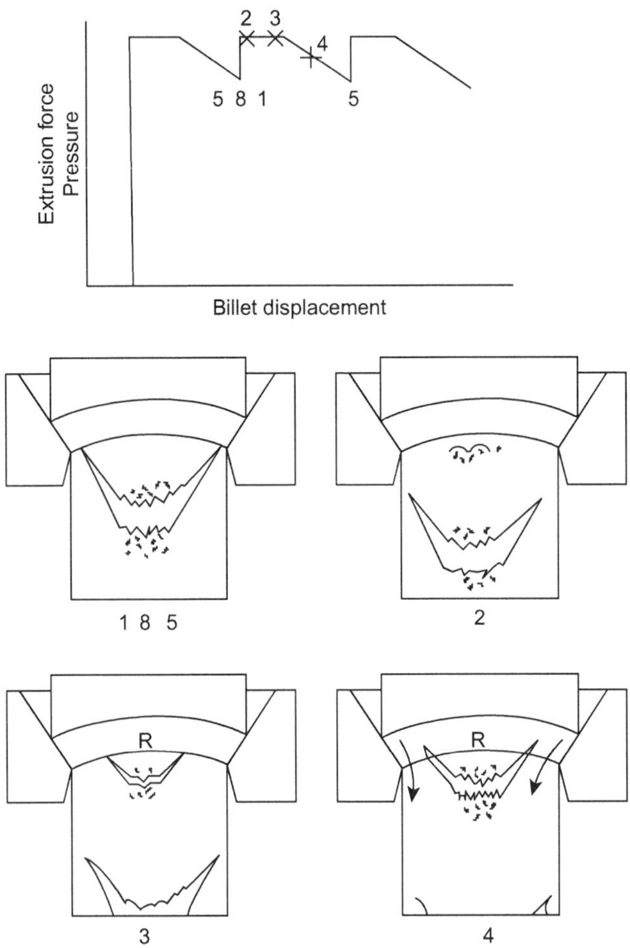

Figure 1.7 Relationship between bar displacement, extrusion force, and central burst shape. *From Pepe, J.J., 1976. Central burst formation during hydrostatic extrusion. Met. Eng. Quart. 16(1), 46–58.*

bar extrusion was often observed in industrial extrusion machines, few researches on Christmas-tree cracking in bar extrusion were performed in laboratory extrusion machines.

Wilcox and Whitton (1959) performed the single pass bar extrusion of copper and aluminum alloys. Although Christmas-tree cracking was observed for the extrusion of aluminum alloys, Christmas-tree cracking was not observed for the extrusion of copper. The relationship between the punch displacement and the extrusion pressure was examined to clarify the mechanism of the formation of the Christmas-tree cracking. Fig. 1.8 shows the

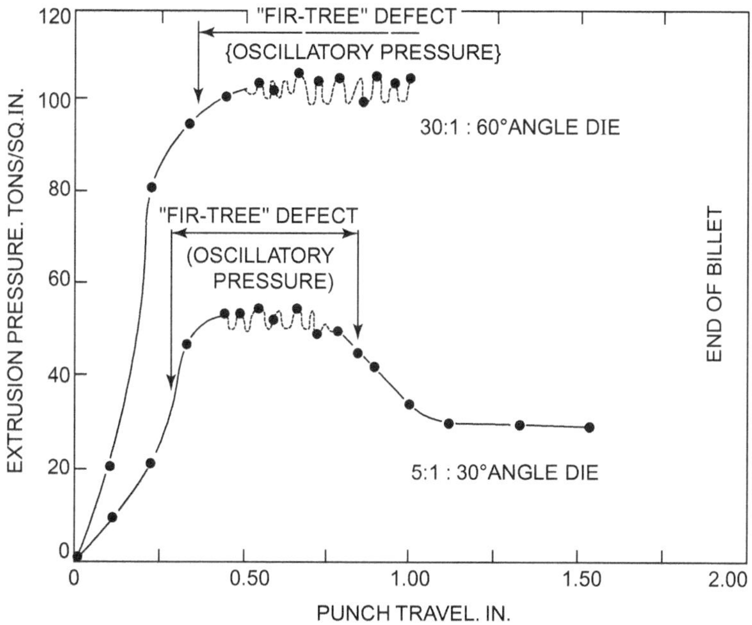

Figure 1.8 Relationship between punch displacement and extrusion pressure. *From Wilcox, R.J., Whitton, P.W., 1959. Further experiments on the cold extrusion of metals using lubrication at slow speed. J. Inst. Met. 88, 145–149.*

relationship between the punch displacement and the extrusion pressure (Wilcox and Whitton, 1959). The mechanism of the formation of Christmas-tree cracking was presumed as follows. First, the extrusion pressure increased when the surface of the bar stuck to the bearing of the die. Next, the extrusion pressure decreased when the surface of the bar in the bearing of the die cracked and the surface of the bar stuck to the bearing of the die was extruded outside the bearing of the die. Then, the extrusion pressure increased again when the surface of the bar stuck to the bearing of the die. Thus, with increasing the punch displacement, the extrusion pressure oscillated. Except for the Christmas-tree cracking, the following defect was mentioned. When the single pass bar extrusion of copper was performed, most of the surface of the bar in the length direction of the bar had a mottled appearance, which was assumed to occur due to a stick-slip phenomenon.

1.8 Surface cracking in cylinder upsetting

Surface cracking in cylinder upsetting, in which the surface in the radial direction fractures during upsetting, is one of the fractures in forging.

Jenner and Dodd (1981) performed a review on cold upsetting and free surface ductility. Several researches on surface cracking in cylinder upsetting are summarized in the following.

Kudo and Aoi (1967) performed the cylinder upsetting of an annealed medium carbon steel called JIS S45C, an equivalent to ISO C45E4. The conventional cylindrical coordinate system was assumed. First, at the surface in the radial direction and at the center in the axial direction, the components of the strain increment $d\varepsilon_\theta$ and $d\varepsilon_z$ were measured. Next, the components of the strain ε_θ and ε_z were calculated along the strain history. Then, the components of the stress σ_θ and σ_z were calculated along the strain history using the Lévy–Mises constitutive equation on the relationship between the stress and the strain increment. The following three types of cracks were observed: the longitudinal crack, the oblique crack, and the mixed crack, which is a mixture of the longitudinal crack and the oblique crack.

Fig. 1.9 shows the schematic representation of the fracture surface of the specimen (Kudo and Aoi, 1967). Fig. 1.9A shows the fracture surface for the oblique crack. The fracture surface of the oblique crack was perpendicular to the θz-plane, and the angle between the fracture surface of

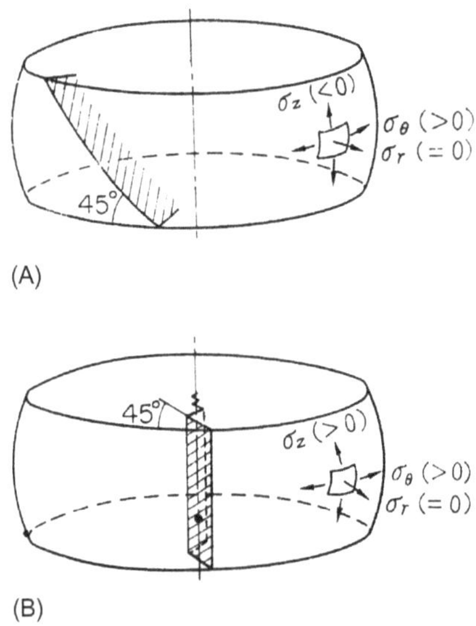

Figure 1.9 Schematic representation of fracture surface of specimen. (A) Fracture surface for oblique crack. (B) Fracture surface for longitudinal crack. *From Kudo, H., Aoi, K., 1967. Effect of compression test condition upon fracturing of a medium carbon steel. J. Jpn. Soc. Technol. Plast. 8(72), 17–27 (in Japanese).*

the oblique crack and the z-axis was equal to 45 degrees. When the oblique crack appeared, σ_θ was the maximum principal stress and σ_z was the minimum principal stress. Hence, the fracture surface of the oblique crack coincided with the surface in which the maximum shear stress yielded. Fig. 1.9B shows the fracture surface for the longitudinal crack. The fracture surface of the longitudinal crack was perpendicular to the $r\theta$-plane, and the angle between the fracture surface of the longitudinal crack and the r-axis was equal to 45 degrees. When the longitudinal crack appeared, σ_θ was the maximum principal stress and σ_r was the minimum principal stress. Hence, the fracture surface of the longitudinal crack coincided with the surface in which the maximum shear stress yielded.

Fig. 1.10 shows the relationship between the axial strain and the circumferential strain for various upsetting conditions (Kudo and Aoi, 1967). The coordinate of the cross mark indicates the fracture strain. The broken line indicates the strain history for homogeneous compression. Hence, the gradient of the broken line is equal to 0.5. The relationships between the axial strain at fracture and the circumferential strain at fracture for various upsetting conditions were expressed by a straight line, the gradient of which was equal to 0.5 and parallel to the broken line. The longitudinal crack appeared when the magnitude of the fracture strain was relatively small, whereas the oblique crack appeared when the magnitude of the fracture strain was relatively large. A lubricated flat die was used and a grooved flat die on the surface of which numerous concentric grooves were engraved, was used to prevent the slip between the cylinder and the die. When the initial height/diameter ratio of a cylinder was specified, the magnitude of the axial strain at fracture and the magnitude of the circumferential strain at fracture obtained using the grooved flat die, were smaller than the magnitude of the axial strain at fracture and the magnitude of the circumferential strain at fracture obtained using the lubricated flat die. When the flat die used was specified, with increasing the initial height/diameter ratio of a cylinder, the magnitude of the axial strain at fracture and the magnitude of the circumferential strain at fracture increased.

Thomason (1969) performed the cylinder upsetting of a spheroidized annealed high-carbon steel that had longitudinal surface defects, which were artificial defects in the form of machined longitudinal grooves. When an experimental condition was specified, the ductility of the cylinder that had the artificial defects was lower than the ductility of the cylinder that had no artificial defects. With increasing the initial height/diameter ratio of a cylinder, the ductility of the cylinder increased. With

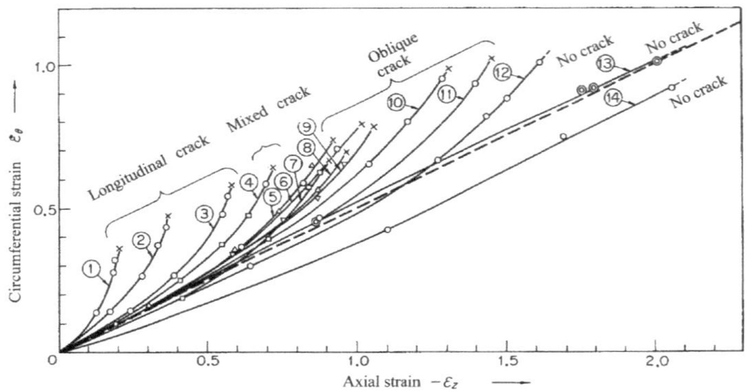

Curve number	Compression method	Initial shape H/D	Curve number	Compression method	Initial shape H/D
1	Groove	6/12	8	Lubrication	12/12
2	Groove	12/12	9	Lubrication	20/12
3	Groove	20/12	10	Cone	24/12
4	No lubrication	24/12	11	Cone	24/12
5	Lubrication	6/12	12	Cone	20/12
6	Cone	20/12	13	Cone	20/12
7	No lubrication	30/12	14	Cone	20/12

Figure 1.10 Relationship between axial strain and circumferential strain for various upsetting conditions. *From Kudo, H., Aoi, K., 1967. Effect of compression test condition upon fracturing of a medium carbon steel. J. Jpn. Soc. Technol. Plast. 8(72), 17–27 (in Japanese).*

decreasing the friction between the platen and the cylinder, the ductility of the cylinder increased. With decreasing the depth of the machined longitudinal grooves, the ductility of the cylinder increased.

Kobayashi (1970) performed the cylinder upsetting and the ring upsetting of an annealed SAE 1040 carbon steel. First, the following results, which had been already confirmed by Kudo and Aoi (1967), were reconfirmed. Whether the axial stress σ_z was compressive or tensile at fracture depended on the experimental condition. The oblique crack appeared when the axial stress σ_z was compressive at fracture, whereas the longitudinal crack appeared when the axial stress σ_z was tensile at fracture. Next, when the experimental condition in upsetting a ring was identical to the experimental condition in upsetting a cylinder, the crack mode appeared in upsetting the ring was almost the same as the crack mode appeared in upsetting the cylinder.

Kuhn and Lee (1971) performed the cylinder upsetting of as-drawn AISI 1045 carbon steel and annealed AISI 1045 carbon steel. The conventional cylindrical coordinate system was assumed. The axial strain ε_z and the circumferential strain ε_θ were measured at the surface in the radial direction and at the center in the axial direction. The radial strain ε_r was calculated using both the axial strain ε_z and the circumferential strain ε_θ from the condition of volume constancy.

Fig. 1.11 shows the relationship between the axial strain, the circumferential strain, the radial strain, and the overall height strain (Kuhn and

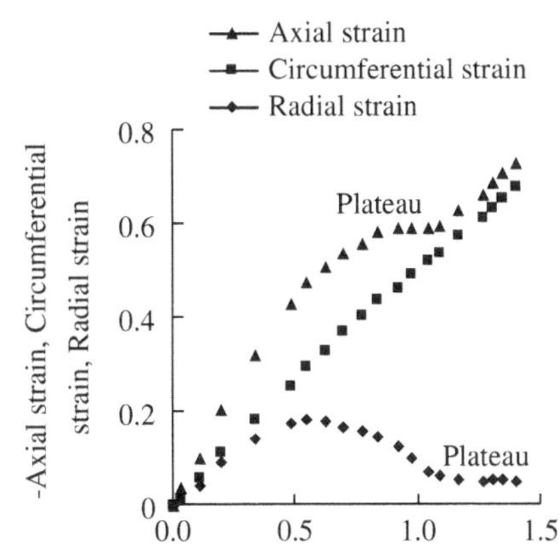

Figure 1.11 Relationship between axial strain, circumferential strain, radial strain, and overall height strain. *From Kuhn, H.A., Lee, P.W., 1971. Strain instability and fracture at the surface of upset cylinders. Metall. Trans. 2(11), 3197–3202.*

Lee, 1971). The overall height strain is defined as $\ln(h/h_0)$, where h_0 denotes the initial height of the cylinder and h denotes the current height of the cylinder. With increasing the magnitude of the overall height strain, the magnitude of the axial strain ε_z and the magnitude of the circumferential strain ε_θ increased smoothly in the early stage. However, in the intermediate stage, with increasing the magnitude of the overall height strain, the magnitude of the axial strain ε_z suddenly reached a plateau and remained constant, whereas the magnitude of the circumferential strain ε_θ continued to increase monotonically, and the magnitude of the radial strain ε_r decreased monotonically. Then, in the final stage, with increasing the magnitude of the overall height strain, the magnitude of the axial strain ε_z began to increase again monotonically and the magnitude of the circumferential strain ε_θ continued to increase monotonically, whereas the magnitude of the radial strain ε_r suddenly reached a plateau and remained constant. At the end of the final stage, the oblique crack appeared; the fracture surface of the oblique crack was perpendicular to the θz-plane, and the angle between the fracture surface of the oblique crack and the z-axis was equal to 45 degrees. The plateau on the magnitude of the axial strain ε_z in the intermediate stage was not observed in previous studies (Kudo and Aoi, 1967; Thomason, 1969; Kobayashi, 1970), because the gauge length for strain measurement in this study was set to be much smaller than the gauge length for strain measurement in the previous studies.

Fig. 1.12 shows the stress state immediately before fracture, the deformation mode immediately before fracture, and the fracture surface in cylinder upsetting and stretch forming (Kuhn and Lee, 1971). The length direction of the groove of the sheet in stretch forming was made to coincide with the x-direction, and the width direction of the groove of the sheet in stretch forming was made to coincide with the y-direction. The thickness direction of the sheet coincided with the z-direction. By superimposing a hydrostatic stress that was equal to the axial stress σ_z on the stress state immediately before fracture in cylinder upsetting, the superimposed stress state was almost the same as the stress state immediately before fracture in stretch forming. Because the radial strain increment $d\varepsilon_r$ was equal to zero immediately before fracture, plane-strain deformation occurred in cylinder upsetting, whereas because the strain increment in the length direction of the groove of the sheet $d\varepsilon_x$ was equal to zero immediately before fracture (Azrin and Backofen, 1970), plane-strain deformation occurred in stretch forming. Hence, the deformation mode

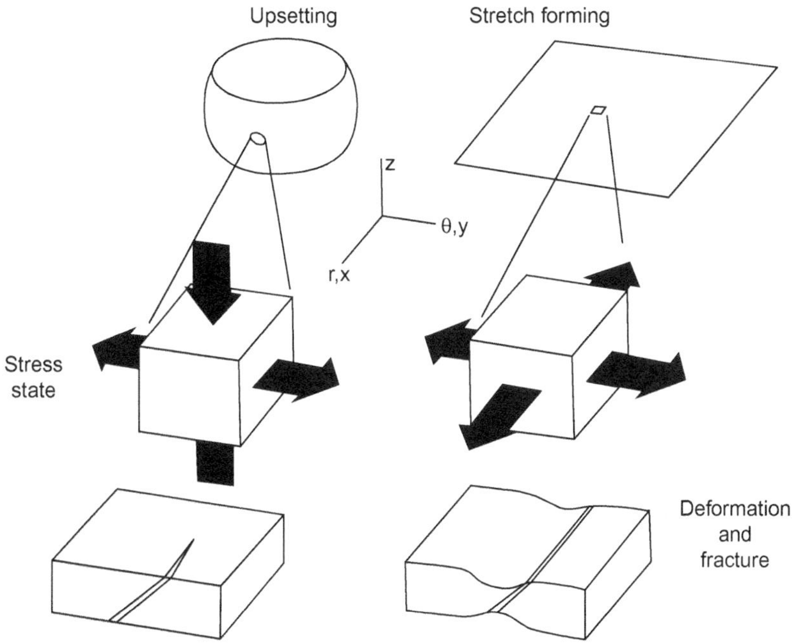

Figure 1.12 Stress state immediately before fracture, deformation mode immediately before fracture, and fracture surface in cylinder upsetting and stretch forming. *From Kuhn, H.A., Lee, P.W., 1971. Strain instability and fracture at the surface of upset cylinders. Metall. Trans. 2(11), 3197–3202.*

immediately before fracture in cylinder upsetting was identical to the deformation mode immediately before fracture in stretch forming. The fracture surface in cylinder upsetting was perpendicular to the θz-plane, and the angle between the fracture surface in cylinder upsetting and the z-axis was equal to 45 degrees. The fracture surface in stretch forming was perpendicular to the yz-plane, and the angle between the fracture surface in stretch forming and the z-axis was equal to 45 degrees (Beaver, 1983). Hence, the similarities between the stress state immediately before fracture, the deformation mode immediately before fracture, and the fracture surface in cylinder upsetting and the stress state immediately before fracture, the deformation mode immediately before fracture, and the fracture surface in stretch forming, were demonstrated.

Lee and Kuhn (1973) performed not only cylinder upsetting but also plate rolling, semicylinder bending, and plane-strain bending using a 1045 carbon steel, a 1020 carbon steel, and a 303 stainless steel. Fig. 1.13 shows the specimens and the surface strain states in cylinder upsetting, plate rolling, semicylinder bending, and plane-strain bending (Lee and Kuhn,

1973). A specimen for plate rolling and a specimen for semicylinder bending were manufactured from cylinders by machining the sides of the cylinders, whereas a specimen for plane-strain bending was manufactured

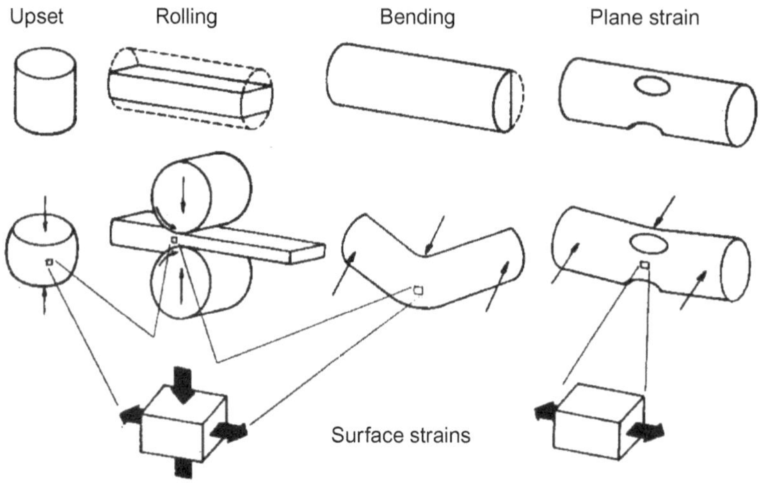

Figure 1.13 Specimens and surface strain states in cylinder upsetting, plate rolling, semicylinder bending, and plane-strain bending. *From Lee, P.W., Kuhn, H.A., 1973. Fracture in cold upset forging—a criterion and model. Metall. Trans. 4(4), 969–974.*

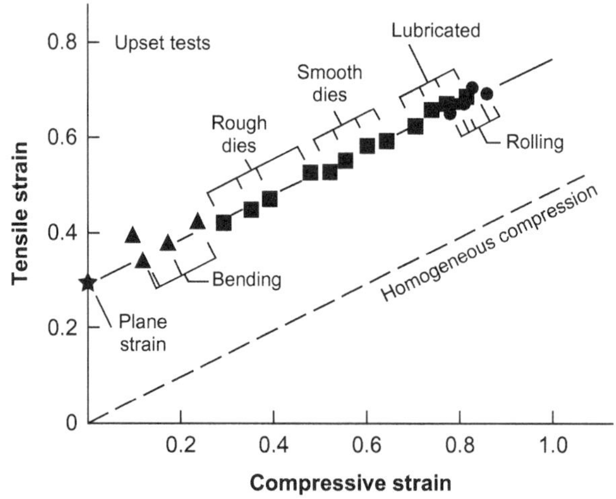

Figure 1.14 Relationship between tensile strain at fracture and compression strain at fracture in cylinder upsetting, plate rolling, semicylinder bending, and plane-strain bending for a 1045 carbon steel. *From Lee, P.W., Kuhn, H.A., 1973. Fracture in cold upset forging—a criterion and model. Metall. Trans. 4(4), 969–974.*

from a cylinder by drilling a hole in the cylinder. Fig. 1.14 shows the relationship between the tensile strain at fracture and the compression strain at fracture in cylinder upsetting, plate rolling, semicylinder bending and plane-strain bending for a 1045 carbon steel (Lee and Kuhn, 1973). The broken line indicates the strain history for homogeneous compression. Hence, the gradient of the broken line is equal to 0.5. The relationships between the axial strain at fracture and the circumferential strain at fracture for various upsetting, rolling, and bending conditions were expressed by a straight line, the gradient of which was equal to 0.5 and parallel to the broken line. Because the compression strain at fracture was equal to zero in plane-strain bending, the coordinate of the tensile strain at fracture in plane-strain bending was identical to the coordinate of the intercept of the line. For a 1045 carbon steel, a1020 carbon steel, and a 303 stainless steel, the relationships between the axial strain at fracture and the circumferential strain at fracture for various upsetting, rolling, and bending conditions were expressed by a straight line, the gradient of which was equal to 0.5. However, the coordinate of the intercept of the line for a 1045 carbon steel, the coordinate of the intercept of the line for a 1020 carbon steel, and the coordinate of the intercept of the line for a 303 stainless steel, were different from each other. The relationships between the axial strain at fracture and the circumferential strain at fracture for various upsetting conditions, which were obtained in Thomason (1969) and Kobayashi (1970), were shown to be expressed by a straight line, the gradient of which was equal to 0.5.

Oh and Kobayashi (1976) performed both the experiment and the simulation of the cylinder upsetting of an aluminum alloy 7075-T6. Fig. 1.15 shows the relationship between the axial strain at fracture and the reduction in height at fracture (Oh and Kobayashi, 1976). The circumferential strain was assumed to be the tensile principal strain, whereas the axial strain was assumed to be the compressive principal strain. The curve indicated by "Simulation" in Fig. 1.15 was calculated as follows. First, with reference to Lee and Kuhn (1973), the principal strains at fracture in the plane of a free surface were assumed to satisfy the equation on the principal strains at fracture which was defined in Oh and Kobayashi (1976) of Section 1.3. Next, the material constant in the equation on the principal strains at fracture was determined by the plane-strain tensile test of a sheet (Clausing, 1970) and the upsetting of a cylinder. Furthermore, the strain distribution during cylinder upsetting was calculated by the axisymmetric rigid-plastic finite-element method (Lee and Kobayashi, 1973).

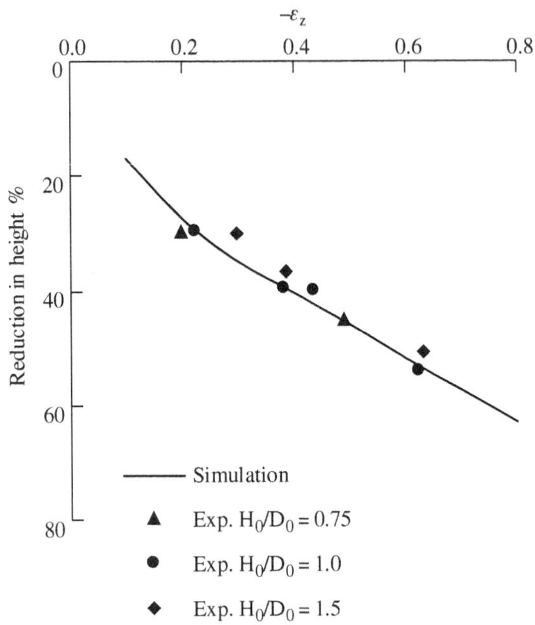

Figure 1.15 Relationship between axial strain at fracture and reduction in height at fracture. *From Oh, S.I., Kobayashi, S., 1976. Workability of aluminum alloy 7075-T6 in upsetting and rolling. Trans. ASME J. Eng. Ind. 98(3), 800–806.*

When the tensile principal strain in the plane of a free surface and the compressive principal strain in the plane of a free surface, which were calculated from the strain distribution during cylinder upsetting, satisfied the equation on the principal strains at fracture, surface cracking was assumed to occur. The reduction in height at which the tensile principal strain and the compressive principal strain satisfied the equation on the principal strains at fracture, agreed with the reduction in height at fracture obtained experimentally.

Sowerby et al. (1984) and Sowerby and Chandrasekaran (1984) performed not only cylinder upsetting but also stepped cylinder upsetting using various kinds of carbon steels. Fig. 1.16 shows the specimens and the tools for the compression test, the collar test, and the punch test (Sowerby et al., 1984). Conventional cylinder upsetting was performed in the compression test, whereas stepped cylinder upsetting was performed in the collar test, and the diameter of the tool was slightly larger than the diameter of the specimen in the punch test. Fig. 1.17 shows the relationship between the axial strain and the circumferential strain in the

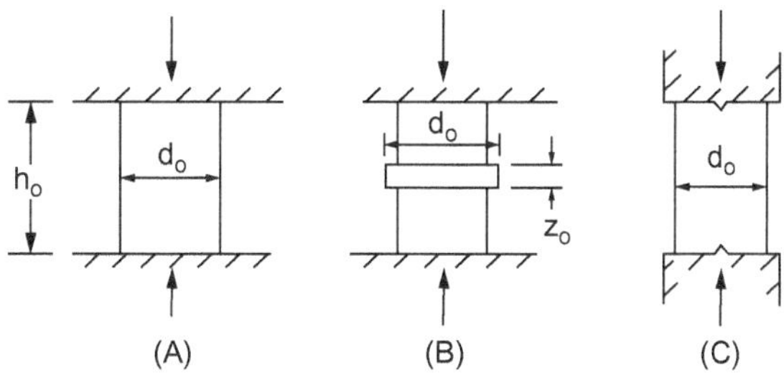

Figure 1.16 Specimens and tools for compression test, collar test, and punch test. (A) Compression test. (B) Collar test. (C) Punch test. *From Sowerby, R., O'Reilly, I., Chandrasekaran, N., Dung, N.L., 1984. Materials testing for cold forging. Trans. ASME J. Eng. Mater. Technol. 106(1), 101–106.*

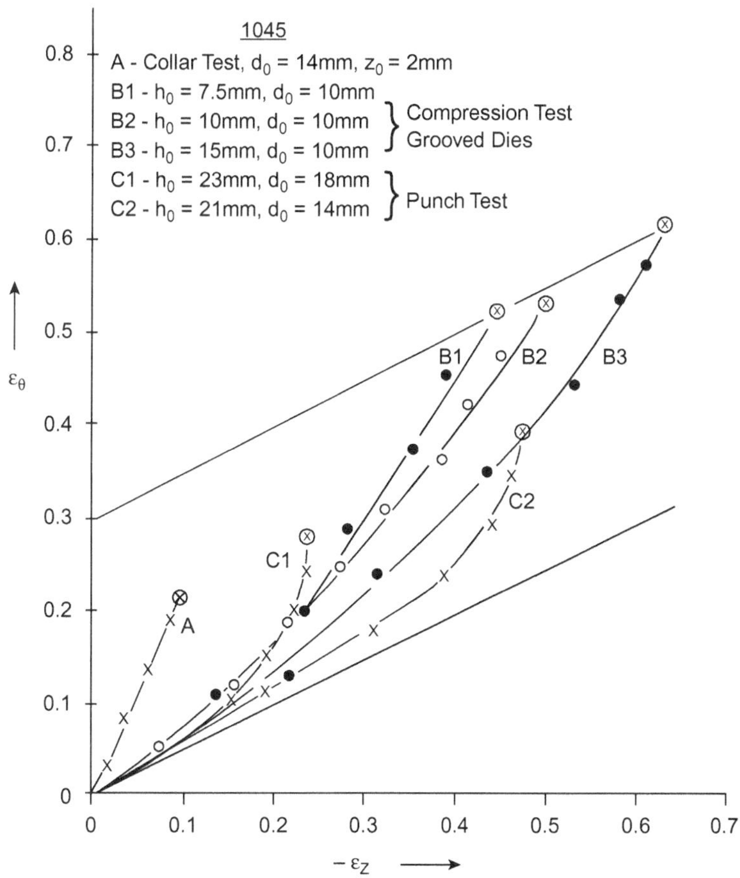

Figure 1.17 Relationship between axial strain and circumferential strain in compression test, collar test, and punch test for an AISI 1045 carbon steel. *From Sowerby, R., O'Reilly, I., Chandrasekaran, N., Dung, N.L., 1984. Materials testing for cold forging. Trans. ASME J. Eng. Mater. Technol. 106(1), 101–106.*

compression test, the collar test, and the punch test for an AISI 1045 carbon steel (Sowerby et al., 1984). The coordinate of the cross mark indicates the fracture strain. The magnitudes of the fracture strains in the collar test and the magnitudes of the fracture strains in the punch test were smaller than the magnitudes of the fracture strains in the compression test. The line, which passes through the origin and the gradient of which is equal to 0.5, indicates the strain history for homogeneous compression. The coordinates of the fracture strains in the compression test were on a line, which was parallel to the line passing through the origin. However, the coordinates of the fracture strains in the collar test and the coordinates of the fracture strains in the punch test were not on the line that passed through the coordinates of the fracture strains in the compression test. In other words, there was no single line that passed through the coordinates of the fracture strains in all the tests: the compression test, the collar test, and the punch test. For an AISI 1045 carbon steel and each one of the other carbon steels experimented, no single line passed through the coordinates of the fracture strains in all the tests: the compression test, the collar test, and the punch test.

1.9 Forming limit in sheet forming

There are two types of forming limits in sheet forming: necking forming limit and fracture forming limit. Necking forming limit is the forming limit at which the necking of a sheet occurs or the unevenness of the thickness of a sheet appears, whereas fracture forming limit is the forming limit at which a sheet fractures or failures. Hence, the necking forming limit differs from the fracture forming limit. Because the subject of this book is ductile fracture, the necking forming limit is not required to be dealt with, and the fracture forming limit should only be dealt with. However, because the necking forming limit is generally considered to be the forming limit in industrial metal-forming, the necking forming limit is described briefly in the following.

The concept of the necking forming limit was proposed in the 1960s (Keeler and Backofen, 1963; Keeler, 1965; Goodwin, 1968). The sheet, in which the scribed circles of the diameter of 0.2 inches were embedded, was loaded in the plane of the sheet in two directions which were orthogonal to each other, until the necking of the sheet occurred. The load was applied so that the strain ratio $\varepsilon_2/\varepsilon_1$ was constant during loading, where ε_1 is the major principal strain in the plane of the sheet, and ε_2 is the

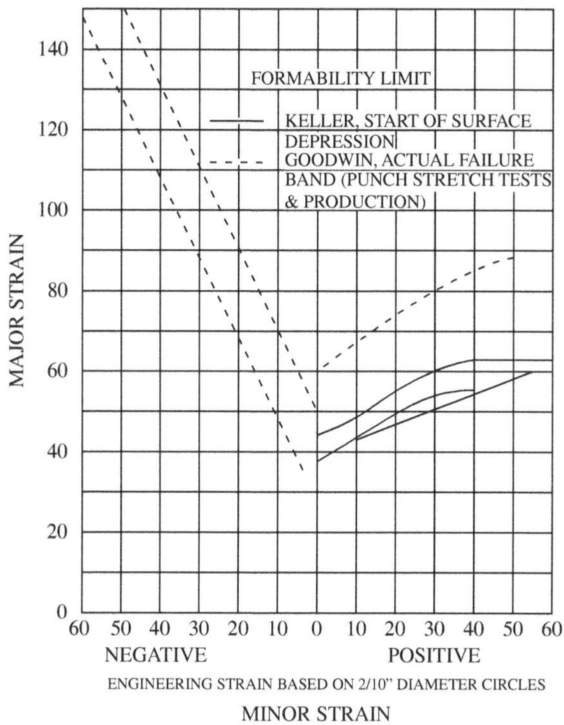

Figure 1.18 Necking forming limit. *From Goodwin, G.M., 1968. Application of strain analysis to sheet metal forming problems in the press shop. La Metallurgia Italiana 60 (8), 767–774.*

minor principal strain in the plane of the sheet. Then, the major principal strain and the minor principal strain in the region where the necking of the sheet occurred, were calculated from the lengths of the major axis and the minor axis of the ellipse which was originally the scribed circle embedded in the sheet.

Fig. 1.18 shows the necking forming limit (Goodwin, 1968). The solid curves were obtained by Keeler (1965), whereas the broken curves were obtained by Goodwin (1968). Because the diameter of the scribed circles was not sufficiently small, the lengths of the major axis and the minor axis of the ellipse were not uniform in the region in which the necking of the sheet occurred. Hence, when the minor principal strain was specified, there were two major principal strains which were on the necking forming limit curves. In other words, when the range of the minor principal strain was specified, there were two necking forming limit curves: lower necking forming limit curve and upper necking forming limit curve. The combination of the major principal strain and the minor principal strain, which lay below the lower necking forming limit curve was considered to be safe for the fracture of the sheet. The combination of the major principal strain and the minor principal strain, which lay above the lower necking forming limit curve and below the upper necking forming limit curve, was considered to be hazardous for the fracture of the sheet.

Because the necking forming limit could be calculated from the thickness of a sheet at necking, the necking forming limit is not necessarily related to the fracture of the sheet. However, since the fracture forming limit should be calculated from the thickness of a sheet at fracture, the fracture forming limit is closely related to the fracture of the sheet.

Fig. 1.19 shows the fracture forming limit (Yoshida et al., 1968). ε_1 is the major principal strain in the plane of the sheet, and ε_2 is the minor principal strain in the plane of the sheet. The experiments on the following three types of tensile tests of the sheet were performed: uniaxial tensile test, plane-strain tensile test, and balanced biaxial tensile test. The strain in the thickness direction of the sheet at the fracture surface, and the strain in the direction parallel to the fracture surface in the plane of the sheet were measured. The strain in the direction perpendicular to the fracture surface in the plane of the sheet was calculated from the condition of volume constancy. The fracture forming limit curve was able to be approximated by a line, the gradient of which was equal to one. Hence, the

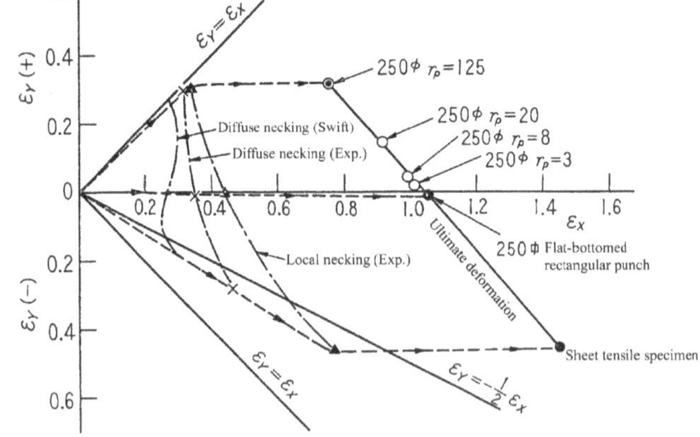

Figure 1.19 Fracture forming limit. *From Yoshida, K., Abe, K., Hosono, K., Takezoe, A., 1968. An experimental study of ultimate ductility and average ductility in steel sheet forming. Rep. Inst. Phys. Chem. Res. 44(3), 128–139 (in Japanese).*

fracture forming limit curve indicated that the sheet fractured when the thickness of the sheet became a certain value. The fact that the sheet fractured when the thickness of the sheet became a certain value was confirmed for rimmed steels and for killed steels. The fracture forming limit curve, the diffuse necking limit curve on which diffuse necking commenced and the local necking limit curve on which local necking commenced, were shown. The strain histories for the uniaxial tensile test, the plane-strain tensile test, and the balanced biaxial tensile test were also demonstrated. When the sheet was subjected to local necking for the uniaxial tensile test, the plane-strain tensile test, and the balanced biaxial tensile test, although the major principal strain increased, the minor principal strain scarcely changed; plane-strain deformation occurred.

Fig. 1.20 shows the necking forming limit and the fracture forming limit for an aluminum–magnesium alloy 5154 (Embury and LeRoy, 1978). Although the gradient of the fracture forming limit curve was equal to one in Fig. 1.19, the gradient of the fracture forming limit curve was much larger than one. When the load was applied so that the strain ratio $\varepsilon_2/\varepsilon_1$ was equal to one during loading, the necking forming limit virtually coincided with the fracture forming limit. For a fine-grained aluminum alloy, Embury and

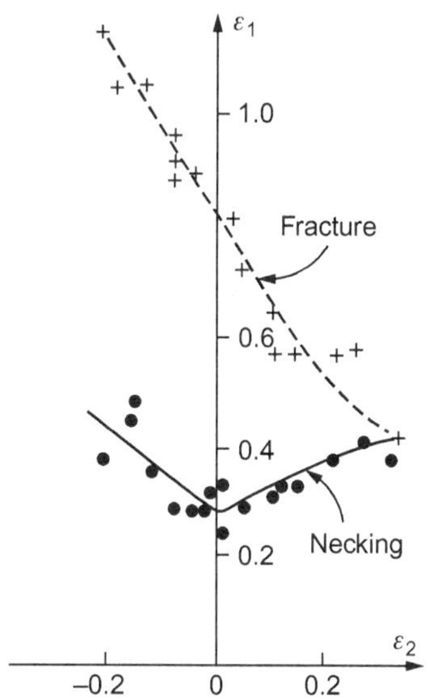

Figure 1.20 Necking forming limit and fracture forming limit for aluminum–magnesium alloy 5154. *From Embury, J. D., LeRoy, G. H., 1978. Failure maps applied to metal deformation processes. In: Taplin, D. M. R., (Ed.), Advances in Research on the Strength and Fracture of Materials. Pergamon Press, New York, U. S. A., pp. 15–42.*

Duncan (1981) showed that the gradient of the fracture forming limit curve was much smaller than one, and that the necking forming limit curve and the fracture forming limit curve sufficiently separated from each other.

1.10 Rupture in shearing

Shearing is a metal-forming process in which a material is divided into two parts by rupture. Hence, the mechanism of the fracture in shearing significantly depends on the ductility of the material. Johnson and Slater (1967) performed a review on the slow and fast blanking of metals at ambient and high temperatures. Several fundamental researches on the rupture in shearing are summarized in the following.

Chang and Swift (1950) performed the experiment on the shearing of metal bars using various kinds of metals. First, the experiment on the shearing without the clearance between the punch and the die was performed. No crack appeared and smooth sheared surfaces were obtained in lead and tin, whereas cracks appeared and uneven fracture surfaces were obtained in aluminum, copper, brass, and mild steel. Fig. 1.21 shows the square grids showing the progression of the fracture in the shearing of mild steel (Chang and Swift, 1950). The square grids were scribed on the faces of a bar before shearing, and no clearance between the punch and the die was assumed. The crack that generated at the punch corner propagated in the material, and the crack that generated at the die corner propagated in the material. However, because these two cracks did not meet, a tongue, which was indicated by diagonal lines in Fig. 1.21D, was formed. Next, the experiment on the shearing in various clearances was performed and an optimum clearance between the punch and the die, which was determined in the consideration of the cleanest surface appeared due to division and the least distortion of the square grids, was obtained in each metal. No clearance between the punch and the die was recommended for lead and tin, which were ductile materials, whereas 5% clearance between the punch and the die was recommended for aluminum, copper, brass, and mild steel, which were harder materials. Fig. 1.22 shows the effect of the clearance on the fracture in the shearing of mild steel (Chang and Swift, 1950). Although the cleanest surface was obtained at 20% clearance, the surface at 10% clearance was much superior geometrically to the surface at 20% clearance, and the optimum clearance between the punch and the die for mild steel was determined exclusively in the consideration of the geometry of the surface.

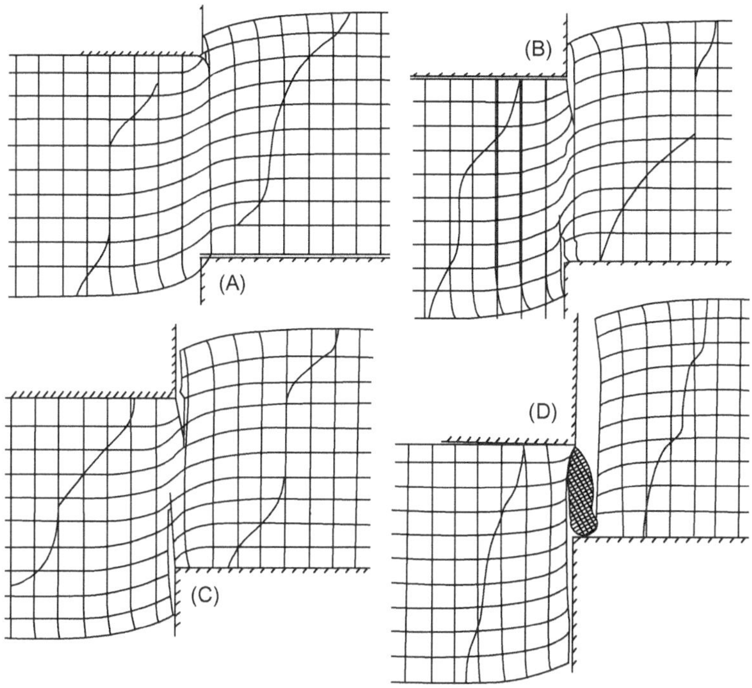

Figure 1.21 Square grids showing progression of fracture in shearing of mild steel. (A) At maximum load (B) 25% penetration, (C) 32% penetration, (D) 60% penetration. *From Chang, T.M., Swift, H.W., 1950. Shearing of metal bars. J. Inst. Met. 78, 119–146.*

Chang (1951) performed the experiment on the shearing of metal blanks using various kinds of metals. The basic modes of the fracture in shearing of metal blanks using circular tools were essentially the same as the basic modes of the fracture in shearing of metal bars using rectangular tools (Chang and Swift, 1950). The following optimum clearances between the punch and the die were recommended in each metal in the consideration of the cleanest surface appeared due to division and the least shearing work: cast iron, 5%–10%; mild steel, 5%–10%; brass, 0%–10%; copper, 0%–10%; zinc, 0%–5%; aluminum, 0%–5%; lead, 0%.

1.11 Fracture and rupture in hole flanging and piercing

The flanging of sheet metal is similar to the bending of sheet metal. The bent part of sheet metal to the whole sheet metal is relatively large, whereas the flanged part of sheet metal to the whole sheet metal is

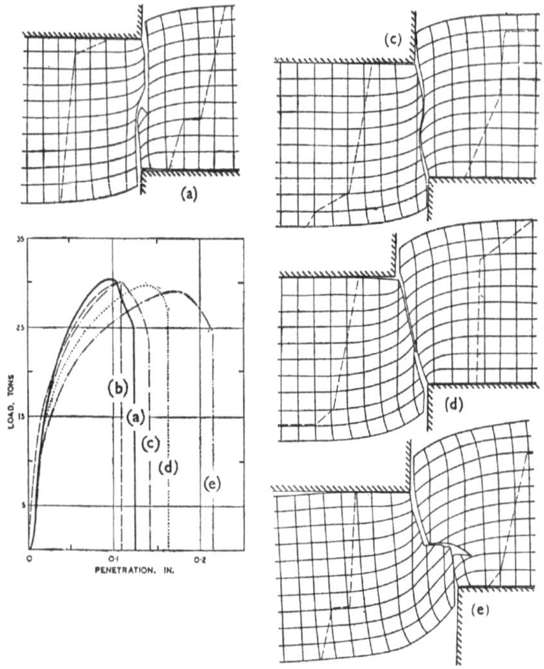

Figure 1.22 Effect of clearance on fracture in shearing of mild steel. (A) Nil (B) 5% (C) 10% (D) 20% (E) 30%. *From Chang, T.M., Swift, H.W., 1950. Shearing of metal bars. J. Inst. Met. 78, 119–146.*

relatively small. Hole flanging is a flanging in which the part of the sheet metal that surrounds a hole drilled in the sheet metal is flanged. The piercing of sheet metal is similar to the blanking of sheet metal. In blanking the part of the sheet metal removed by the punch is utilized and the remaining part of the sheet metal is scrapped, whereas in piercing the part of the sheet metal removed by the punch is scrapped and the remaining part of the sheet metal is utilized. Johnson et al. (1980) performed a review on the piercing and the hole flanging of sheet metals. There are three types of deformation modes in hole flanging and piercing (Chitkara and Johnson, 1974): lip formation, petal formation, and plug formation. Lip, which is formed when the part of the sheet metal that surrounds a hole is flanged, yields when the sheet metal does not fracture, whereas petal, which is formed when the part of the sheet metal that surrounds a hole is flanged, yields when the sheet metal fractures. Plug, which is formed when the sheet metal is pierced, yields when a part of the sheet metal is removed by the punch. Several researches on fracture and rupture in hole flanging and piercing are summarized in the following.

Richards (1955a,b,c,d,e) performed a series of studies on the production of the component of the sheet metal which had a flanged part. First, Richards (1955a) obtained theoretically the relationship between the height of the flanged part, the punch diameter, and the initial hole diameter, on the assumption that the thickness of the flanged part coincided with the initial thickness of the sheet metal. Next, Richards (1955b) performed the experiment on hole flanging using a mild steel. When the punch diameter was specified, with decreasing the initial hole diameter, the height of the flanged part increased. The minimum of the initial hole diameter at which the flanged part did not fracture and lip was formed, was obtained for various punch diameters. With increasing the ratio of the initial thickness of the sheet metal to the punch diameter, the ratio of the minimum of the initial hole diameter at which the flanged part did not fracture to the punch diameter decreased. Moreover, Richards (1955c) performed the experiment on hole flanging using a tinplate and an aluminum. The following result, which was confirmed in Richards (1955b), was reconfirmed. With increasing the ratio of the initial thickness of the sheet metal to the punch diameter, the ratio of the minimum of the initial hole diameter at which the flanged part did not fracture to the punch diameter decreased. Finally, Richards (1955d,e) performed the experiment on hole flanging for various die diameters, punch diameters, initial hole diameters, and initial thicknesses of sheet metals using a mild steel and a tinplate. The relationships between the initial hole diameter and the height of the flanged part were obtained schematically for various die diameters and initial thicknesses of sheet metals.

Johnson et al. (1973) performed the hole flanging and the piercing of circular sheet metals using conically headed cylindrical punches for various sheet metals. Fig. 1.23 shows the relationship between the punch penetration and the punch load for various initial hole diameters. With increasing the initial hole diameter, the punch penetration at which the sheet metal fractured and petal was formed, increased. With increasing the initial hole diameter further, petal came not to be formed and lip came to be formed, in other words, the sheet metal came not to fracture. Petal was formed, when the punch the cone angle of which was relatively small was used and when the initial hole diameter of the sheet metal was zero or relatively small. Plug was formed, when the punch the cone angle of which

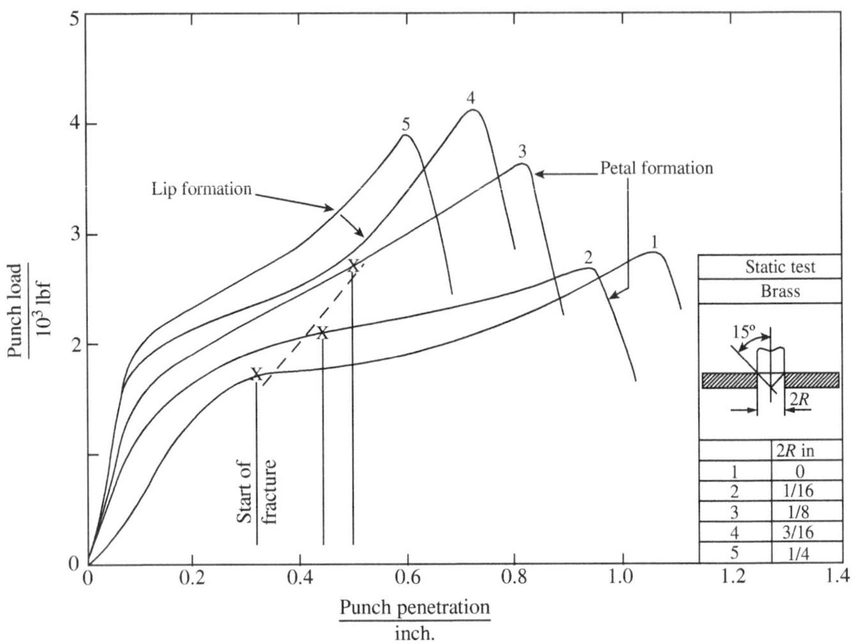

Figure 1.23 Relationship between punch penetration and punch load for various initial hole diameters. *From Johnson, W., Chitkara, N.R., Ibrahim, A.H., Dasgupta, A.K., 1973. Hole flanging and punching of circular plates with conically headed cylindrical punches. J. Strain Anal. 8(3), 228–241.*

was relatively large was used and when the initial hole diameter of the sheet metal was zero or relatively small.

1.12 Fracture in tensile test

Tensile test is a fundamental material test to obtain not only the relationship between the tensile stress and the tensile strain of the material but also the ductility of the material. Furthermore, tensile test is useful to obtain the connection between the formation of shear bands and the material fracture. Hence, tensile test is indispensable to predict ductile fracture in metal-forming processes. Several researches on fracture in tensile test are summarized in the following.

Bridgman (1952) showed experimentally in his famous book that the true strain at fracture in the tensile direction in the tensile test increased, with increasing the hydrostatic pressure in which the tensile test was performed. Moreover, Bridgman (1952) derived analytically the stress distribution at the neck of the specimen in the tensile test, which has been utilized as an approximate stress distribution at the neck of the specimen in the tensile test.

Rosenfield and Hahn (1966) performed the tensile test of a bar using plane carbon steels in various ambient low temperatures and various strain rates, and obtained an empirical stress–strain relationship, an empirical brittle fracture criterion, and an empirical ductile fracture criterion for various ambient low temperatures and various strain rates. The empirical ductile fracture criterion was derived from the combination of the void coalescence criterion proposed by McClintock (1968a) and the relationship between the void diameter at fracture and the equivalent stress at fracture, which was derived with reference to the criterion of crack growth for brittle fracture due to Irwin (1957). Fig. 1.24 shows the strain

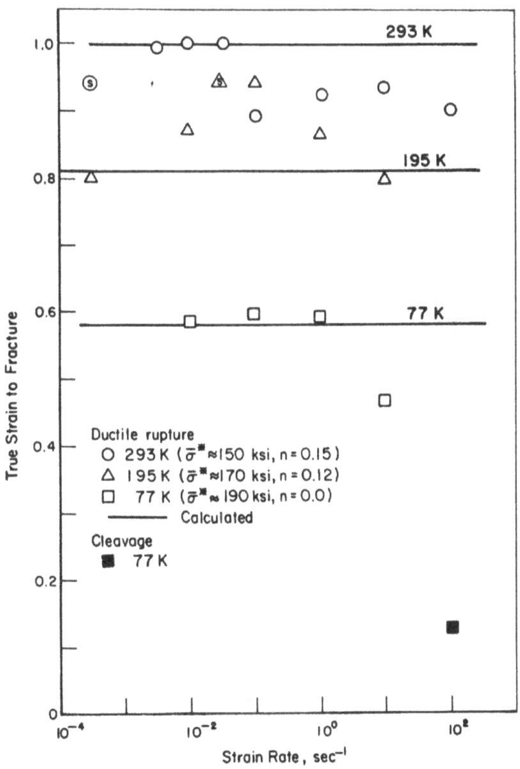

Figure 1.24 Strain to fracture in various ambient low temperatures and various strain rates for X-52 pipe steel. *From Rosenfield, A.R., Hahn, G.T., 1966. Numerical descriptions of the ambient low-temperature, and high-strain rate flow and fracture behavior of plane carbon steel. Trans. ASM 59, 962–980.*

to fracture in various ambient low temperatures and various strain rates for X-52 pipe steel (Rosenfield and Hahn, 1966).

Clausing (1970) performed the axisymmetric tensile test and the plane-strain tensile test using seven structural steels and compared the ductility in the plane-strain tensile test with the ductility in the axisymmetric tensile test in each structural steel. Fig. 1.25 shows the plane-strain tensile specimen (Clausing, 1970). The thick parts on both sides of the thin part were thick enough to cause the elastic deformation in the thick parts and to constrain the plastic deformation in the thin part. The ratio B/L was large enough so that approximate plane-strain deformation occurred in

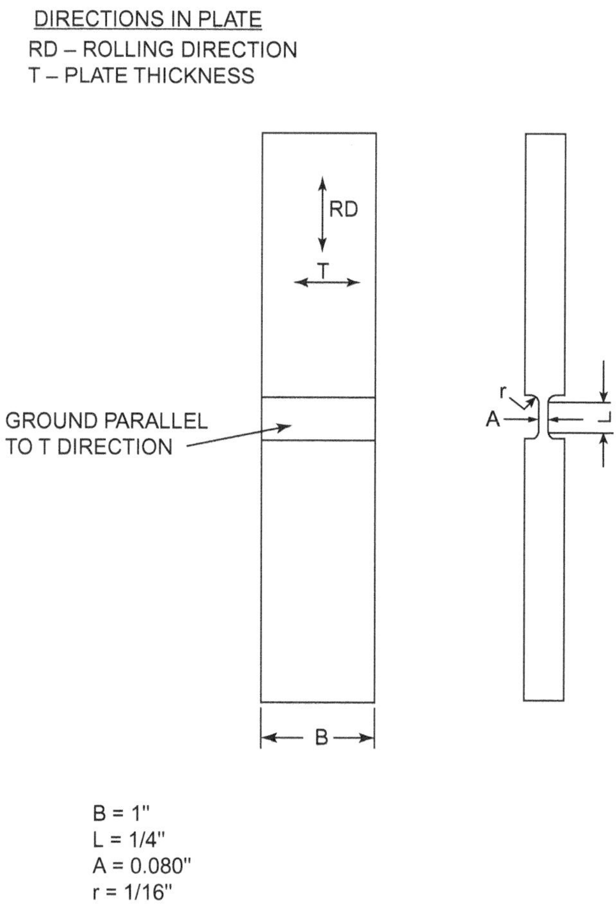

Figure 1.25 Plane-strain tensile specimen. *From Clausing, D.P., 1970. Effect of plastic strain state on ductility and toughness. Int. J. Fract. Mech. 6(1), 71–85.*

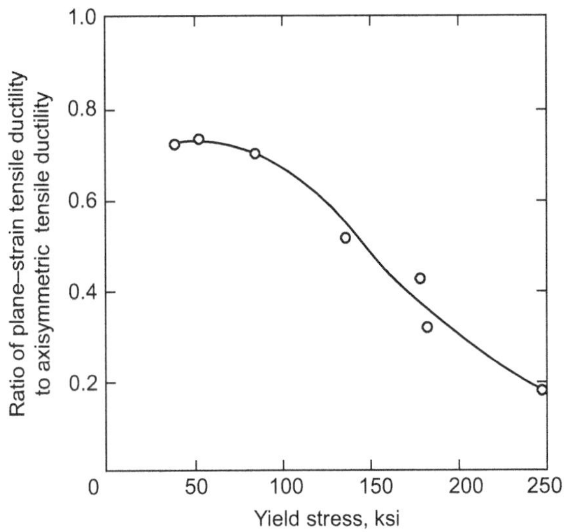

Figure 1.26 Relationship between ratio of plane-strain tensile ductility to axisymmetric tensile ductility and yield stress. *From Clausing, D.P., 1970. Effect of plastic strain state on ductility and toughness. Int. J. Fract. Mech. 6(1), 71–85.*

the thin part, whereas the ratio L/A was large enough so that approximate uniform deformation occurred in the thin part. The true strain at fracture in the tensile direction in the axisymmetric tensile test, which was defined as axisymmetric tensile ductility, and the true strain at fracture in the tensile direction in the plane-strain tensile test, which was defined as plane-strain tensile ductility, were measured, and the ratio of the plane-strain tensile ductility to the axisymmetric tensile ductility was calculated. Fig. 1.26 shows the relationship between the ratio of the plane-strain tensile ductility to the axisymmetric tensile ductility and the yield stress (Clausing, 1970). The ratio of the plane-strain tensile ductility to the axisymmetric tensile ductility was smaller than one in the seven structural steels. Furthermore, with increasing the yield stress, the ratio of the plane-strain tensile ductility to the axisymmetric tensile ductility decreased monotonically.

Mackenzie et al. (1977) performed the tensile test of four high strength steels using notched round tensile specimens and obtained the effect of the triaxiality of stress state on the equivalent plastic strain required to initiate ductile fracture. Fig. 1.27 shows the relationship between the equivalent plastic strain at fracture initiation and the triaxiality of the stress state for HY130 (Mackenzie et al., 1977). The equivalent plastic strain at

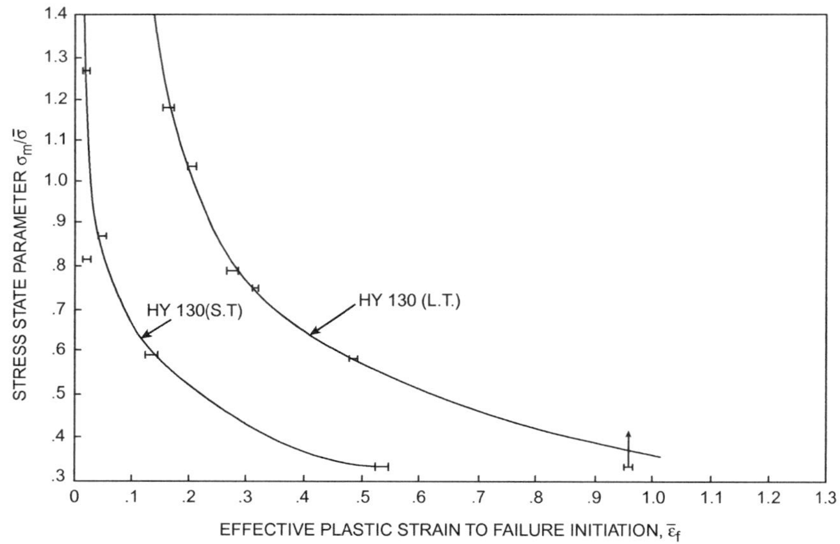

Figure 1.27 Relationship between equivalent plastic strain at fracture initiation and triaxiality of stress state for HY130. *From Mackenzie, A.C., Hancock, J.W., Brown, D.K., 1977. On the influence of state of stress on ductile failure initiation in high strength steels. Eng. Fract. Mech. 9(1), 167–188.*

fracture initiation was calculated from $2\ln(d_0/d)$, where d_0 was the initial diameter at the notch root and d was the diameter at fracture initiation at the notch root. The triaxiality of the stress state was defined as $\sigma_m/\overline{\sigma}$, where σ_m was the mean normal stress and $\overline{\sigma}$ was the equivalent stress. The mean normal stress σ_m was calculated from the approximate stress distribution at the neck of the specimen in the tensile test (Bridgman, 1952) using the initial diameter at the notch root and the initial radius of the notch root. The equivalent stress $\overline{\sigma}$ was calculated using the equivalent plastic strain at fracture initiation $2\ln(d_0/d)$. Experimental points were indicated by the capital letter H. With increasing the triaxiality of the stress state, the equivalent plastic strain at fracture initiation deceased. Specimens were machined from a thick plate. When the tensile direction of the specimen was parallel to the width direction of the thick plate, the specimen was called LT specimen, whereas when the tensile direction of the specimen was parallel to the thickness direction of the thick plate, the specimen was called ST specimen. When the triaxiality of the stress state was specified, the equivalent plastic strain at fracture initiation obtained using the LT specimen was larger than the equivalent plastic strain at fracture initiation obtained using the ST specimen. Fig. 1.28 shows the relationship between the equivalent plastic strain at fracture initiation and the triaxiality of the stress state for all the materials (Mackenzie et al., 1977). Experimental points were omitted for clarity. Q1 was equivalent to HY80 in strength, and Marrel was equivalent to HY110 in strength, whereas ESR was equivalent to HY130 in specification. HY130, Q1, Marrel, and ESR were high strength steels. When the triaxiality of the stress state was specified, the equivalent plastic strains at fracture initiation obtained using the LT specimen were larger than the equivalent plastic strains at fracture initiation obtained using the ST specimen for HY130, Q1, and ESR. EN8 was a medium strength steel, Swedish iron was a low strength iron, whereas L64 and L65 were high strength aluminum–copper alloys. The relationships between the equivalent plastic strain at fracture initiation and the triaxiality of the stress state for EN8, Swedish iron, L64, and L65 were appended using broken lines in Fig. 1.28. With increasing the triaxiality of the stress state, the equivalent plastic strain at fracture initiation deceased for all the materials. The effect of the triaxiality of the stress state on the equivalent plastic strain at fracture initiation for the high strength steels, was larger than the effect of the triaxiality of the stress state on the equivalent plastic strain at fracture initiation for the other materials.

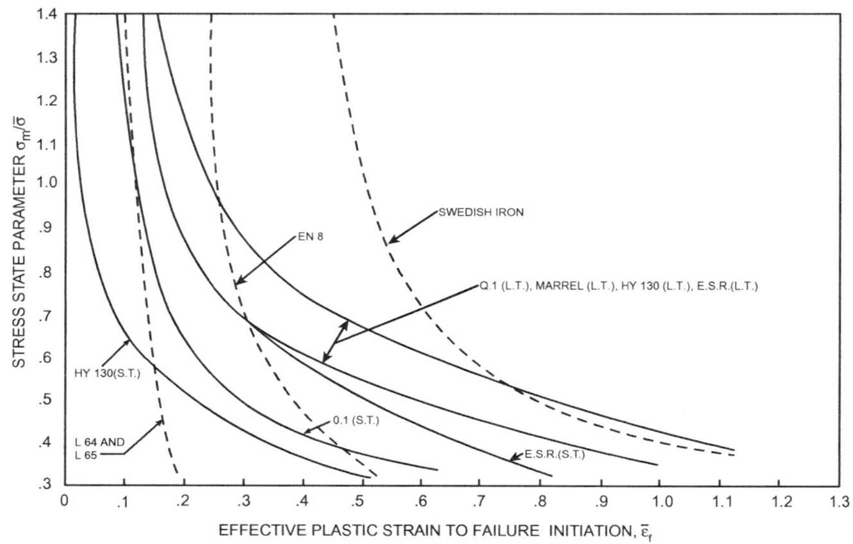

Figure 1.28 Relationship between equivalent plastic strain at fracture initiation and triaxiality of stress state for all materials. *From Mackenzie, A.C., Hancock, J.W., Brown, D.K., 1977. On the influence of state of stress on ductile failure initiation in high strength steels. Eng. Fract. Mech. 9(1), 167–188.*

Johnson and Cook (1985) performed the torsion test, the tensile Hopkinson bar test, and the quasistatic tensile test, using oxygen-free high thermal conductivity copper, Armco iron, and 4340 steel, which was medium carbon and low alloy steel, and obtained the general expression for the strain at fracture. The strain at fracture was assumed to depend on strain rate, temperature, and stress triaxiality, and was assumed to be expressed by the product of three sets of blankets using five material constants. The expression in the first set of blankets, which represented the effect of stress triaxiality, was identical to the expression proposed by Hancock and Mackenzie (1976). The expression in the second set of blankets represented the effect of strain rate, whereas the expression of the third set of blankets represented the effect of temperature. Fig. 1.29 shows the relationship between the stress triaxiality and the equivalent plastic strain at fracture (Johnson and Cook, 1985). The stress triaxiality in the quasistatic tensile test was calculated using a commercial elastic–plastic

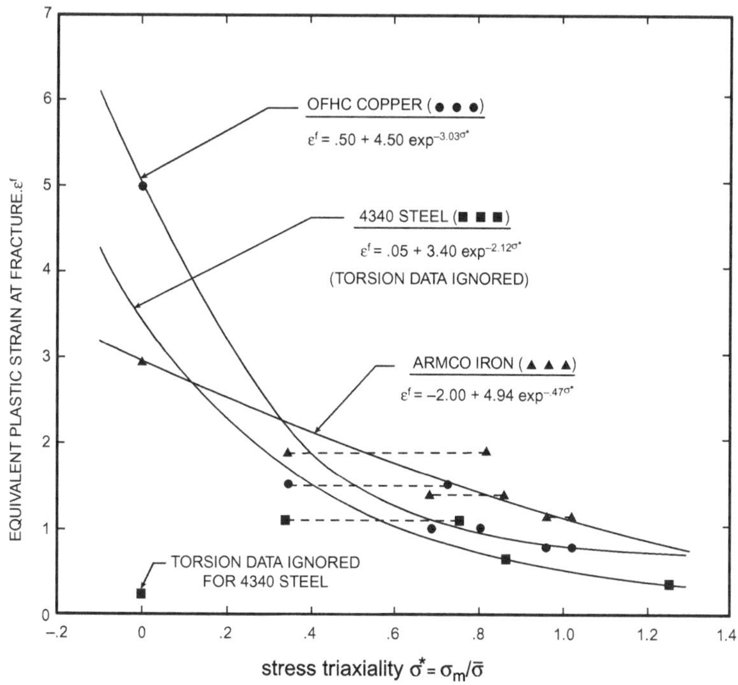

Figure 1.29 Relationship between stress triaxiality and equivalent plastic strain at fracture. *From Johnson, G.R., Cook, W.H., 1985. Fracture characteristics of three metals subjected to various strains, strain rates, temperatures and pressures. Eng. Fract. Mech. 21(1), 31–48.*

Table 1.3 Strain at which shear bands are first observed and angle between shear bands and tensile axis or compression axis.

	Observed	Predicted Deformation theory	Flow theory
Tension			
$\varepsilon_{1,\mathrm{crit}}$	+ 0.034	+ 0.084	+ 0.184
θ	± (38 ± 2) degrees	± 42.55 degrees	± 45 degrees
Compression			
$\varepsilon_{1,\mathrm{crit}}$	− 0.034	− 0.085	− 0.184
θ	± (55 ± 2) degrees	± 47.45 degrees	± 45 degrees

From Anand, L., Spitzig, W.A., 1980. Initiation of localized shear bands in plane strain. J. Mech. Phys. Solids 28(2), 113–128.

finite-element software, whereas the stress triaxiality in the torsion test was assumed to be zero. The strain to fracture was more dependent on the stress triaxiality, whereas the strain to fracture was less dependent on the strain rate and the temperature.

Anand and Spitzig (1980) performed the plane-strain tensile test and the plane-strain compression test using a maraging steel, observed shear bands using an optical microscope, and compared the results obtained experimentally with the results obtained analytically with reference to Hill and Hutchinson (1975) using the Prandtl–Reuss constitutive equation and the Hencky constitutive equation. Table 1.3 shows the strain at which the shear bands are first observed and the angle between the shear bands and the tensile axis or the compression axis (Anand and Spitzig, 1980). In the plane-strain tensile test, the strain at which the shear bands were first observed was slightly larger than the strain at which diffuse necking commenced, whereas the strain at which the shear bands were first observed was much smaller than the strain at which the material fractured.

1.13 Fracture in shear test

Shear test is a fundamental material test to obtain the relationship between the shear stress and the shear strain of the material and also the ductility of the material. Hence, shear test is indispensable to predict ductile fracture in metal-forming processes. Several researches on fracture in shear test are summarized in the following.

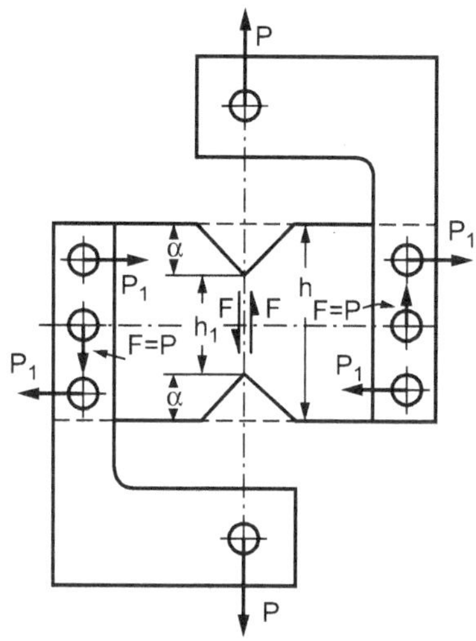

Figure 1.30 Iosipescu simple shear method. *From Iosipescu, N., 1967. New accurate procedure for single shear testing of metals. J. Mater. 2(3), 537–566.*

Iosipescu (1967) proposed an accurate method for simple shear testing of materials. Fig. 1.30 shows the Iosipescu simple shear method (Iosipescu, 1967). The specimen was combined with the two surrounding devices. The test section at which fracture was evaluated, was the section at the center in the horizontal direction of the specimen. The two surrounding devices were designed so that, in the test section, only shear stresses were produced and no normal stresses were produced; no bending moment was generated. Furthermore, the specimen that had V-shaped notches was designed to fracture the specimen in the test section at which no bending moment was generated, and to obtain the uniform distribution of the shear stresses in the test section, which was experimentally confirmed by means of a qualitative photoelastic study. The Iosipescu simple shear method was standardized in ASTM and used for the evaluation of the fracture of composite materials.

Miyauchi (1984) proposed a planar simple shear test in sheet materials, and the relationship between the shear stress and the shear strain of the material was obtained using the planar simple shear test. Fig. 1.31 shows the conventional planar simple shear test (Miyauchi, 1984). Although the direction of shear deformation coincided with the tensile direction initially, the direction of shear deformation did not coincide with the tensile direction during the shear test because the direction of shear deformation changed; simple shear deformation occurred initially, simple shear

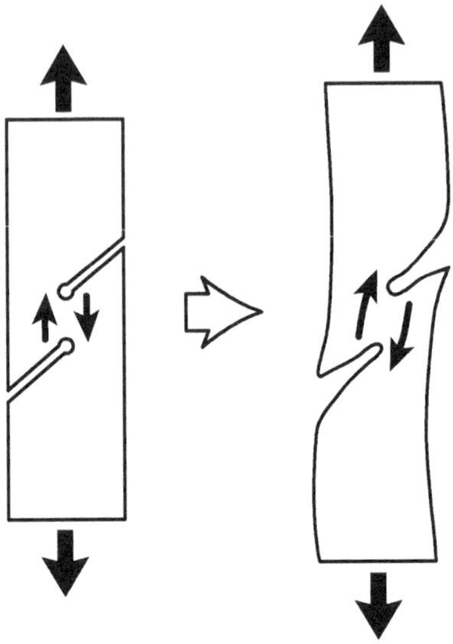

Figure 1.31 Conventional planar simple shear test. *From Miyauchi, K., 1984. A proposal of a planar simple shear test in sheet metals. Sci. Pap. Inst. Phys. Chem. Res. 78(3), 27–40.*

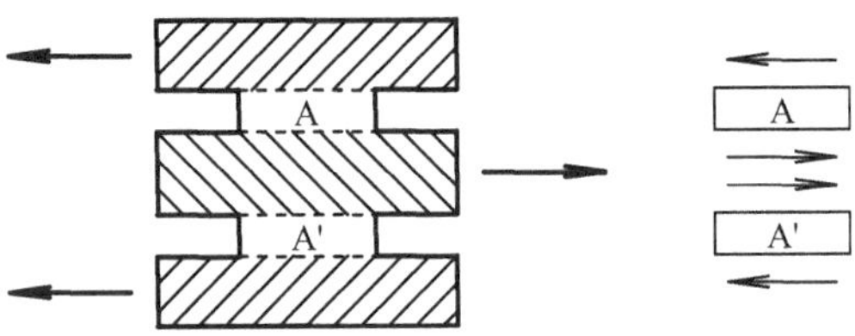

Figure 1.32 Miyauchi planar simple shear test. *From Miyauchi, K., 1984. A proposal of a planar simple shear test in sheet metals. Sci. Pap. Inst. Phys. Chem. Res. 78(3), 27–40.*

deformation did not occur during the shear test. Hence, the relationship between the shear stress and the shear strain of the material was not able to be obtained. Fig. 1.32 shows the Miyauchi planar simple shear test (Miyauchi, 1984). The parts in which diagonal lines were drawn moved in the horizontal direction during the shear test. Hence, the direction of shear deformation coincided with the tensile direction during the shear test; simple shear deformation occurred during the shear test. Therefore, the relationship between the shear stress and the shear strain of the material was able to be obtained, and the relationships between the shear stress and the shear strain for Ti-killed mild steel, rephosphorized high strength steel, and dual phase high strength steel were obtained. The Miyauchi planar simple shear test, which was proposed to obtain the relationship between the shear stress and the shear strain of the material, could be applied to obtain the ductility of the material.

1.14 Fracture in other material tests

Several material tests other than tensile test and shear test have been designed to obtain the ductility of the material and not the relationship between the stress and the strain of the material. These material tests are indispensable to clarify the mechanism of ductile fracture in metal-forming processes. Several researches on fracture in other material tests are summarized in the following.

Bao and Wierzbicki (2004) proposed several material tests and obtained the relationship between the equivalent plastic strain to fracture and the stress triaxiality using 2024-T351 aluminum alloy. First, in the conventional upsetting of a cylinder (Kudo and Aoi, 1967), the friction between the cylinder and the die was indispensable to fracture the surface of the cylinder. However, in general, the phenomenon of the friction between the specimen and the tool was not clarified. Hence, to remove the friction between the specimen and the tool, a specimen, which was similar to a circumferentially notched tensile specimen (Mackenzie et al., 1977), was proposed and uniaxially compressed to fracture the surface of the notch root of the specimen. Next, a specimen for the combined test of the tensile test and the shear test was proposed by changing the shape of the test section of the specimen for the shear test, which was similar to the Iosipescu simple shear test specimen combined with the two surrounding devices (Iosipescu, 1967). Fig. 1.33 shows the specimen for the combined test of the tensile test and the shear test (Bao and Wierzbicki, 2004). An average stress triaxiality, which was defined as the average of the stress triaxiality during deformation, was calculated using a commercial finite-element software. The average stress triaxiality was approximately zero when the specimen for the shear test was employed, whereas the average stress triaxiality was approximately 0.1 when the specimen for the combined test of the tensile test and the shear test was employed. Fig. 1.34 shows the relationship between the average stress triaxiality and the equivalent plastic strain to fracture (Bao and Wierzbicki, 2004). Shear fracture occurred when the average stress triaxiality was negative, whereas fracture occurred due to void formation when the average stress triaxiality was larger than 0.4.

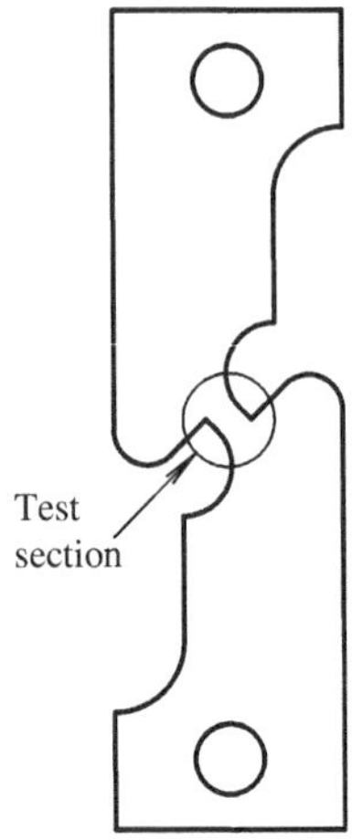

Figure 1.33 Specimen for combined test of tensile test and shear test. *From Bao, Y., Wierzbicki, T., 2004. On fracture locus in the equivalent strain and stress triaxiality space. Int. J. Mech. Sci. 46(1), 81–98.*

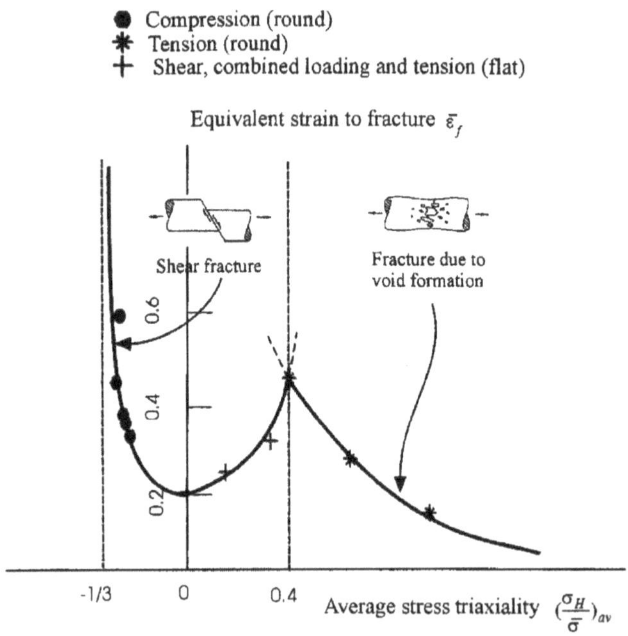

Figure 1.34 Relationship between average stress triaxiality and equivalent strain to fracture. *From Bao, Y., Wierzbicki, T., 2004. On fracture locus in the equivalent strain and stress triaxiality space. Int. J. Mech. Sci. 46(1), 81–98.*

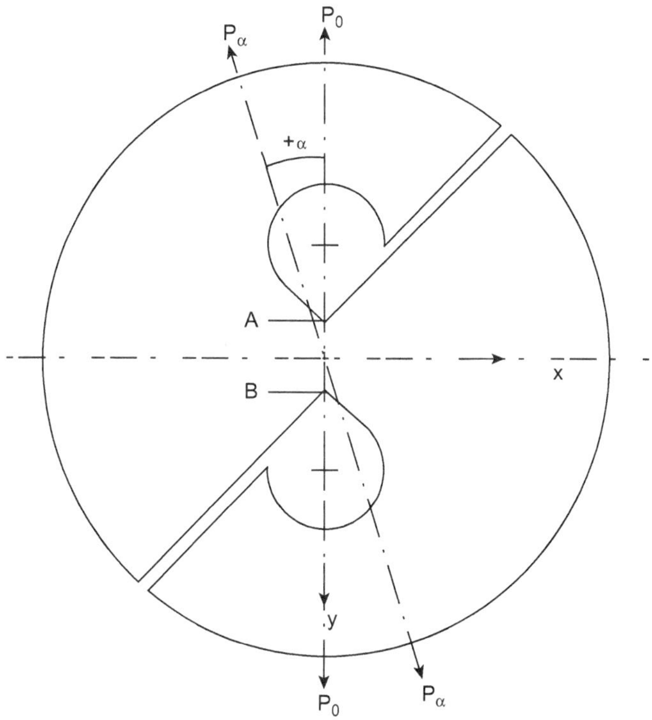

Figure 1.35 Arcan specimen. *From Arcan, M., Hashin, Z., Voloshin, A., 1978. A method to produce uniform plane-stress states with applications of fiber-reinforced materials. Exp. Mech. 18(4), 141–146.*

Arcan et al. (1978) proposed a specially designed plane specimen to produce a uniform plane-stress state and obtained the relationship between the shear stress and the shear strain of fiber-reinforced materials. Fig. 1.35 shows the Arcan specimen (Arcan et al., 1978). The specimen was plane circular with two cutouts. The angle between the vertical direction and the length direction of the rectilinear parts of the two cutouts was 45 degrees. The test section of the specimen was AB, and the stress state on AB was proved to be approximately uniform by a photoelastic method and a strain-gage method. When the loading direction was identical to the vertical direction, a pure shear state was obtained. To obtain a uniform plane-stress state on AB, the magnitude of the angle between the vertical direction and the loading direction was restricted to be smaller than 45 degrees; the maximum principal stress and the minimum principal stress were restricted to be different in codes.

Mohr and Henn (2007) proposed a newly designed flat specimen to investigate the onset of fracture in metals at a low stress triaxiality and an

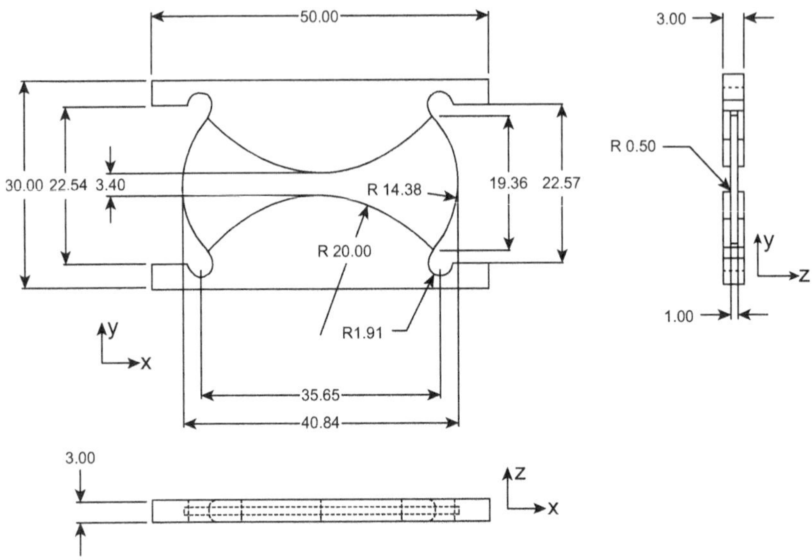

Figure 1.36 Newly designed flat specimen. *Mohr, D., Henn, S., 2007. Calibration of stress-triaxiality dependent crack formation criteria: a new hybrid experimental–numerical method. Exp. Mech. 47(6), 805–820.*

intermediate stress triaxiality, and obtained the relationship between the stress triaxiality and the maximum principal strain, which was calculated using a commercial finite-element software. Fig. 1.36 shows the newly designed flat specimen (Mohr and Henn, 2007). The test section of the specimen was the center of the specimen. The thickness of the two fan-shaped parts of the specimen was made to be thinner than the thickness of the other parts of the specimen so that a crack initiated in the test section of the specimen and did not initiate at the boundary of the specimen. The loading direction was made to be arbitrary, with reference to the Arcan specimen (Arcan et al., 1978), although the loading direction was restricted in the Arcan specimen. When the loading direction was identical to the horizontal direction, a pure shear state was obtained, in which the stress triaxiality was equal to zero. When the loading direction was identical to the vertical direction, a state of transversely constrained plane-stress compression was obtained, in which the stress triaxiality was equal to $-1/\sqrt{3}$, or a state of transversely constrained plane-stress tension was obtained, in which the stress triaxiality was equal to $+1/\sqrt{3}$. Hence, the stress triaxiality was able to be changed from $-1/\sqrt{3}$ to $+1/\sqrt{3}$.

CHAPTER 2

Macroscopic ductile fracture criteria

2.1 Introduction

When ductile fracture occurs in a material after plastic deformation, void nucleation, void growth, and void coalescence occur microscopically in the material (Dodd and Bai, 1987); the material deteriorates during plastic deformation due to voids. Hence, the deterioration of the material should be considered in the simulation of the plastic deformation of the material, in other words, the deterioration of the material should be coupled with the plastic deformation of the material. However, to perform the coupled simulation of the plastic deformation of the material and the deterioration of the material is not necessarily practical (Fischer et al., 1995). Hence, the uncoupled simulation of the plastic deformation of the material and the deterioration of the material is frequently performed, in other words, the simulation of the plastic deformation of the material in which the deterioration of the material is not considered, is repeatedly performed. In summary, there are two types of simulations to predict macroscopic ductile fracture: the uncoupled simulation of the plastic deformation of the material and the deterioration of the material and the coupled simulation of the plastic deformation of the material and the deterioration of the material.

There are two types of macroscopic ductile fracture criteria that are used in the uncoupled simulation of the plastic deformation of the material and the deterioration of the material. One type of macroscopic ductile fracture criterion is defined using the definite integral of an integrand, which is a function of stress such as the equivalent stress, the maximum principal stress, and the mean normal stress, through the equivalent plastic strain along the strain history. Because both stress and strain are not microscopic but macroscopic physical quantities, the macroscopic ductile fracture criterion defined using both stress and strain does not necessarily have a definite physical meaning microscopically. The Freudenthal fracture criterion (1950), the Cockcroft and Latham fracture criterion (1968),

DOI: https://doi.org/10.1016/B978-0-12-814772-6.00002-3

the Brozzo et al. fracture criterion (1972), and the Oyane fracture criterion (1972), belong to this type of macroscopic ductile fracture criterion.

In the other type of macroscopic ductile fracture criterion, the following two deformation modes are assumed: sound deformation mode and void nucleating deformation mode. When the energy required for the sound deformation mode is smaller than the energy required for the void nucleating deformation mode, sound deformation is assumed to occur. However, when the energy required for the sound deformation mode is larger than the energy required for the void nucleating deformation mode, void nucleating deformation is assumed to occur. In other words, a void is assumed to nucleate in the material and macroscopic ductile fracture is assumed to occur. The idea that when the energy required for a deformation mode is a minimum the deformation mode actually occurs, is based on the upper bound theorem (Prager and Hodge, 1951; Avitzur, 1968a). This type of macroscopic ductile fracture criterion is principally used by Avitzur (1980). In this type of macroscopic ductile fracture criterion, in general, a rigid, perfectly plastic material is assumed; no elastic deformation of the material is assumed and no strain hardening of the material is assumed. Moreover, in this type of macroscopic ductile fracture criterion, in general, ductile fracture criterion depends on forming conditions and does not depend on the ductility of the material. In the upper bound method, since the velocity field which represents the flow of the material is assumed, flow modes (not deformation modes) are assumed.

There are two types of macroscopic ductile fracture criteria which are used in the coupled simulation of the plastic deformation of the material and the deterioration of the material. In one type of macroscopic ductile fracture criterion, the decrease in the material density due to void nucleation and void growth is considered; the increase in the void volume fraction of the material due to void nucleation and void growth is considered. In this type of macroscopic ductile fracture criterion, the Gurson yield function (1977) is principally used. Because a macroscopic ductile fracture criterion is not included in the Gurson yield function, a macroscopic ductile fracture criterion should be defined additionally. For instance, the material is assumed to fracture when the void volume fraction of the material becomes larger than a certain void volume fraction. Since the material density, which is related to the void volume fraction of the material, is able to be measured macroscopically by the Archimedes method, the material density calculated from the simulation is able to be compared with the material density obtained in the experiment.

In the other type of macroscopic ductile fracture criterion, the decrease in the flow stress of the material due to the deterioration of the material is considered. In this type of macroscopic ductile fracture criterion, the Lemaitre damage model (1985) is principally used. Because the decrease in the flow stress of the material occurs due to the damage of the material, the Lemaitre damage model is one of the damage models in damage mechanics. The material is assumed to be undamaged when the damage variable D is equal to zero, whereas the material is assumed to be ruptured when the damage variable D is equal to one. Saanouni (2012) describes the damage mechanics in metal forming. Because a macroscopic ductile fracture criterion is included in the Lemaitre damage model, it is not necessary to define a macroscopic ductile fracture criterion additionally. Although the damage variable D is able to be measured microscopically by the observation of micrographic pictures (Lemaitre and Dufailly, 1987), it is not necessarily easy to compare the damage variable calculated from the simulation with the damage variable obtained in the experiment.

In the following sections, classical papers on macroscopic ductile fracture criteria are introduced for each of the metal-forming processes, that is, rolling, forging, and sheet forming, among others.

2.2 Alligatoring and central burst in strip rolling

Zhu and Avitzur (1988) performed the analysis of strip rolling by the upper bound method and obtained the criterion for the prevention of split ends, that is, alligatoring. The following two flow modes were assumed: sound flow mode and split end flow mode. Fig. 2.1 shows the velocity field for the sound flow mode (Zhu and Avitzur, 1988). Γ_1 and Γ_2 were velocity discontinuity lines and a single rotational triangle ΔABK, the angular velocity of which was equal to the angular velocity of the roll, was assumed. The x-coordinate of the point K was optimized by minimizing the energy dissipated on the velocity discontinuity lines. If the point K was made to coincide with the point E and the velocity discontinuity line Γ_1 was removed, the velocity field for the split end flow mode was obtained. Fig. 2.2 shows the velocity field for the split end flow mode (Zhu and Avitzur, 1988). Because the angular velocity of the strip on the line BE was equal to the angular velocity of the roll, split ends were formed. When the energy required for the sound flow mode was larger than the energy required for the split end flow mode, split ends were assumed to occur. An approximate inequality, which represented

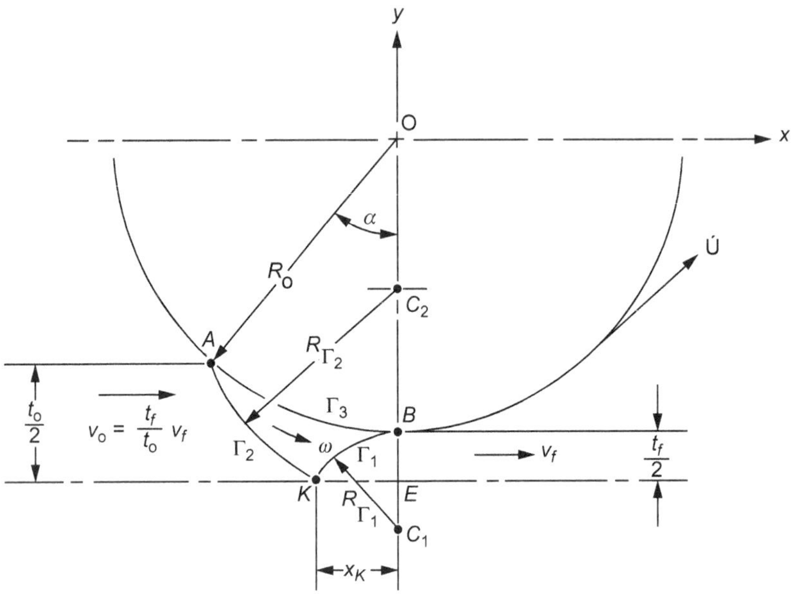

Figure 2.1 Velocity field for sound flow mode. *From Zhu, Y.D., Avitzur, B., 1988. Criteria for the prevention of split ends. Trans. ASME J. Eng. Ind. 110(2), 162–172.*

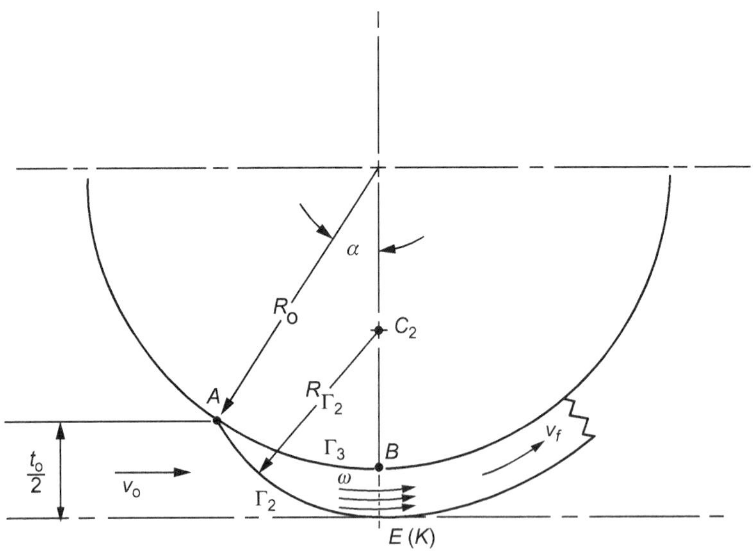

Figure 2.2 Velocity field for split end flow mode. *From Zhu, Y.D., Avitzur, B., 1988. Criteria for the prevention of split ends. Trans. ASME J. Eng. Ind. 110(2), 162–172.*

the criterion of the occurrence of the strip end, was obtained. The approximate inequality indicated that split ends had a tendency to occur with increasing the final thickness of the strip and with decreasing the roll radius, when the initial thickness of the strip was specified.

Avitzur et al. (1988) performed the analysis of strip rolling by the upper bound method and obtained the criterion for the prevention of central bursts during rolling. The following two flow modes were assumed: sound flow mode and central burst flow mode. Fig. 2.3 shows the velocity field for the sound flow mode and the central burst flow mode (Avitzur et al., 1988). The rigid body region, which was indicated by Zone II, was bounded by three velocity discontinuity lines Γ_1, Γ_2, and Γ_3. The location of the apex of the rigid body region, the coordinates of which were $(-X, \epsilon)$, was optimized by minimizing the energy dissipated on the velocity discontinuity lines and due to the front tension and the back tension. When ϵ was equal to zero the sound flow mode was obtained, whereas when ϵ was not equal to zero the central burst flow

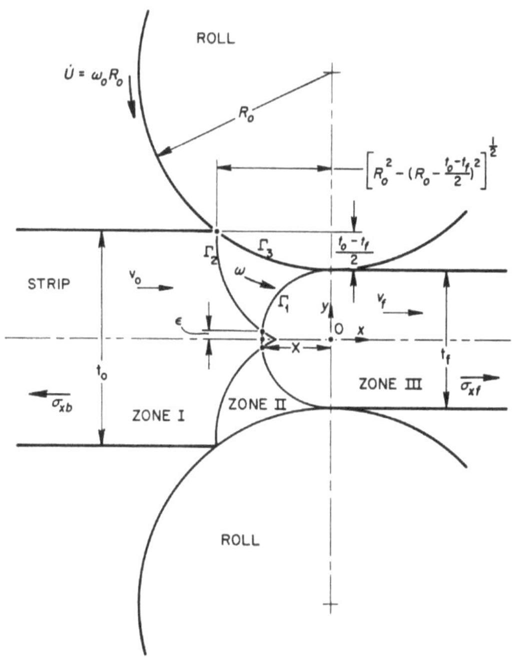

Figure 2.3 Velocity field for sound flow mode and central burst flow mode. *From Avitzur, B., Van Tyne, C.J., Turczyn, S., 1988. The prevention of central bursts during rolling. Trans. ASME J. Eng. Ind. 110(2), 173–178.*

mode was obtained, because the other two rigid body regions, that is, the region I and the region III, contacted with each other when ϵ was not equal to zero. When the initial thickness of the strip was specified, central bursts had a tendency to occur with increasing the final thickness of the strip, with decreasing the roll radius, with increasing the front tension, and with increasing the back tension.

2.3 Central burst in wire drawing and bar extrusion

Avitzur (1968b) performed the analysis of central bursting defects in wire drawing and bar extrusion by the upper bound method and obtained the criterion for the prevention of central bursts during wire drawing and bar extrusion. The following two flow modes were assumed: flow mode with central bursts and flow mode without central bursts. Fig. 2.4 shows the velocity field for the flow mode with central bursts (Avitzur, 1968b). Zone I and Zone III were rigid bodies. Zone II was bounded by the conical surface of the die Γ_3, a cylindrical surface Γ_4, and two toroidal surfaces Γ_1 and Γ_2. Zone I and Zone III were adjacent over the surface Γ_5. The velocity of the rigid body of Zone I was v_o, whereas the velocity of the rigid body of Zone III was v_f. Because the velocity v_f was larger than the velocity v_o due to the incompressibility condition, Zone III was moved away from Zone I along the surface Γ_5 and thus the central burst

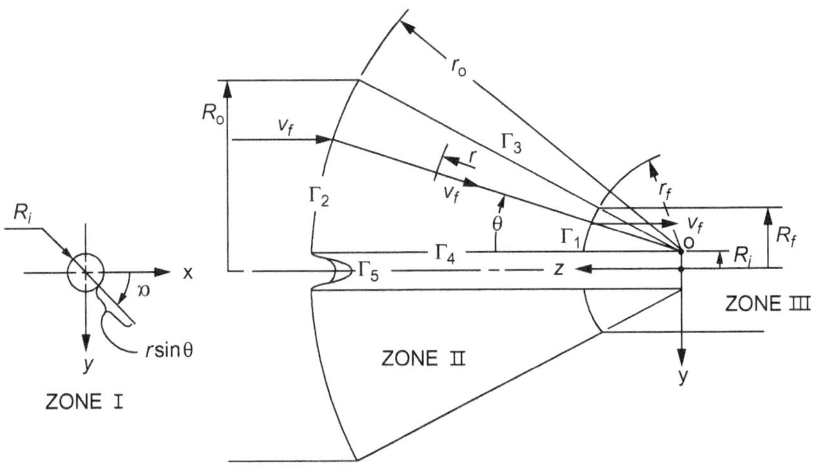

Figure 2.4 Velocity field for flow mode with central bursts. *From Avitzur, B., 1968. Analysis of central bursting defects in extrusion and wire drawing. Trans. ASME J. Eng. Ind. 90(1), 79–91.*

was introduced. The energy was dissipated in the material of Zone II, on the velocity discontinuity lines of Γ_1, Γ_2, Γ_3, and Γ_4, and due to the front tension and the back tension. When the front tension required to perform wire drawing decreased with increasing R_i/R_f from zero, the central burst was assumed to occur, whereas when the front tension required to perform wire drawing increased with increasing R_i/R_f from zero, the central burst was not assumed to occur. Fig. 2.5 shows the criterion for the central burst (Avitzur, 1968b). The central burst occurred when the coordinate, which represented a drawing condition, was below the solids lines. The central burst had a tendency to occur with increasing the die angle, with decreasing the reduction in the area, and with increasing the ratio of the frictional stress to the shear yield stress m. A dead zone, which was also a defect and was obtained in the analysis performed by Avitzur (1965) using the upper bound method, was formed when the coordinate, which represented a drawing condition, was below the dashed lines.

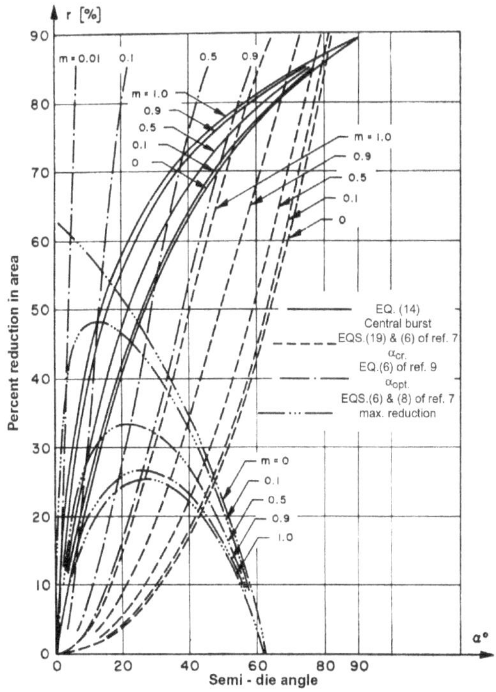

Figure 2.5 Criterion for central burst. *From Avitzur, B., 1968. Analysis of central bursting defects in extrusion and wire drawing. Trans. ASME J. Eng. Ind. 90(1), 79–91.*

Zimerman and Avitzur (1970) performed the analysis of the effect of strain hardening on central bursting defects in drawing and extrusion by the upper bound method and obtained the effect of strain hardening on the criterion for the prevention of central bursts during drawing and extrusion. A rigid-plastic linear strain-hardening material was assumed. The equivalent stress was modified along the flow of the material to calculate the energy dissipated in the material of Zone II, on the velocity discontinuity lines of Γ_1, Γ_2, Γ_3, and Γ_4, and due to the front tension and the back tension. When the front tension required to perform wire drawing decreased with increasing R_i/R_f from zero, the central burst was assumed to occur. Fig. 2.6 shows the criterion for the central burst for strain-hardening materials (Zimerman and Avitzur, 1970). The central burst occurred when the coordinate, which represented a drawing condition, was on the right side of the solids lines. The central burst had a tendency not to occur with increasing the strain-hardening coefficient β. Both the central burst and the dead zone obtained in the analysis

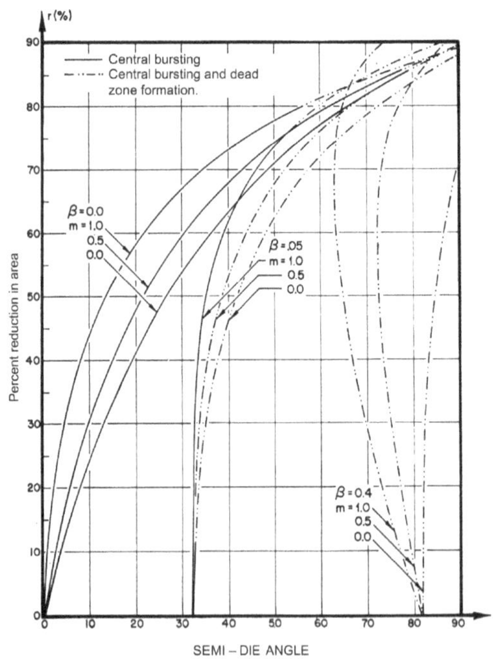

Figure 2.6 Criterion for central burst for strain-hardening materials. *From Zimerman, Z., Avitzur, B., 1970. Analysis of the effect of strain hardening on central bursting defects in drawing and extrusion. Trans. ASME J. Eng. Ind. 92(1), 135–145.*

performed by Avitzur (1965) appeared when the coordinate, which represented a drawing condition, was on the right side of the two-dot chain lines. However, when the dead zone appeared, the central burst did not necessarily occur because the die angle decreased substantially.

Chen et al. (1979) performed the simulation of axisymmetric extrusion and drawing by the rigid-plastic finite-element method and Oh et al. (1979) evaluated central bursts in axisymmetric extrusion and drawing using the following two ductile fracture criteria: the McClintock fracture criterion (1968a) and the Cockcroft and Latham fracture criterion (1968). When the hyperbolic sine function in the McClintock fracture criterion was approximated by a linear function and the Cockcroft and Latham fracture criterion was expressed in nondimensional form, the modified McClintock fracture criterion and the modified Cockcroft and Latham fracture criterion were proved to be virtually the same. Moreover, the value of the material constant in the modified McClintock fracture criterion was shown to be the double of the value of the material constant in the modified Cockcroft and Latham fracture criterion. The fracture strain calculated using the modified McClintock fracture criterion and the fracture strain calculated using the modified Cockcroft and Latham fracture criterion were almost the same in cylinder upsetting and in uniaxial tensile test. Fig. 2.7 shows the fracture strain in uniaxial tensile test for various

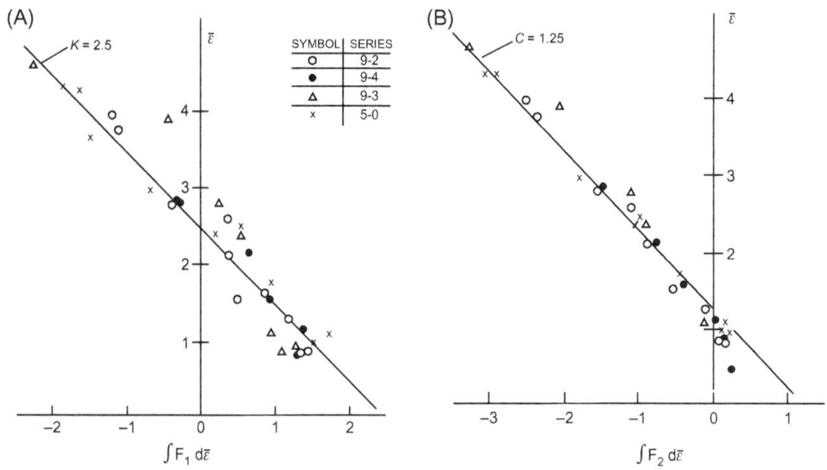

Figure 2.7 Fracture strain in uniaxial tensile test for various mean normal stresses. (A) McClintock. (B) Cockcroft and Latham. *From Oh, S.I., Chen, C.C., Kobayashi, S., 1979. Ductile fracture in axisymmetric extrusion and drawing. Part 2. Workability in extrusion and drawing. Trans. ASME J. Eng. Ind. 101(1), 36–44.*

mean normal stresses (Oh et al., 1979). The value of the material constant in the modified McClintock fracture criterion was assumed to be 2.5, whereas the value of the material constant in the modified Cockcroft and Latham fracture criterion was assumed to be 1.25. The fracture strain obtained in the experiment performed by Bridgman (1952) was also demonstrated. To determine the horizontal coordinate of the fracture strain obtained in the experiment, several assumptions were made. The fracture strains calculated from the modified McClintock fracture criterion and the modified Cockcroft and Latham fracture criterion agreed with the fracture strain obtained in the experiment. In the simulation of extrusion and drawing, the presence or absence of central burst calculated from the simulation using the value of the material constant obtained from the tensile test, differed from the presence or absence of central burst calculated from the simulation using the value of the material constant obtained from the compression test. The presence or absence of central burst calculated from the simulation using the value of the material constant obtained from the tensile test or the compression test, differed from the presence or absence of central burst obtained in the experiment.

Avitzur and Choi (1986) performed the analysis of central bursting defects in plane-strain drawing and extrusion by the upper bound method and obtained the criterion for the prevention of central bursts during plane-strain drawing and extrusion. The following two flow modes were assumed: proportional flow mode in which central bursts were not assumed and central bursting flow mode. Fig. 2.8 shows the velocity field

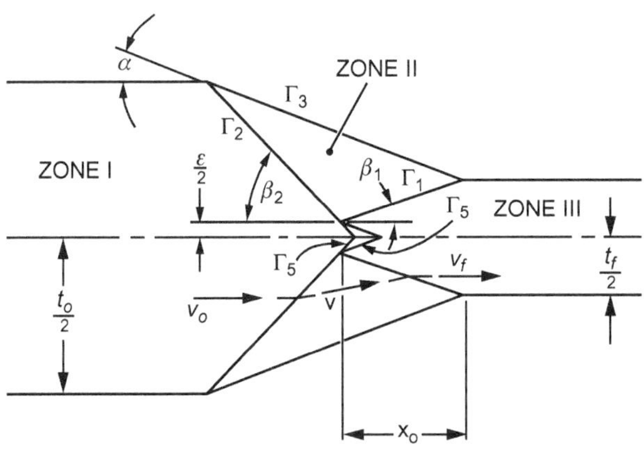

Figure 2.8 Velocity field for central bursting flow mode. *From Avitzur, B., Choi, J.C., 1986. Analysis of central bursting defects in plane strain drawing and extrusion. Trans. ASME J. Eng. Ind. 108(4), 317–321.*

for the central bursting flow mode (Avitzur and Choi, 1986). Zone I, Zone II, and Zone III were rigid bodies. Zone II was bounded by three velocity discontinuity lines Γ_1, Γ_2, and Γ_3. Because the velocity of the rigid body of Zone III v_f was larger than the velocity of the rigid body of Zone I v_0 due to the incompressibility condition, Zone III was moved away from Zone I along the surface Γ_5 and thus the central burst was introduced. First, the height of the central burst ϵ was set to be zero, and the x-coordinate of the intersecting point of Γ_1 and Γ_2, that is, $-X_0$ was optimized by minimizing the energy dissipated on the velocity discontinuity lines of Γ_1, Γ_2, and Γ_3, and due to the front tension and the back tension. Next, when the dissipated energy decreased with increasing ϵ from zero, the central burst was assumed to occur. The central burst had a tendency to occur, with increasing the die angle and with decreasing the reduction in thickness. With increasing the ratio of the frictional stress to the shear yield stress m, the central burst had a tendency to occur in drawing, whereas the central burst had a tendency not to occur in extrusion. The central burst had a tendency to occur, with increasing the back tension in drawing and with increasing the front tension in extrusion.

Aravas (1986) performed the simulation of bar extrusion by the elastic–plastic finite-element method using the Gurson yield function (1977) and discussed the occurrence of central bursts. The friction between the material and the die was not assumed and the reduction in the area in each die was assumed to be 0.25. First, a single pass extrusion was simulated. When the die angle was equal to 5 degrees, no void growth occurred along the axis of symmetry, whereas when the die angle was equal to 15 degrees, void growth occurred along the axis of symmetry. Next, a multipass extrusion was simulated. The extrusion condition of the second pass was set to be the same as the extrusion condition of the first pass. When the die angle was equal to 15 degrees, further void growth occurred along the axis of symmetry. Zimerman et al. (1971) showed experimentally that central burst defects occurred after several passes of a multipass extrusion when the die angle was equal to 15 degrees and the reduction in the area in each die was equal to 0.25. Zimerman et al. (1971) also showed experimentally that central burst defects did not appear after several passes of a multipass extrusion when the die angle was equal to 5 degrees and the reduction in the area in each die was equal to 0.25. Hence, the elastic–plastic finite-element method using the Gurson yield function was shown to be effective to predict whether or not central bursts occurred during axisymmetric extrusion.

Moritoki (1991) obtained the criterion for central bursting defects in plane-stain drawing and extrusion in terms of the collapse of the unique solution. There are two types of multiplicities: a statical multiplicity and a kinematical multiplicity. The statical multiplicity prescribed diffuse necking, whereas the kinematical multiplicity, which was dealt with in this study, corresponded to local necking. The criterion obtained was expressed using a simple equation, in which variables were the tensile strain in forming direction, the strain-hardening exponent, the shear yield stress and the mean normal stress, which was calculated from the slip-line field theory (Johnson et al., 1982). The central burst had a tendency to occur in drawing, with decreasing the strain-hardening exponent, with increasing the frictional stress between the material and the die, and with increasing the back tension. The central burst had a tendency to occur in extrusion, with decreasing the strain-hardening exponent, with decreasing the frictional stress between the material and the die, and with decreasing the front compression. Fig. 2.9 shows the criterion for central burst (Moritoki, 1991). The presence or absence of central burst obtained in

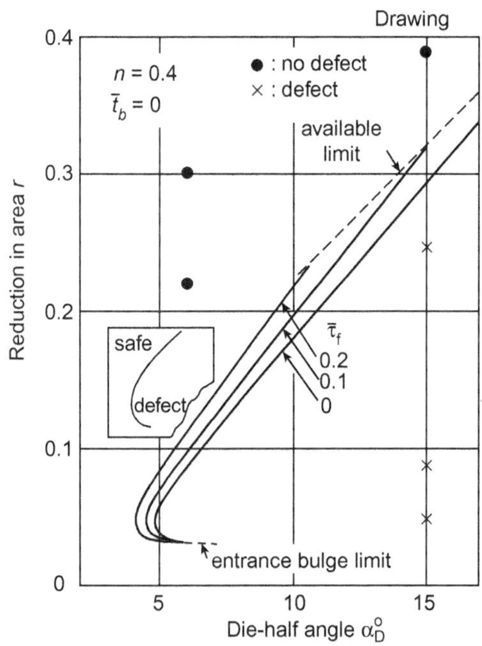

Figure 2.9 Criterion for central burst. *From Moritoki, H., 1991. The criterion for central bursting and its occurrence in drawing and extrusion under plane strain. Int. J. Plast. 7 (7), 713–731.*

the experiment performed by Orbegozo (1968) was also shown. The presence or absence of central burst calculated from the analysis agreed with the presence or absence of central burst obtained in the experiment, although the analytical condition of plane-strain drawing was different from the experimental condition of axisymmetric drawing.

Reddy et al. (1996) proposed a simple criterion for the prediction of central burst in extrusion and confirmed the validity of the simple criterion by the simulation using the rigid-plastic finite-element method of mixed formulation (Bathe, 1996). The simple criterion was that central burst occurred when the mean normal stress at a point on the center axis became tensile. The presence or absence of central burst calculated from the simulation agreed with the presence or absence of central burst obtained in the experiment performed by Hoffmanner (1971). Fig. 2.10 shows the criterion for central burst (Reddy et al., 1996). The presence or absence of central burst calculated using the simple criterion, in other words, the mean normal stress criterion agreed approximately with the presence or absence of central burst calculated from the analysis performed by Avitzur (1968b). Reddy et al. (2000) applied the simple criterion for

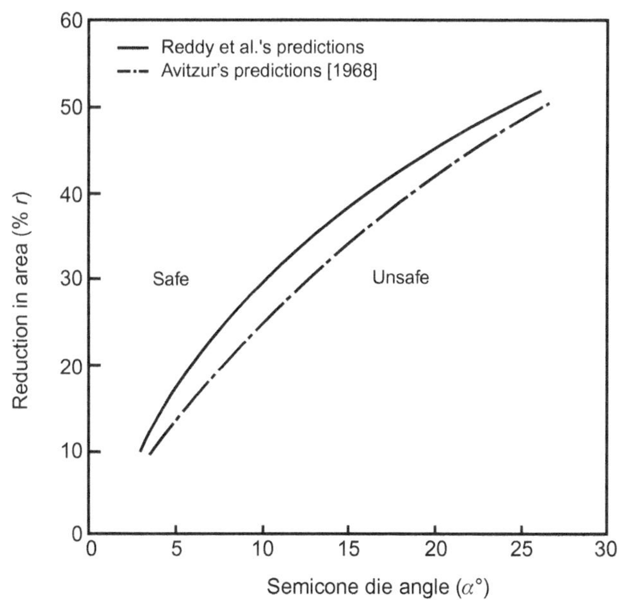

Figure 2.10 Criterion for central burst. *From Reddy, N.V., Dixit, P.M., Lal, G.K., 1996. Central bursting and optimal die profile for axisymmetric extrusion. Trans. ASME J. Manuf. Sci. Eng. 118(4), 579–584.*

the prediction of central burst in drawing. The presence or absence of central burst calculated from the simulation almost agreed with the presence or absence of central burst obtained in the experiment performed by Orbegozo (1968).

Komori (1999a) developed an in-house rigid-plastic finite-element software, with reference to the conventional rigid-plastic finite-element software (Kobayashi et al., 1989), by which the behavior of crack propagation after ductile fracture was able to be simulated. A node separation method was developed to propagate cracks. The phenomenon that central bursts, that is, chevron cracks appeared periodically in the axial direction of the material in multipass wire drawing, was simulated using the developed in-house rigid-plastic finite-element software. The Gurson yield function (1977) was used. When the void volume fraction of the material became larger than a certain void volume fraction and the axial stress of the material became larger than a certain axial stress, the material was assumed to fracture. Because the number of finite elements used in the simulation was not necessarily sufficient, the direction of crack propagation calculated from the simulation was not necessarily appropriate. Komori (2003) performed the simulation of chevron cracks in multipass wire drawing using a large number of finite elements, and obtained the effects of various kinds of ductile fracture criteria on the formation and evolution of chevron cracks. Both the Gurson yield function and the Mises yield function were used. When the Gurson yield function was used, the material was assumed to fracture on the condition that the void volume fraction of the material became larger than a certain void volume fraction. When the Mises yield function was used, the uncoupled simulation of the plastic deformation of the material and the deterioration of the material was performed. Fig. 2.11 shows the finite-element meshes after ductile fracture calculated using the Gurson yield function (Komori, 2003). The shapes of chevron cracks calculated using the Gurson yield

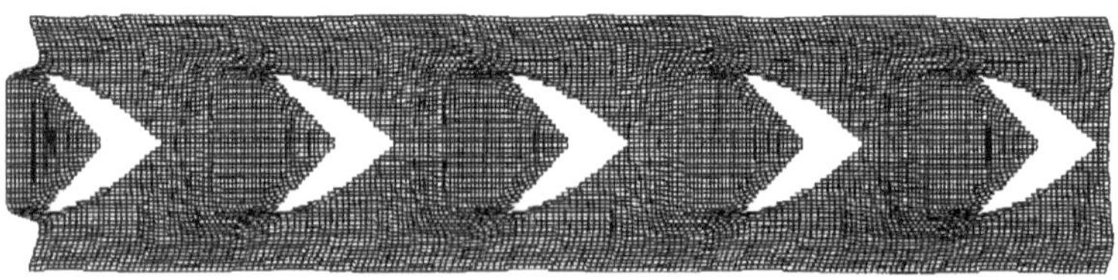

Figure 2.11 Finite-element meshes after ductile fracture calculated using the Gurson yield function. *From Komori, K., 2003. Effect of ductile fracture criteria on chevron crack formation and evolution in drawing. Int. J. Mech. Sci. 45(1), 141–160.*

function and the Oyane fracture criterion (1972) agreed well with the shape of chevron cracks obtained in the experiment performed by the author. The shapes of chevron cracks calculated using the Cockcroft and Latham fracture criterion (1968) and the Brozzo et al. fracture criterion (1972) somewhat agreed with the shape of chevron cracks obtained experimentally. However, the shape of chevron cracks calculated using the Freudenthal fracture criterion (1950) did not agree with the shape of chevron cracks obtained experimentally.

Saanouni et al. (2004) performed the simulation of chevron cracks in single pass bar extrusion by the damage mechanics, and obtained the effects of various forming conditions on the occurrence of chevron cracks. The damage model proposed by Mariage et al. (2002) was used. An element deletion method was employed to propagate cracks. The presence or absence of central burst calculated from the simulation almost agreed with the presence or absence of central burst obtained in the experiment performed by Zimerman et al. (1971). Fig. 2.12 shows the finite-element meshes and the value of the damage variable D after ductile fracture (Saanouni et al., 2004). The material was undamaged when D was equal to zero, whereas the material was ruptured when D was equal to one.

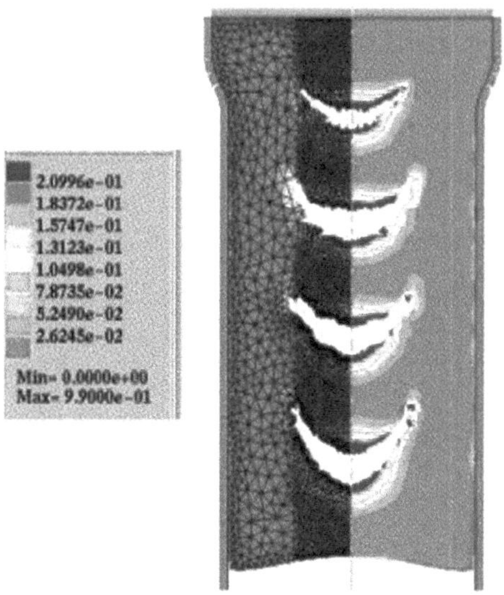

Figure 2.12 Finite-element meshes and value of damage variable D after ductile fracture. *From Saanouni, K., Mariage, J.F., Cherouat, A., Lestriez, P., 2004. Numerical prediction of discontinuous central bursting in axisymmetric forward extrusion by continuum damage mechanics. Comput. Struct. 82(27), 2309–2332.*

The central burst had a tendency to occur, with decreasing the reduction in the area, with increasing the die angle, with decreasing the ductility of the material, and with decreasing the coefficient of friction between the material and the die.

McAllen and Phelan (2005) performed the simulation of central bursts in single pass bar extrusion, and obtained the effects of various kinds of ductile fracture criteria on the development of cracks. An element deletion method was employed to propagate cracks. The following three kinds of ductile fracture criteria were used: the Cockcroft and Latham fracture criterion (1968), the Oyane fracture criterion (1972), and the Chaouadi et al. fracture criterion (1994), which was derived from the void growth model proposed by Rice and Tracey (1969). A commercial elastic–plastic finite-element software into which a user subroutine was incorporated, was used. The presence or absence of central burst calculated using the Cockcroft and Latham fracture criterion and the Oyane fracture criterion almost agreed with the presence or absence of central burst obtained in the experiment performed by Ko and Kim (2000). Fig. 2.13 shows the finite-element meshes after ductile fracture for various kinds of ductile fracture criteria (McAllen and Phelan, 2005). McAllen and Phelan (2007) performed the simulation of central bursts in multipass wire drawing. The Gurson yield function (1977) modified by Tvergaard (1981) was used, whereas the Chaouadi et al. fracture criterion modified by the authors was developed and used. The presence or absence of central burst calculated using the modified Chaouadi et al. fracture criterion

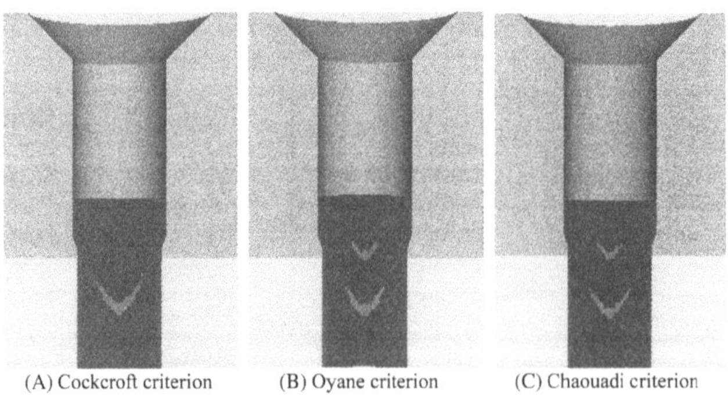

Figure 2.13 Finite-element meshes after ductile fracture for various kinds of ductile fracture criteria. *From McAllen, P., Phelan, P., 2005. A method for the prediction of ductile fracture by central bursts in axisymmetric extrusion. Proc. Inst. Mech. Eng. C: J. Mech. Eng. Sci. 219(3), 237–250.*

almost agreed with the presence or absence of central burst obtained in the experiment performed by Orbegozo (1968). When central burst occurred, the mean die pressure fluctuated periodically with increasing the drawing length. The wavelength of the fluctuation of the mean die pressure in the drawing direction was identical to the distance between two neighboring central bursts in the drawing direction.

2.4 Christmas-tree cracking in bar extrusion

Clift et al. (1990) performed the simulation of axisymmetric bar extrusion by the elastic–plastic finite-element method using the Mises yield function and predicted the location of fracture initiation for each ductile fracture criterion. The ductile fracture criteria used were the McClintock et al. fracture criterion (1966), the McClintock fracture criterion (1968a), the Oyane fracture criterion (1972), the Ghosh fracture criterion (1976), the Freudenthal fracture criterion (1950), the Cockcroft and Latham fracture criterion (1968), the Brozzo et al. fracture criterion (1972), the Norris et al. fracture criterion (1978), and the Atkins fracture criterion (1981). The experiment of axisymmetric bar extrusion was performed using 60–40 brass and 7075 aluminum alloy, and the fir-tree cracking, which is also called the Christmas-tree cracking, appeared on the surfaces of the specimens of both 60–40 brass and 7075 aluminum alloy. In the cases of the Oyane fracture criterion, the Ghosh fracture criterion, the Freudenthal fracture criterion, the Cockcroft and Latham fracture criterion, the Brozzo et al. fracture criterion, and the Atkins fracture criterion, the following results were obtained. For both 60–40 brass and 7075 aluminum alloy, the location of the fir-tree cracking obtained experimentally agreed with the location of the node in the finite-element mesh at which the cumulative value of each ductile fracture criterion was at maximum.

Hambli and Badie-Levet (2000) performed the simulation of axisymmetric bar extrusion using a commercial elastic–plastic finite-element software into which a user subroutine for the Lemaitre damage model (1985) was incorporated. A fir-tree cracking appeared on the surface of the specimen in the experiment of axisymmetric bar extrusion using a plasticine, whereas the value of the damage variable D was largest on the surface of a bar in the simulation. Hence, a surface cracking was confirmed to appear both in the experiment and in the simulation of axisymmetric bar extrusion.

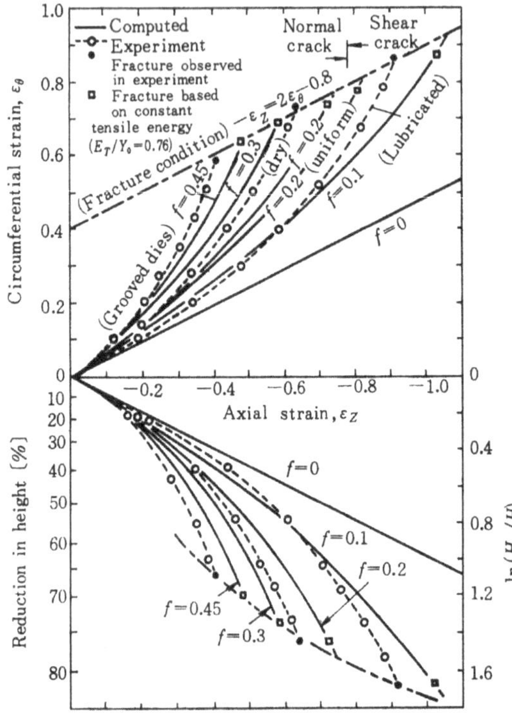

Figure 2.14 Relationship between axial strain and circumferential strain at fracture in cylinder upsetting. *From Kobayashi, S., Lee, C.H., Shah, S.N., 1973. Analysis of rigid-plastic deformation problems by the matrix method. Journal of the Japan Society for Technology of Plasticity 14 (153), 770–778 (in Japanese).*

2.5 Surface cracking in cylinder upsetting

Kobayashi and Lee (1973) performed the simulation of cylinder upsetting by the rigid-plastic finite-element method in various ratios of the frictional stress to the shear yield stress between the material and the die, and predicted surface cracking in cylinder upsetting using the Cockcroft and Latham fracture criterion (1968). Fig. 2.14 shows the relationship between the axial strain and the circumferential strain at fracture in cylinder upsetting (Kobayashi et al., 1973). The experiment was performed using an annealed SAE 1040 steel. The material was assumed to fracture at the surface in the radial direction and at the center in the axial direction. When the cumulative value of the Cockcroft and Latham fracture criterion at fracture was specified, the relationships between the axial strain at fracture and the circumferential strain at fracture calculated from the simulation were expressed by a straight line, the gradient of which was equal to 0.5. When the axial stress at fracture was tensile in the simulation, the longitudinal crack, which was indicated by Kudo and Aoi (1967), appeared in the experiment, whereas when the axial stress at fracture was compressive in the simulation, the oblique crack, which was also indicated by Kudo and Aoi, appeared in the experiment.

Sowerby et al. (1985) performed the simulation of cylinder upsetting and also stepped cylinder upsetting by the rigid-plastic finite-element method, and predicted surface cracking in cylinder upsetting and stepped cylinder upsetting using the McClintock fracture criterion (1968a). Although the McClintock fracture criterion considered neither void nucleation nor void coalescence, and only considered void growth, the cumulative value of the McClintock fracture criterion was assumed to be the damage accumulated. The accumulated damage was largest in the finite-element on the surface of the specimen when the reduction in height of the specimen was specified; surface cracking was predicted to occur. Fig. 2.15 shows the relationship between the axial strain and the circumferential strain at fracture in cylinder upsetting and stepped cylinder upsetting (Sowerby et al., 1985). The relationships between the axial strain and the circumferential strain at fracture obtained experimentally

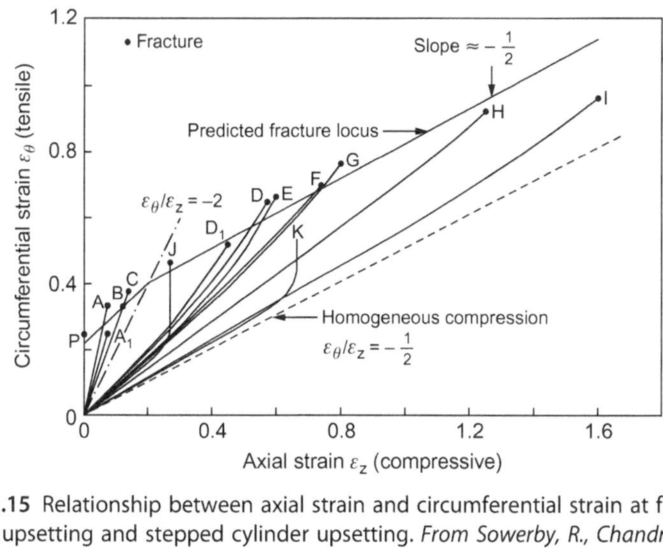

Figure 2.15 Relationship between axial strain and circumferential strain at fracture in cylinder upsetting and stepped cylinder upsetting. *From Sowerby, R., Chandrasekaran, N., Dung, N.L., Mahrenholtz, O., 1985. The prediction of damage accumulation during upsetting tests based on McClintock's model. Forschung im Ingenieurwesen 51(5), 147–150.*

using AISI 1045 carbon steel were also shown. The relationships between the axial strain and the circumferential strain during cylinder upsetting and stepped cylinder upsetting were virtually linear both in the experiment and in the simulation. Hence, the average value of the accumulated damage at fracture obtained by the experiment was able to be calculated, and used to obtain the relationship between the axial strain and the circumferential strain at fracture calculated from the simulation.

Clift et al. (1990) performed the simulation of cylinder upsetting by the elastic–plastic finite-element method using the Mises yield function and predicted the location of fracture initiation in cylinder upsetting for each ductile fracture criterion, which was shown in Section 2.5. The experiment of cylinder upsetting using four kinds of cylinders, the initial height/diameter ratios of which were different from each other, was performed using 60–40 brass, and surface cracking appeared in each kind of cylinders. In the cases of the McClintock fracture criterion (1968a), the Freudenthal fracture criterion (1950), and the Brozzo et al. fracture criterion (1972), the following results were obtained. For the four kinds of cylinders, the initial height/diameter ratios of which were different from each other, the location of the surface cracking obtained experimentally agreed with the location of the node in the finite-element mesh at which the cumulative value of each ductile fracture criterion was at maximum.

Bontcheva and Iankov (1991) performed the simulation of cylinder upsetting by the thermal rigid-plastic finite-element method using the Lemaitre damage model (1985) and predicted the location of fracture initiation in cylinder upsetting for various initial height/diameter ratios of cylinders and various ratios of the frictional stress to the shear yield stress between the material and the die. The material fractured at the center in the radial direction and at the center in the axial direction for low initial height/diameter ratio of a cylinder, whereas the material fractured at the surface in the radial direction and at the center in the axial direction for high initial height/diameter ratio of a cylinder. With increasing the ratio of the frictional stress to the shear yield stress between the material and the die, the reduction in height at which the material fractured decreased.

Moritoki (1993) obtained the criterion for surface cracking in cylinder upsetting by a method based on the sufficient condition for the collapse of the unique solution. In the experiment of cylinder upsetting by Kudo and Aoi (1967), the following two types of surface cracking appeared: oblique crack and longitudinal crack or perpendicular crack. In the analysis, when the stability required for the oblique deformation mode was

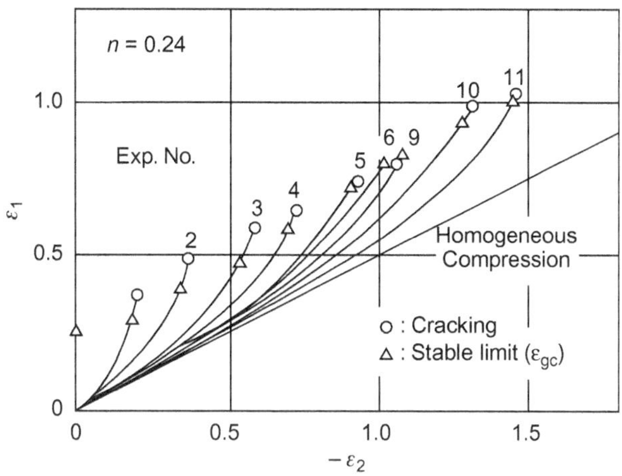

Figure 2.16 Relationship between axial strain and circumferential strain at fracture in cylinder upsetting. *From Moritoki, H., 1993. Free surface ductility in upsetting. Int. J. Plast. 9(4), 507–523.*

smaller than the stability required for the perpendicular deformation mode, the oblique crack was assumed to occur, whereas when the stability required for the perpendicular deformation mode was smaller than the stability required for the oblique deformation mode, the perpendicular crack was assumed to occur. Fig. 2.16 shows the relationship between the axial strain and the circumferential strain at fracture in cylinder upsetting (Moritoki, 1993). The circle marks indicated the fracture strains obtained experimentally by Kudo and Aoi (1967), whereas the triangle marks indicated the fracture strains calculated analytically by the author.

The perpendicular crack appeared both in the experiment and in the analysis in the cases of numbers 1, 2, and 3, whereas the oblique crack appeared both in the experiment and in the analysis in the cases of numbers 5, 6, 9, 10, and 11. Furthermore, both the perpendicular crack and the oblique crack appeared both in the experiment and in the analysis in the case of number 4.

Gouveia et al. (1996) performed the simulation of cylinder upsetting, stepped cylinder upsetting, tapered cylinder upsetting, and hollow cylinder upsetting by the rigid-plastic finite-element method, and predicted the location of fracture initiation in each cylinder upsetting for each ductile fracture criterion. The ductile fracture criteria used were the Oyane fracture criterion (1972), the Freudenthal fracture criterion (1950), the Cockcroft and Latham fracture criterion (1968), and the Brozzo et al. fracture criterion (1972). The experiment of each cylinder upsetting was

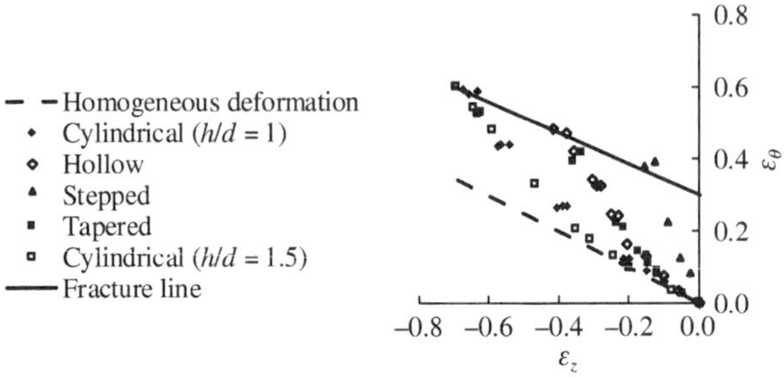

Figure 2.17 Relationship between axial strain and circumferential strain at fracture in cylinder upsetting, stepped cylinder upsetting, tapered cylinder upsetting, and hollow cylinder upsetting. *From Gouveia, B.P.P.A., Rodrigues, J.M.C., Martins, P.A.F., 1996. Fracture predicting in bulk metal forming. Int. J. Mech. Sci. 38(4), 361–372.*

performed using a lead alloy. Fig. 2.17 shows the relationship between the axial strain and the circumferential strain at fracture in cylinder upsetting, stepped cylinder upsetting, tapered cylinder upsetting, and hollow cylinder upsetting (Gouveia et al., 1996). The relationships between the axial strain at fracture and the circumferential strain at fracture obtained in the experiment were expressed by a straight line, the gradient of which was approximately equal to the gradient of the line that showed the relationship between the axial strain and the circumferential strain for homogeneous deformation. In each cylinder upsetting, when the surface cracking occurred experimentally at a location, the cumulative value of each ductile fracture criterion at the location was calculated both in the experiment and in the simulation. In each ductile fracture criterion, the deviation between the cumulative value of each cylinder upsetting and the average of the cumulative values of all cylinder upsettings was calculated both in the experiment and in the simulation. The deviation was large in the Freudenthal fracture criterion, whereas the deviations were small in the Oyane fracture criterion, the Cockcroft and Latham fracture criterion, and the Brozzo et al. fracture criterion. In an tapered cylinder upsetting, the location of fracture initiation calculated from the simulation using the Brozzo et al. fracture criterion differed from the location of fracture initiation obtained in the experiment, whereas the cumulative value of the Cockcroft and Latham fracture criterion temporarily decreased during upsetting. Hence, the Oyane fracture criterion was considered to be the most appropriate fracture criterion.

Kim et al. (1999) performed the simulation of cylinder upsetting and cylinder heading by the rigid-viscoplastic finite-element method, and predicted the location of fracture initiation in cylinder upsetting and cylinder heading using the Freudenthal fracture criterion (1950) and the Cockcroft and Latham fracture criterion (1968). The experiments of cylinder upsetting and cylinder heading were performed using various kinds of aluminum alloys. In cylinder upsetting, the cumulative value of the Cockcroft and Latham fracture criterion was a maximum at the surface in the radial direction and at the center in the axial direction, whereas the cumulative value of the Freudenthal fracture criterion was a maximum at the surface in the radial direction and at the end in the axial direction. Since the material fractured at the surface in the radial direction and at the center in the axial direction, the prediction of the location of fracture initiation by the Cockcroft and Latham fracture criterion was more reasonable than the prediction of the location of fracture initiation by the Freudenthal fracture criterion. In cylinder heading, the prediction of the location of fracture initiation by the Cockcroft and Latham fracture criterion was also shown to be more reasonable than the prediction of the location of fracture initiation by the Freudenthal fracture criterion. The cumulative value of the Cockcroft and Latham fracture criterion at fracture in cylinder upsetting was approximately 10% smaller than the cumulative value of the Cockcroft and Latham fracture criterion at fracture in cylinder heading.

Gänser et al. (2001) performed the simulation of cylinder upsetting using a commercial elastic–plastic finite-element software, and evaluated two types of ductile fracture criteria by the comparison of the fracture strains calculated from the simulation and the fracture strains obtained in the experiment performed by Kudo and Aoi (1967). The two types of ductile fracture criteria were the Gunawardena et al. fracture criterion (1993) and the Atkins fracture criterion (1981). The Gunawardena et al. fracture criterion was derived from the relationship between the fracture strain and the mean normal stress proposed by Hancock and Mackenzie (1976). Fig. 2.18 shows the relationship between the axial strain and the circumferential strain at fracture in cylinder upsetting in the case of the Gunawardena et al. fracture criterion (Gänser et al., 2001). The mathematical structures of the Gunawardena et al. fracture criterion and the Atkins fracture criterion were different from each other. However, the relationships between the axial strain and the circumferential strain at fracture in the cases of the Gunawardena et al. fracture criterion and the Atkins fracture criterion agreed with the relationship between the axial

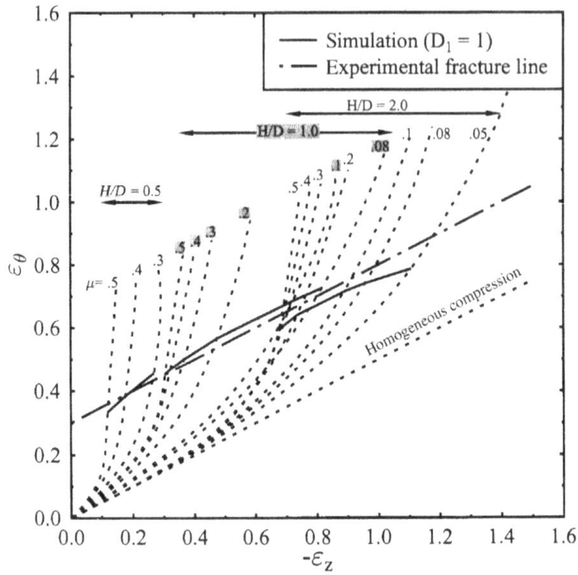

Figure 2.18 Relationship between axial strain and circumferential strain at fracture in cylinder upsetting in case of Gunawardena et al. fracture criterion. *From Gänser, H.P., Atkins, A.G., Kolednik, O., Fischer, F.D., Richard, O., 2001. Upsetting of cylinders: a comparison of two different damage indicators. Trans. ASME J. Eng. Mater. Technol. 123(1), 94–99.*

strain and the circumferential strain at fracture obtained experimentally by Kudo and Aoi.

Ragab (2002) obtained the relationship between the axial strain and the circumferential strain at fracture in cylinder upsetting using the semiempirical solution in cylinder upsetting due to Dadras (1981) and the Gurson yield function (1977) modified by Tvergaard (1981). The components of the stress were calculated using the Dadras semiempirical solution, whereas the components of the strain increment and the void volume fraction increment were calculated using the flow rule associated with the Gurson–Tvergaard yield function. The material was assumed to fracture at the surface in the radial direction and at the center in the axial direction. When strain localization occurred, that is, the circumferential strain increment divided by the axial strain increment became infinite, the material was assumed to fracture. Fig. 2.19 shows the relationship between the axial strain and the circumferential strain at fracture in cylinder upsetting (Ragab, 2002). The analytical results were compared with the experimental results on cold-drawn 1045 steel obtained by Kuhn et al. (1973).

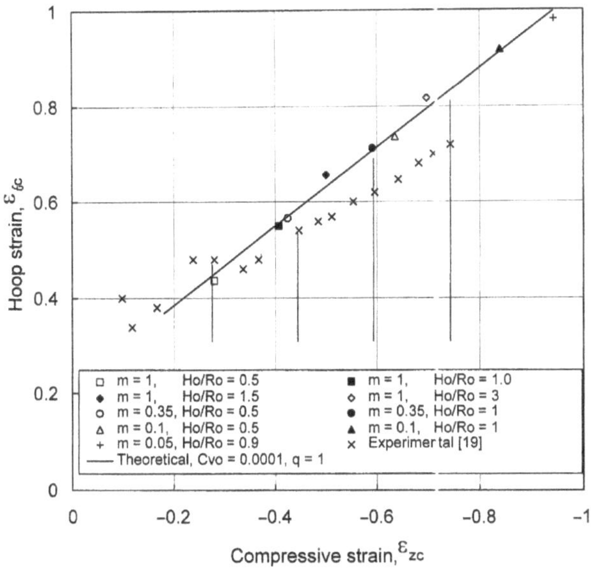

Figure 2.19 Relationship between axial strain and circumferential strain at fracture in cylinder upsetting. *From Ragab, A.R., 2002. Fracture limit curve in upset forging of cylinders. Mater. Sci. Eng. A 334(1–2), 114–119.*

2.6 Forming limit in sheet forming

Necking forming limit is the forming limit at which the necking of a sheet occurs or the unevenness of the thickness of a sheet appears. A large number of researches on the analysis or the simulation of the necking forming limit have been performed. Swift (1952) analyzed the onset of diffuse necking. However, since diffuse necking forming limit was not related to fracture forming limit, the analysis of the diffuse necking forming limit is not dealt with. Hill (1952) analyzed the onset of local necking, which was obtained when the strain ratio $\varepsilon_2/\varepsilon_1$ was smaller than zero and which was not obtained when the strain ratio $\varepsilon_2/\varepsilon_1$ was larger than zero, where ε_1 is the major principal strain in the plane of the sheet, and ε_2 is the minor principal strain in the plane of the sheet. Marciniak and Kuczyński (1967) simulated the onset of local necking when the strain ratio $\varepsilon_2/\varepsilon_1$ was larger than zero in the assumption of the initial unevenness of the thickness of a sheet. Stören and Rice (1975) analyzed the onset of local necking when the strain ratio $\varepsilon_2/\varepsilon_1$ was not only smaller than zero but also larger than zero using the total strain theory, that is, the deformation theory and not the incremental strain theory, that is, the flow theory. However, since local necking forming limit was scarcely related to fracture forming limit, the analysis or the simulation of the local necking forming limit is not dealt with.

The local necking forming limit is not necessarily applicable to industrial metal-forming, because the local necking forming limit strongly depends on the strain history to which the sheet is subjected. Hence, Arrieux et al. (1982) proposed a local necking forming limit using the principal stress in the plane of the sheet, and showed that the local necking forming limit on the principal stress scarcely depended on the strain history to which the sheet was subjected. However, the simulation of the local necking forming limit on the principal stress is also not dealt with, because the local necking forming limit on the principal stress was scarcely related to fracture forming limit.

Glover et al. (1977) illustrated a variety of possible criteria for fracture forming limit in sheet forming both in the principal strain space and in the principal stress space. The criteria illustrated were the maximum principal strain criterion, the maximum principal stress criterion, the maximum equivalent stress criterion, the maximum mean normal stress criterion, the maximum thickness strain criterion, the maximum shear stress criterion, and the Cockcroft and Latham fracture criterion (1968).

Ghosh (1976) proposed a criterion for fracture forming limit in sheet forming under biaxial loading, and compared the fracture forming limit calculated from the analysis with the fracture forming limit obtained in the experiment. The fracture surface was assumed to coincide with the surface in which the maximum shear stress yielded. Hence, the angle between the direction of the maximum principal stress in the plane of the sheet and the normal direction of the fracture surface was assumed to be 45 degrees, and the angle between the thickness direction of the sheet and the normal direction of the fracture surface was also assumed to be 45 degrees. The criterion was derived from the void coalescence due to shear along the fracture surface, and was a function of the stress ratio σ_2/σ_1, where σ_1 is the major principal stress in the plane of the sheet, and σ_2 is the minor principal stress in the plane of the sheet. Fig. 2.20 shows the fracture forming limit for low carbon aluminum-killed steel (Ghosh, 1976). The fracture forming limits obtained in the experiment were denoted by black circles, whereas the fracture forming limit calculated from the analysis was denoted by a solid line. The number of material constants in the criterion was one, and the unique material constant was determined such that the fracture forming limit calculated from the

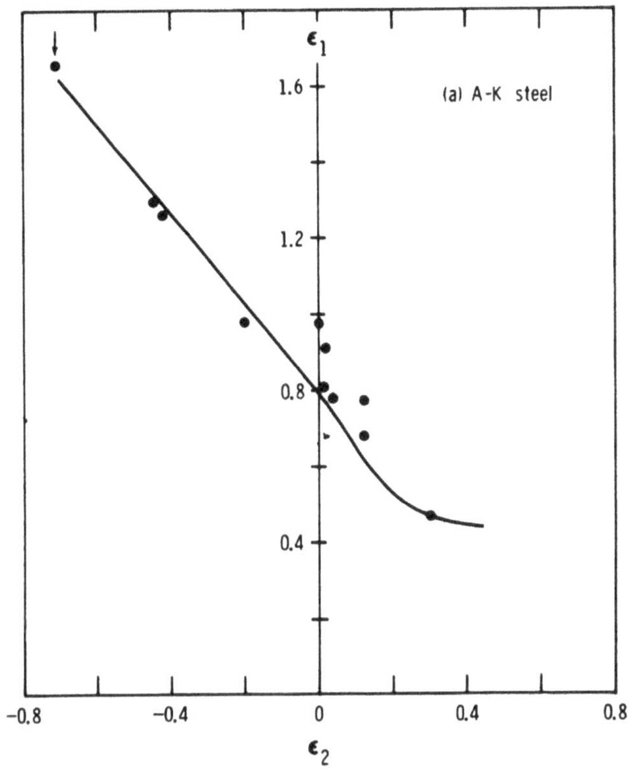

Figure 2.20 Fracture forming limit for low carbon aluminum-killed steel. *From Ghosh, A.K., 1976. A criterion for ductile fracture in sheets under biaxial loading. Metall. Trans. A 7(4), 523–533.*

analysis in uniaxial tension agreed with the fracture forming limit obtained in the experiment in uniaxial tension.

Chu and Needleman (1980) obtained the necking forming limit and also the fracture forming limit in punch stretching of a sheet by the elastic–plastic finite-element method using the Gurson yield function (1977). Fig. 2.21 shows the necking forming limit and the fracture forming limit (Chu and Needleman, 1980). The necking forming limit was assumed when the local necking of a sheet preceded the attainment of the maximum void volume fraction, whereas the fracture forming limit was assumed when the attainment of the maximum void volume fraction preceded the local necking of a sheet. The solid line indicated the necking forming limit, which was obtained when the strain ratio $\varepsilon_2/\varepsilon_1$ was close to one, and at which the local necking of a sheet appeared. The cross

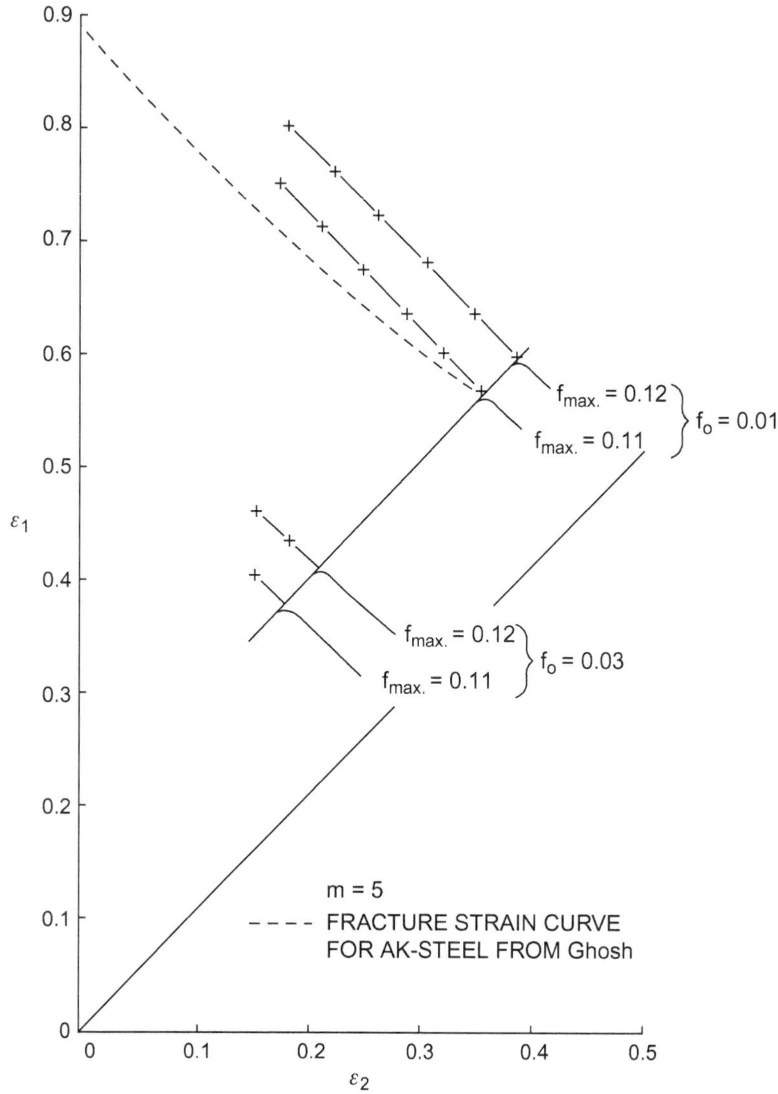

Figure 2.21 Necking forming limit and fracture forming limit. *From Chu, C.C., Needleman, A., 1980. Void nucleation effects in biaxially stretched sheets. Trans. ASME J. Eng. Mater. Technol. 102(3), 249–256.*

marks indicated the fracture forming limit, which was obtained when the strain ratio $\varepsilon_2/\varepsilon_1$ was close to zero, and at which the maximum void volume fraction was assumed to be a certain value. The broken line indicated the fracture forming limit obtained experimentally by Ghosh (1976).

Atkins (1981) proposed a criterion for fracture forming limit in sheet forming such that the major and minor principal strains at fracture in the plane of the sheet calculated from the analysis agreed with the major and minor principal strains at fracture in the plane of the sheet obtained experimentally and shown in Embury and LeRoy (1978). The Atkins fracture criterion was derived from the modification of the Norris et al. fracture criterion (1978).

Chow et al. (1997) obtained the necking forming limit and also the fracture forming limit in punch stretching of a sheet by the damage mechanics. The diffuse necking, the local necking, and the fracture were predicted in a unified approach. However, the damage variable D for the local necking in the case of positive strain ratio, that is, $\varepsilon_2/\varepsilon_1 > 0$ was made to differ from the damage variable D for the local necking in the case of negative strain ratio, that is, $\varepsilon_2/\varepsilon_1 < 0$. The experiment of punch stretching was performed using aluminum alloy 6111-T4, and the necking forming limit calculated from the analysis agreed with the necking forming limit obtained experimentally.

Jain et al. (1999) obtained the fracture forming limit in punch stretching of a sheet using aluminum alloy AA6111-T4, and evaluated 16 kinds of ductile fracture criteria in comparison between the fracture forming limit calculated from the analysis and the fracture forming limit obtained experimentally. The Freudenthal fracture criterion (1950), the Cockcroft and Latham fracture criterion (1968), the Hancock and Mackenzie fracture criterion (1976), the Ghosh fracture criterion (1976), the Bressan and Williams fracture criterion (1983), the Oyane fracture criterion (1972), and the Brozzo et al. fracture criterion (1972) were included in the sixteen kinds of ductile fracture criteria. Fig. 2.22 shows the fracture forming limit for aluminum alloy AA6111-T4 (Jain et al., 1999). The horizontal axis shows the strain ratio $\varepsilon_2/\varepsilon_1$ and not the minor principal strain in the plane of the sheet ε_2. The fracture forming limit calculated from the analysis using the Tresca fracture cirterion, in which the material was assumed to fracture when the maximum principal stress reached a certain value and which was identical to the Bressan and Williams fracture criterion, agreed with the fracture forming limit obtained experimentally.

Takuda et al. (2000) obtained the necking forming limit and also the fracture forming limit in punch stretching of a sheet using two kinds of aluminum alloys A1100-O and A5182-O, and predicted both the necking forming limit and the fracture forming limit in punch stretching by the rigid-plastic finite-element method incorporating the Oyane fracture

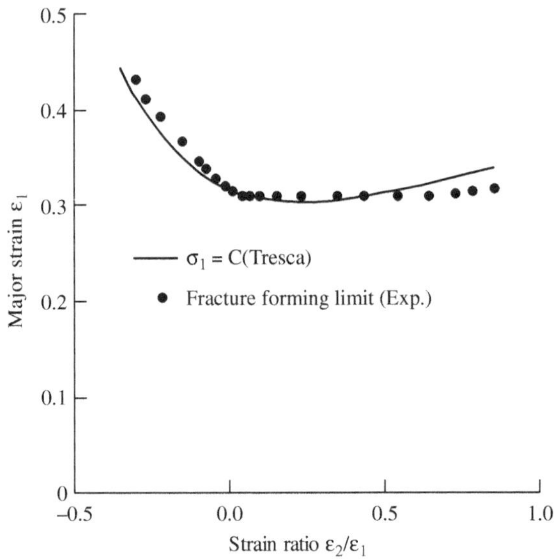

Figure 2.22 Fracture forming limit for aluminum alloy AA6111-T4. *From Jain, M., Allin, J., Lloyd, D.J., 1999. Fracture limit prediction using ductile fracture criteria for forming of an automotive aluminum sheet. Int. J. Mech. Sci. 41(10), 1273–1288.*

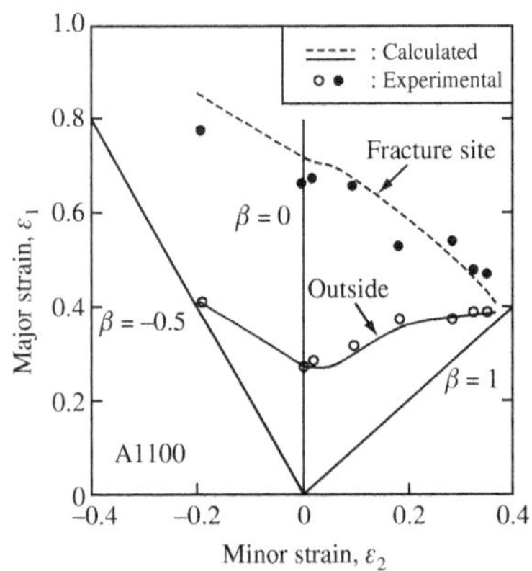

Figure 2.23 Necking forming limit and fracture forming limit for aluminum alloy A1100-O. *From Takuda, H., Mori, K., Takakura, N., Yamaguchi, K., 2000. Finite element analysis of limit strains in biaxial stretching of sheet metals allowing for ductile fracture. Int. J. Mech. Sci. 42(4), 785–798.*

criterion (1972). Fig. 2.23 shows the necking forming limit and the fracture forming limit for aluminum alloy A1100-O (Takuda et al., 2000). The solid line denoted the necking forming limit, which was calculated from strain components in the finite element near the fracture site, whereas the dashed line denoted the fracture forming limit, which was calculated from strain components in the finite element at the fracture site. The two material constants in the Oyane fracture criterion were determined by two types of material tests, that is, uniaxial tensile test and plane-strain tensile test.

Lei et al. (2002) obtained the fracture forming limit in tube bulging using STKM-11A carbon steel, identified the two material constants in the Oyane fracture criterion (1972), performed the simulations of three kinds of hydroforming processes by the rigid-plastic finite-element method incorporating the Oyane fracture criterion, and predicted the fracture initiation site and the forming limit in the hydroforming processes.

Han and Kim (2003) obtained the necking forming limit and also the fracture forming limit in punch stretching of a sheet using four kinds of steels. In the fracture forming limit diagram, in which the vertical axis denoted the major principal strain in the plane of the sheet ε_1 and the horizontal axis denoted the minor principal strain in the plane of the sheet ε_2, the fracture forming limit obtained experimentally of the material having relatively higher ductility was approximated by a straight line. However, the fracture forming limit obtained experimentally of the material having relatively lower ductility was approximated by neither a straight line nor curved lines calculated from the simulation using several kinds of ductile fracture criteria. The ductile fracture criteria used were the Freudenthal fracture criterion (1950), the Cockcroft and Latham fracture criterion (1968), the Brozzo et al. fracture criterion (1972), and the Oyane fracture criterion (1972). Hence, an empirical fracture criterion, in which the Cockcroft and Latham fracture criterion and the Bressan and Williams fracture criterion (1983) were combined, was proposed. Fig. 2.24 shows the fracture forming limit for the four kinds of steels obtained by the experiment and calculated from the simulation using the empirical fracture criterion (Han and Kim, 2003).

Lee et al. (2004) performed the simulation of punch stretching of a sheet using commercial elastic–plastic finite-element softwares and also the elementary analysis of punch stretching of a sheet for a rigid-plastic material, and obtained the fracture forming limit in punch stretching. The experiment of punch stretching was also performed using mild steel. Fig. 2.25 shows the fracture forming limit for the mild steel (Lee et al., 2004). When the coefficient of friction between the punch and the material was assumed to be zero, the material was subjected to balanced biaxial

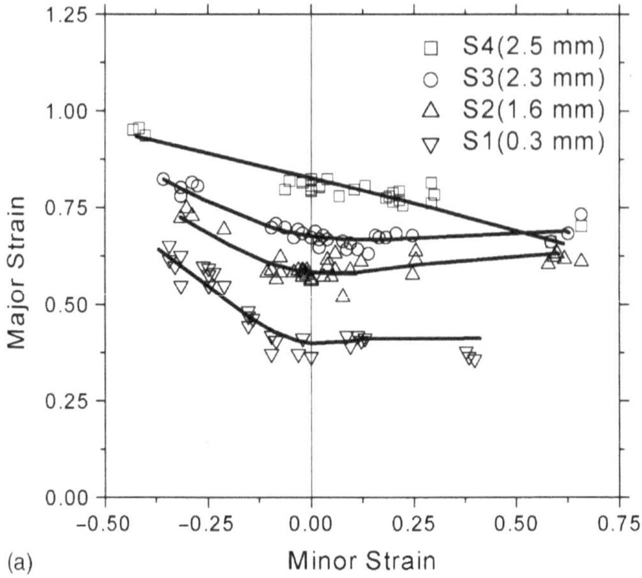

Figure 2.24 Fracture forming limit for four kinds of steels obtained by experiment and calculated from simulation using empirical fracture criterion. *From Han, H.N., Kim, K.H., 2003. A ductile fracture criterion in sheet metal forming process. J. Mater. Process. Technol. 142(1), 231–238.*

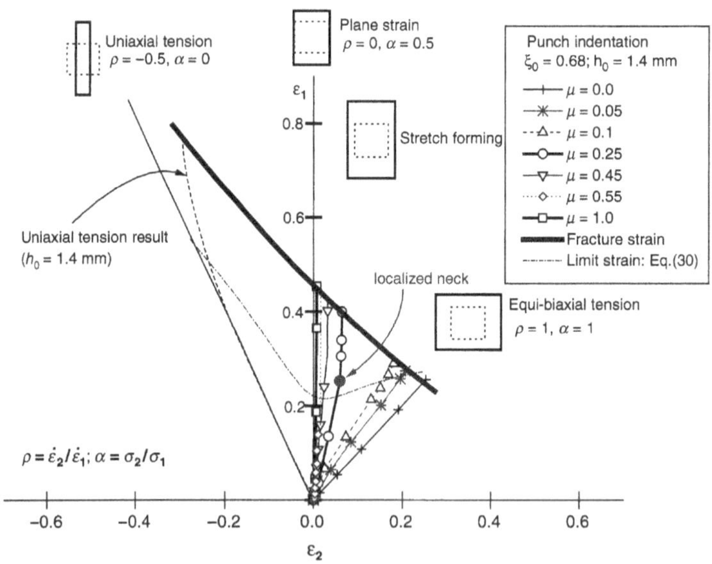

Figure 2.25 Fracture forming limit for mild steel. *From Lee, Y.W., Woertz, J.C., Wierzbicki, T., 2004. Fracture prediction of thin plates under hemi-spherical punch with calibration and experimental verification. Int. J. Mech. Sci. 46(5), 751–781.*

tension, whereas when the coefficient of friction between the punch and the material was assumed to be one, the material was subjected to plane-strain tension. The material was assumed to fracture when the definite integral of the stress triaxiality $\left(\sigma_m/\overline{\sigma}\right)$, where σ_m was the mean normal stress and $\overline{\sigma}$ was the equivalent stress, through the equivalent plastic strain along the strain history became greater than a certain value.

2.7 Rupture in shearing

Taupin et al. (1996) performed the simulation of axisymmetric shearing using a commercial rigid-plastic finite-element software into which a user subroutine was incorporated, and compared the material shape at rupture calculated from the simulation with the material shape at rupture obtained in the experiment. The experiment was performed using low carbon steel and high strength steel. An element deletion method was used to perform the simulation of shearing in which a material was divided into two parts. When the damage value calculated using the McClintock fracture criterion (1968a) became larger than a certain value in an element in the current step of the simulation, the element was deleted in the next step of the simulation. The certain value, which was a material constant, was determined such that the material shape at rupture calculated from the simulation agreed with the material shape at rupture obtained in the experiment. Fig. 2.26 shows the material shape at rupture for low carbon steel (Taupin et al., 1996). The clearance between the punch and the die was assumed to be 15%.

Brokken et al. (1998, 2000) performed the simulation of plane-strain shearing using a commercial elastic–plastic finite-element software into which the user subroutine for an arbitrary Lagrange Euler method (Donea, 1983) combined with remeshing was incorporated. Both an arbitrary Lagrange Euler method, in which the topology of finite-element mesh was preserved, and remeshing, in which the topology of finite-element mesh was not preserved, were used to maintain the quality of finite-element mesh despite the drastic deformation during shearing. An operator split method was used to perform the simulation of shearing in which a material was divided into two parts. Several monitoring points were spaced radially around the node of crack tip in each step of the simulation. When the damage value calculated using the void growth model proposed by Rice and Tracey (1969) became larger than a certain value at one of the radially spaced monitoring points in the current step of the

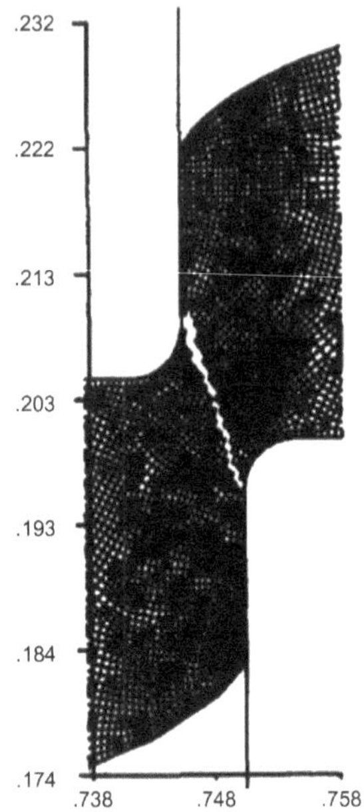

Figure 2.26 Material shape at rupture for low carbon steel. *From Taupin, E., Breitling, J., Wu, W.T., Altan, T., 1996. Material fracture and burr formation in blanking results of FEM simulations and comparison with experiments. J. Mater. Process. Technol. 59(1–2), 68–78.*

simulation, the radially spaced monitoring point was made to coincide with the node of crack tip in the next step of the simulation. Fig. 2.27 shows the finite-element meshes and the damage value at rupture (Brokken et al., 1998). The clearance between the punch and the die was assumed to be 5%.

Goijaerts et al. (2000, 2001) performed the simulations of axisymmetric shearing and tensile test, performed the experiments of axisymmetric shearing and tensile test using five kinds of materials, and evaluated ductile fracture criteria. The simulation of axisymmetric shearing was performed using a commercial elastic–plastic finite-element software into which the user subroutine for an arbitrary Lagrange Euler method (Donea, 1983) combined with remeshing was incorporated, whereas the simulation of tensile test was performed using a commercial three-dimensional elastic–plastic finite-element software. The materials used were ferritic stainless steel, austenitic stainless steel, low carbon steel, brass, and aluminum alloy. The tensile tests of sheets were performed using specimens that were subjected to sheet rolling, to obtain the stress–strain relationship

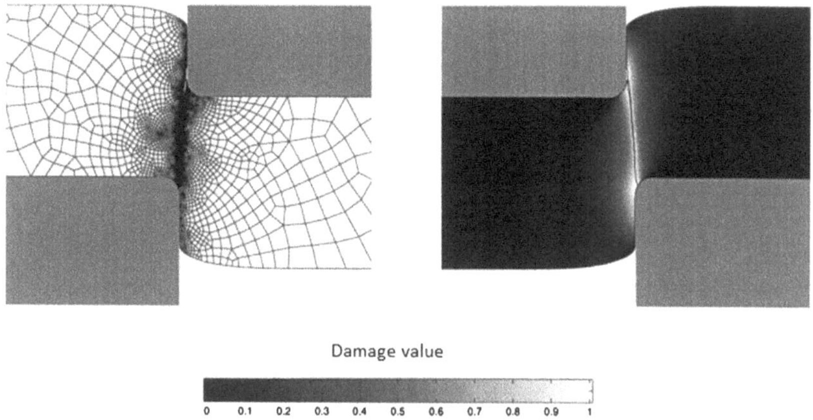

Figure 2.27 Finite-element meshes and damage value at rupture. *From Brokken, D., Brekelmans, W.A.M., Baaijens, F.P.T., 1998. Numerical modelling of the metal blanking process. J. Mater. Process. Technol. 83(1–3), 192–199.*

for large equivalent plastic strain to which the material was subjected during shearing.

The experiment of axisymmetric shearing was performed for a specified clearance between the punch and the die to obtain the punch displacement at ductile fracture initiation. The simulation of axisymmetric shearing was performed for the specified clearance between the punch and the die such that the punch displacement coincided with the punch displacement at ductile fracture initiation obtained experimentally, and a material constant in the ductile fracture criterion derived from the void growth model proposed by Rice and Tracey (1969) was calculated. The simulation of axisymmetric shearing was performed using the material constant for the remaining clearances between the punch and the die, and the punch displacement at ductile fracture initiation was calculated, which was consistent with the punch displacement at ductile fracture initiation obtained experimentally.

The experiment of the tensile test of a sheet was performed, and the minimum thickness of the sheet at rupture was measured. The simulation of the tensile test of a sheet was performed such that the minimum thickness of the sheet coincided with the minimum thickness of the sheet at rupture measured experimentally, and a material constant in the proposed ductile fracture criterion was calculated. The simulation of axisymmetric shearing was performed using the material constant for various clearances between the punch and the die, and the punch displacement at ductile

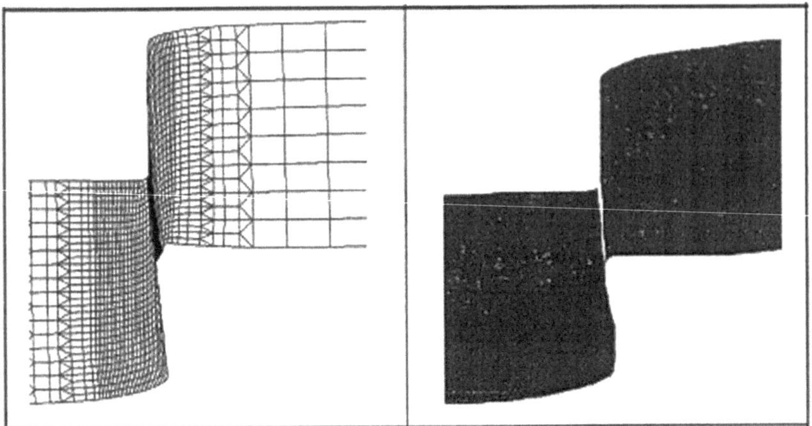

-c- Punch penetration = 70%	-d- Experimental profile (penetration = 70%)

Figure 2.28 Material shape at rupture. *From Hambli, R., 2001. Finite element model fracture prediction during sheet-metal blanking processes. Eng. Fract. Mech. 68(3), 365–378.*

fracture initiation was calculated, which was almost consistent with the punch displacement at ductile fracture initiation obtained experimentally.

Hambli and Potiron (2000) and Hambli (2001a) performed the simulation of axisymmetric shearing using a commercial elastic–plastic finite-element software into which a user subroutine for the Lemaitre damage model (1985) was incorporated. The experiment of axisymmetric shearing was performed using 1060 high carbon steel. The crack was assumed to initiate when the damage variable D reached a critical value, whereas the material was assumed to be completely damaged when D was equal to one, that is, the element stiffness matrix was equal to zero. Fig. 2.28 shows the material shape at rupture (Hambli, 2001a). The clearance between the punch and the die was assumed to be equal to 10%. Because the curvature radii of the corner parts of the punch and the die increased when the surfaces of the punch and the die wore, the simulation of the effect of the curvature radii of the corner parts of the punch and the die on the material shape at rupture was performed. With increasing the curvature radii of the corner parts of the punch and the die, the magnitude of the burr increased.

Hambli (2001b) performed the simulations of axisymmetric shearing by the elastic–plastic finite-element method using the Lemaitre damage model (1985) and also the Gurson yield function (1977). In the simulation using the Gurson yield function, the material was assumed to be completely damaged when the void volume fraction of the material was equal to one. At the punch displacement at which the material was ruptured in the experiment, the element in which the material was assumed to be completely damaged was located inside the clearance between the punch and the die in the simulation using the Lemaitre damage model. However, at the punch displacement at which the material was ruptured in the experiment, the element in which the material was assumed to be completely damaged was located outside the clearance between the punch and the die in the simulation using the Gurson yield function. Hence, it was concluded that the simulation using the Gurson yield function was not able to predict the rupture in shearing. However, the result that the simulation using the Gurson yield function was able to predict the rupture in shearing was demonstrated in subsequent researches.

Komori (2001, 2005) performed the simulation of axisymmetric shearing by the rigid-plastic finite-element method. A node separation method was used to perform the simulation of shearing in which a material was divided into two parts. In order to perform simulations precisely, a method of controlling the punch displacement such that an element was fractured in each step of the simulation, and a method of separating nodes at fracture, were proposed. The effect of various ductile fracture criteria on crack initiation and crack propagation during shearing was shown. The Mises yield function was used basically when various ductile fracture criteria were used. In the simulation using the Gurson yield function (1977), the evolution equation, which denoted the change in the void volume fraction, of either Eq. (6.8), in which A_0 was a material constant, or Eq. (6.9), in which A_1 was a material constant, was used. The material was assumed to fracture when the void volume fraction of the material was equal to 5% in the simulation using the Gurson yield function. The experiments of tensile test and axisymmetric shearing were performed using copper. The simulation of tensile test (Komori, 2002) was performed, and a material constant in each ductile fracture criterion was determined such that the reduction in the area calculated from the simulation agreed with the reduction in the area obtained in the experiment. Fig. 2.29 shows the material shape at rupture (Komori, 2005). The clearance between the punch and the die was assumed to be equal to 7%. The material shapes at rupture calculated using the Gurson yield function and the evolution equation of Eq. (6.8), the Brozzo et al. fracture criterion (1972), and the Oyane fracture criterion (1972), agreed with the material shape at rupture obtained in the experiment. The material shapes at

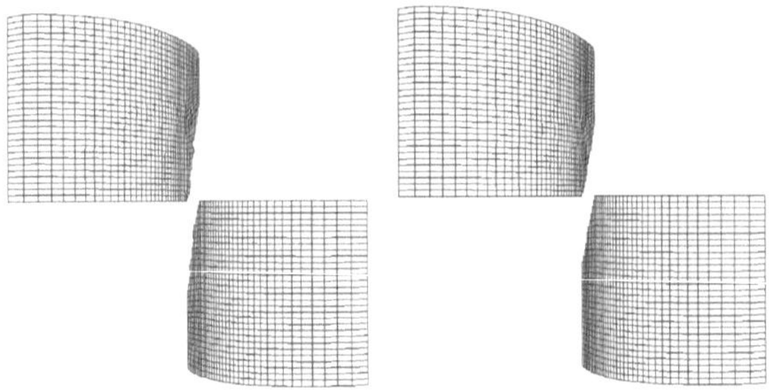

Figure 2.29 Material shape at rupture. (A) Gurson ($A_0 \neq 0$), (B) Gurson ($A_1 \neq 0$). *From Komori, K., 2005. Ductile fracture criteria for simulating shear by node separation method. Theoret. Appl. Fract. Mech. 43(1) 101–114.*

rupture calculated using the Gurson yield function and the evolution equation of Eq. (6.9), and the Cockcroft and Latham fracture criterion (1968), agreed approximately with the material shape at rupture obtained experimentally. However, the material shape at rupture calculated using the Freudenthal fracture criterion (1950) did not agree with the material shape at rupture obtained in the experiment.

2.8 Fracture in hole expanding

Takuda et al. (1999) obtained the hole expanding ratio in hole expanding using a mild steel and a high strength steel, and predicted the hole expanding ratio in hole expanding by the axisymmetric rigid-plastic finite-element method incorporating the Oyane fracture criterion (1972). The two material constants in the Oyane fracture criterion were determined by two types of material tests, that is, uniaxial tensile test and plane-strain tensile test. The experiments and the simulations of hole expanding using a conical-headed punch, a hemispherical-headed punch, and several kinds of flat-headed punches the curvature radii of the corner parts of which were different from each other, were performed. Fig. 2.30 shows the hole expanding ratios for various kinds of punches (Takuda et al., 1999). The material at the hole edge fractured when the conical-headed punch or the hemispherical-headed punch was used, whereas the material at the hole edge did not necessarily fracture, for example, the material which was in contact with the corner part of the punch fractured when a flat-headed punch was used.

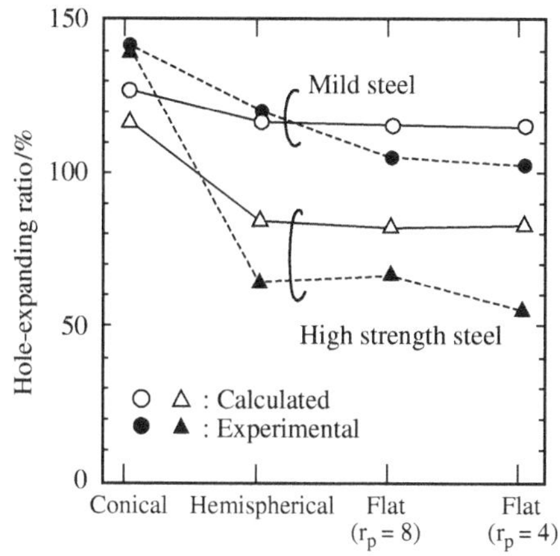

Figure 2.30 Hole expanding ratios for various kinds of punches. *From Takuda, H., Mori, K., Fujimoto, H., Hatta, N., 1999. Prediction of forming limit in bore-expanding of sheet metals using ductile fracture criterion. J. Mater. Process. Technol. 92–93, 433–438.*

Ko et al. (2007) obtained the hole expanding ratio in hole expanding using a high strength steel and a conical-headed punch, and predicted the hole expanding ratio in hole expanding using a commercial three-dimensional elastic–plastic finite-element software into which a user subroutine for various kinds of ductile fracture criteria was incorporated. The material at the upper hole edge fractured in the experiment. Material constants in various kinds of ductile fracture criteria were determined by the uniaxial tensile test and the plane-strain dome test. The hole expanding ratios calculated from the simulations using the Oyane fracture criterion (1972), the Brozzo et al. fracture criterion (1972), the Cockcroft and Latham fracture criterion (1968), and the ductile fracture criterion derived from the void growth model proposed by Rice and Tracey (1969), slightly differed from the hole expanding ratio obtained in the experiment. Hence, a new ductile fracture criterion was proposed, and the hole expanding ratio calculated from the simulation using the new ductile fracture criterion agreed with the hole expanding ratio obtained experimentally.

2.9 Fracture in tensile test

Hancock and Mackenzie (1976) obtained the relationship between the equivalent plastic strain at fracture initiation and the triaxiality of the stress state analytically, which was derived from the void growth model

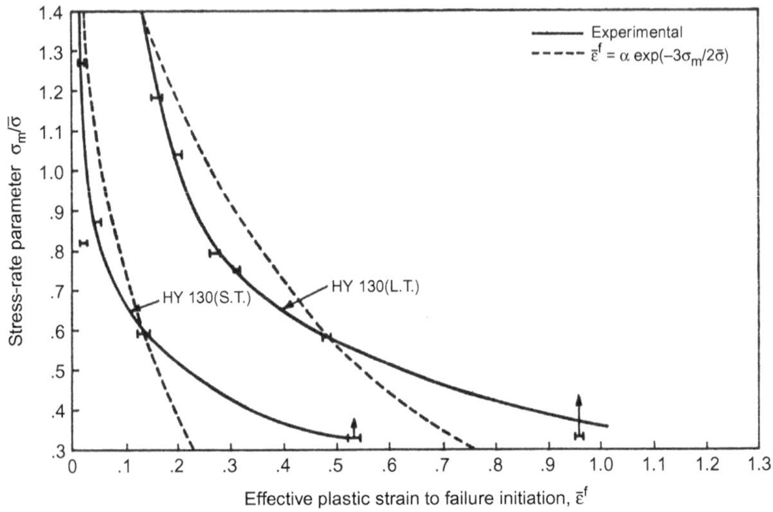

Figure 2.31 Relationship between equivalent plastic strain at fracture initiation and triaxiality of stress state for HY130. *From Hancock, J.W., Mackenzie, A.C., 1976. On the mechanisms of ductile failure in high-strength steels subjected to multi-axial stress-states. J. Mech. Phys. Solids 24(2–3), 147–169.*

proposed by Rice and Tracey (1969) and the assumption that the equivalent plastic strain at fracture initiation was inversely proportional to the hole growth rate. The tensile test using notched round tensile specimens was performed. Fig. 2.31 shows the relationship between the equivalent plastic strain at fracture initiation and the triaxiality of the stress state for HY130 (Hancock and Mackenzie, 1976). A material constant in the relationship between the equivalent plastic strain at fracture initiation and the triaxiality of the stress state was determined such that the relationship calculated from the analysis agreed with the relationship obtained experimentally for a specified circumferentially notched tensile specimen. A relationship similar to the relationship between the equivalent plastic strain at fracture initiation and the triaxiality of the stress state, was derived from the void growth model proposed by McClintock (1968a).

Tvergaard and Needleman (1984) performed the simulation of the tensile test of a bar by the axisymmetric elastic–plastic finite-element method using the Gurson yield function (1977) modified by Tvergaard (1981), and obtained the cup-and-cone fracture. The crack propagated in the radial direction of the bar at the central part of the bar, whereas the crack propagated in both the radial direction and the axial direction of the bar near the surface of the bar. Fig. 2.32 shows the finite-element mesh

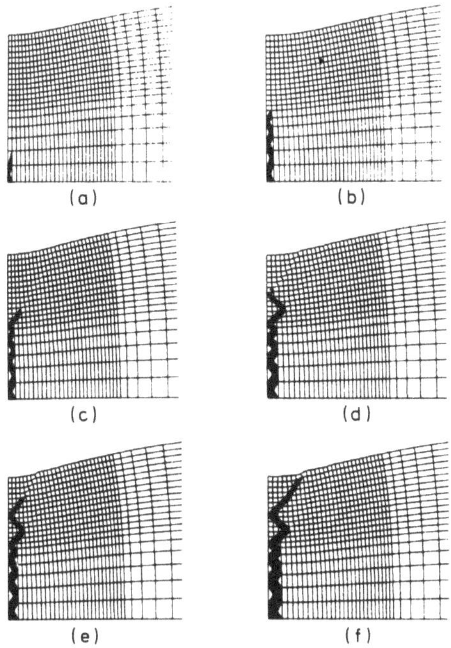

Figure 2.32 Finite-element mesh and fractured elements. *From Tvergaard, V., Needleman, A., 1984. Analysis of the cup-cone fracture in a round tensile bar. Acta Metall. 32(1), 157–169.*

and the fractured elements (Tvergaard and Needleman, 1984). The fractured element represented the crack, and the crack shape depended on the finite-element mesh. When the number of finite elements was not sufficient, the crack propagated in the radial direction of the bar, and no cup-and-cone fracture was obtained. The void volume fraction above which the void volume fraction increased drastically was assumed to be 15%, whereas the void volume fraction at fracture at which the flow stress of the fractured element was equal to zero was assumed to be 25%.

Needleman and Tvergaard (1984) performed the simulations of the axisymmetric tensile test of a notched bar and the plane-strain tensile test of a notched sheet by the elastic–plastic finite-element method using the Gurson yield function (1977) modified by Tvergaard (1981), and compared the simulation result with the experimental result obtained by Hancock and Brown (1983). The void volume fraction above which the void volume fraction increased drastically was assumed to be 15%, whereas the void volume fraction at fracture at which the flow stress of the fractured element was equal to zero was assumed to be 25%.

The material initially fractured at the center in the radial direction of the notched bar both in the simulation and in the experiment. The material initially fractured at the center or at the surface in the width direction of the notched sheet in the simulation, whereas the material initially fractured at the center, at the surface, or both at the center and at the surface in the width direction of the notched sheet in the experiment. The location of fracture initiation in the notched sheet calculated from the simulation almost agreed with the location of fracture initiation in the notched sheet obtained in the experiment.

Besson et al. (2001) performed the simulation of the tensile test of a bar by the axisymmetric elastic–plastic finite-element method using the Rousselier yield function (1987) and the Gurson yield function (1977) modified by Tvergaard (1981), and obtained the cup-and-cone fracture. The localization analysis was performed using the Rice condition for bifurcation (Rice, 1977; Rice and Rudnicki, 1980), and the simulation was performed beyond the localization. Although axisymmetry was assumed, line symmetry was not assumed so that the entire material was discretized by the finite-element mesh. Fig. 2.33 shows the finite-element mesh and the void volume fraction calculated using the Rousselier yield function (Besson et al., 2001). The black part indicated the element the

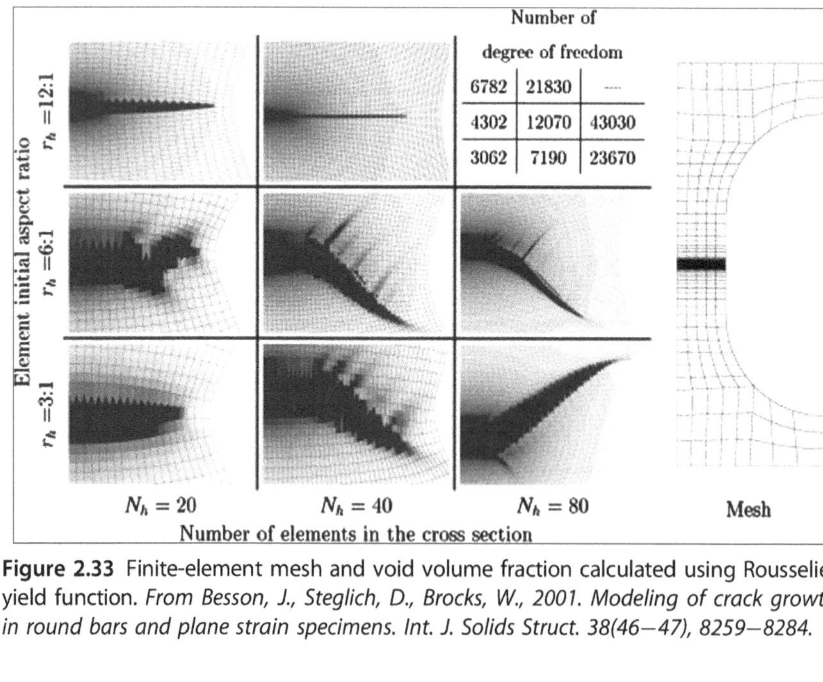

Figure 2.33 Finite-element mesh and void volume fraction calculated using Rousselier yield function. *From Besson, J., Steglich, D., Brocks, W., 2001. Modeling of crack growth in round bars and plane strain specimens. Int. J. Solids Struct. 38(46–47), 8259–8284.*

void volume fraction of which was larger than 10%, and the black part was assumed to be fractured. The shape of the fractured part depended on the finite-element mesh. The cup-and-cone fracture was easily formed when the simulation was performed using the Rousselier yield function, whereas the cup-and-cone fracture was not easily formed when the simulation was performed using the Gurson yield function.

Besson et al. (2003) performed the simulation of the tensile test of a plate by the plane-strain elastic–plastic finite-element method using the Rousselier yield function (1987) and the Gurson yield function (1977) modified by Tvergaard (1981), and obtained slant fracture. The localization analysis was performed using the Rice condition for bifurcation (Rice, 1977; Rice and Rudnicki, 1980), and the simulation was performed beyond the localization. Although vertical line symmetry was not assumed, horizontal line symmetry was assumed so that the right half of the material was discretized by the finite-element mesh. Fig. 2.34 shows the finite-element mesh and the void volume fraction calculated using the Rousselier yield function (Besson et al., 2003). The black part indicated the element the void volume fraction of which was larger than 10%, and the black part was assumed to be fractured. The slant fracture was easily

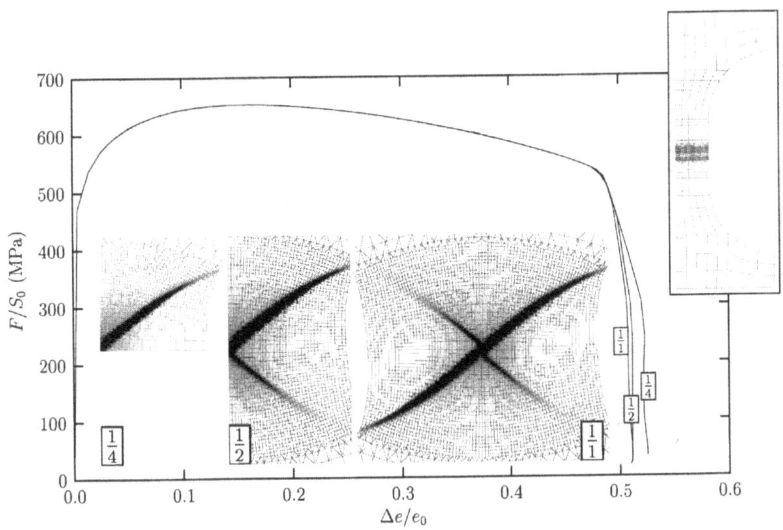

Figure 2.34 Finite-element mesh and void volume fraction calculated using Rousselier yield function. *From Besson, J., Steglich, D., Brocks, W., 2003. Modeling of plane strain ductile rupture. Int. J. Plast. 19(10), 1517–1541.*

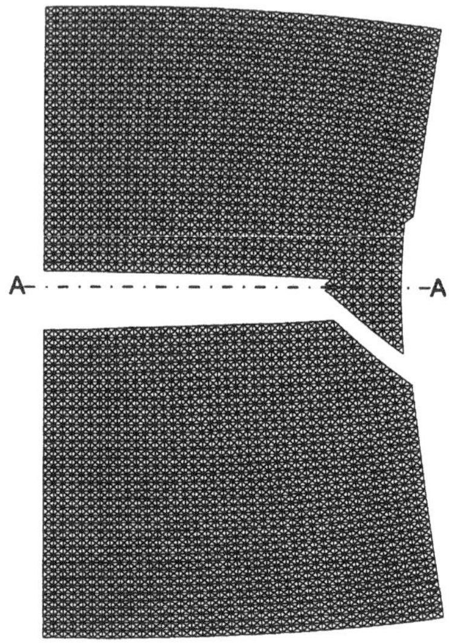

Figure 2.35 Finite-element mesh at rupture. *From Scheider, I., Brocks, W., 2003. Simulation of cup–cone fracture using the cohesive model. Eng. Fract. Mech. 70(14), 1943–1961.*

formed when the simulation was performed using either the Rousselier yield function or the Gurson yield function.

Scheider and Brocks (2003) performed the simulation of the tensile test of a bar by the axisymmetric elastic–plastic finite-element method using the cohesive zone model (Barenblatt, 1962). In the cohesive zone model, two neighboring elements were connected by an interface element at a specified traction, which was a function of the distance between the two sides of the interface element which also belonged to the two neighboring elements. The interface element suffered not only the normal traction, which was a function of the distance in the normal direction between the two sides of the interface element, but also the tangential traction, which was a function of the distance in the tangential direction between the two sides of the interface element. Because the interface element fractured when the interface element suffered a specified energy due to the normal traction or the tangential traction, a crack propagated along the boundary of the two neighboring elements. Fig. 2.35 shows the finite-element mesh at rupture (Scheider and Brocks, 2003). Although axisymmetry was assumed, line symmetry was not assumed so that the entire material was discretized by the finite-element mesh.

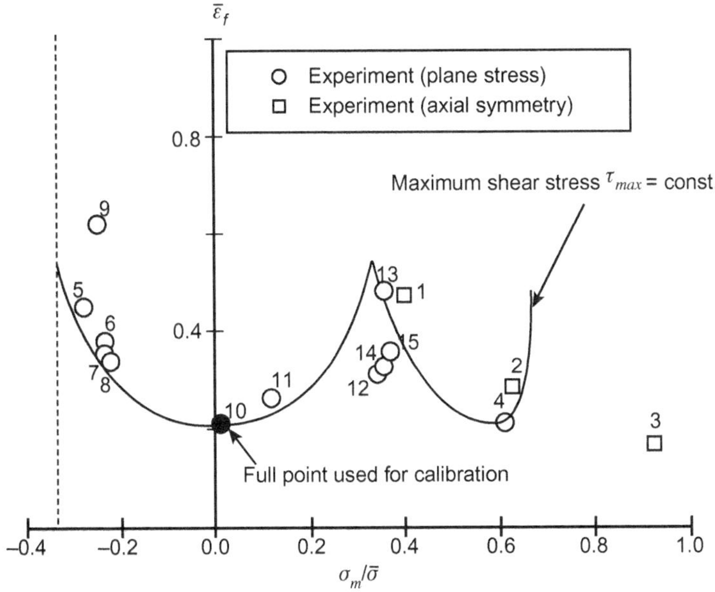

Figure 2.36 Relationship between equivalent plastic strain to fracture and stress triaxiality in case of constant maximum shear stress fracture criterion. *From Wierzbicki, T., Bao, Y., Lee, Y.W., Bai, Y., 2005. Calibration and evaluation of seven fracture models. Int. J. Mech. Sci. 47(4–5), 719–743.*

2.10 Fracture in other material tests

Wierzbicki et al. (2005) evaluated seven kinds of ductile fracture criteria in plane-stress state using the relationship between the equivalent plastic strain to fracture and the stress triaxiality obtained by 15 kinds of material tests using 2024-T351 aluminum alloy. The constant equivalent plastic strain fracture criterion, the constant plastic strain in the thickness direction fracture criterion, the constant maximum shear stress fracture criterion, the Johnson and Cook fracture criterion (1985), and the Cockcroft and Latham fracture criterion (1968) were included in the seven kinds of ductile fracture criteria. The relationship between the equivalent plastic strain to fracture and the stress triaxiality obtained by the experiment, was obtained with the aid of the simulation using a commercial elastic–plastic finite-element software. Fig. 2.36 shows the relationship between the equivalent plastic strain to fracture and the stress triaxiality in the case of the constant maximum shear stress fracture criterion (Wierzbicki et al., 2005). The relationship calculated using the constant maximum shear stress fracture criterion agreed most with the relationship obtained experimentally of the relationships calculated using the seven kinds of ductile fracture criteria.

Bai and Wierzbicki (2010) revisited the Mohr–Coulomb fracture criterion, which had been widely used for relatively brittle materials and in which the equivalent plastic strain to fracture depended on the stress triaxiality and also the Lode parameter, to use the Mohr–Coulomb fracture criterion for ductile materials. Because the Mohr–Coulomb fracture criterion had eight material constants, the effect of each material constant on the equivalent plastic strain to fracture was shown. The constant maximum shear stress fracture criterion was intrinsically included in the Mohr–Coulomb fracture criterion; the Mohr–Coulomb fracture criterion was identical to the constant maximum shear stress fracture criterion when the specified material constant was assumed to be zero. The Mohr–Coulomb fracture criterion was evaluated using the relationship between the equivalent plastic strain to fracture and the stress triaxiality obtained by fifteen kinds of material tests using 2024-T351 aluminum alloy (Wierzbicki et al., 2005), and the validity of the Mohr–Coulomb fracture criterion in plane-stress state was demonstrated.

CHAPTER 3

Microscopic ductile fracture phenomena

3.1 Introduction

In Chapter 1, Macroscopic ductile fracture phenomena, which were observed experimentally using an optical microscope (not an electron microscope), were mainly described to utilize the observed phenomena in Chapter 2, Macroscopic ductile fracture criteria.

Microscopic ductile fracture phenomena are divided into the following three processes: void nucleation, void growth, and void coalescence (Dodd and Bai, 1987). There are a number of review papers on microscopic ductile fracture phenomena (Rosenfield, 1968; Goods and Brown, 1979; Wilsdorf, 1983; Van Stone et al., 1985; Garrison and Moody, 1987; Thompson, 1987).

Tipper (1949) performed the tensile test using mild steel, observed sectioned tensile specimens using an optical microscope, and showed that voids nucleated around inclusions. Puttick (1959) performed the tensile test using tough pitch high conductivity copper, observed sectioned tensile specimens using an optical microscope, and showed that voids nucleated due to either the separation of inclusions from the matrix or the cracking of inclusions, grew during the tensile test, and eventually coalesced to form a crack which yielded final rupture.

Rogers (1960) performed the tensile test using oxygen-free high thermal conductivity copper, observed the longitudinal section of the tensile specimens using both an optical microscope and an electron microscope. Voids nucleated at grain boundaries in the hydrogen-treated copper and within the grain in the nitrogen-treated copper approximately when necking appeared, grew particularly in the necked region with increasing the plastic deformation, and eventually coalesced to form a crack at the center of the neck.

In this chapter, Microscopic ductile fracture phenomena, which are observed experimentally using an electron microscope (not an optical

Ductile Fracture in Metal Forming
DOI: https://doi.org/10.1016/B978-0-12-814772-6.00003-5

microscope), are mainly described to utilize the observed phenomena in Chapter 4, Microscopic ductile fracture criteria.

Microscopic ductile fracture phenomena are divided into void nucleation, void growth, and void coalescence. Although it is not necessarily easy to classify research on microscopic ductile fracture phenomena into a research on void nucleation, a research on void growth, and a research on void coalescence, research on microscopic ductile fracture phenomena is classified into these three researches for convenience.

In the following sections, classical papers on microscopic ductile fracture phenomena are introduced for each of the following processes: void nucleation, void growth, and void coalescence.

3.2 Microscopic phenomena of void nucleation

Gurland and Plateau (1963) performed the experiment of bending thin plates using an aluminum casting alloy, Armco iron, and a ferrite–pearlite steel, and observed the polished convex surface of the bent thin plates using an optical microscope. In the aluminum casting alloy, cracks were observed in the silicon inclusion. In Armco iron, cracks were observed at the interface between the matrix and the inclusion. In the ferrite–pearlite steel, cracks were observed at the ferrite–pearlite interface and in the ferrite matrix. An attempt was made to relate the fracture strain to the inclusion volume fraction in uniaxial tension on the basis of the change in the inclusion shape due to uniaxial tension. The relationship between the inclusion volume fraction and the fracture strain obtained analytically agreed with the relationship between the summation of the void volume fraction and the inclusion volume fraction and the fracture strain obtained experimentally by Edelson and Baldwin (1962). The relationship obtained analytically under the assumption of the soft inclusion, the flow stress of which was equal to the flow stress of the matrix, hardly differed from the relationship obtained analytically under the assumption of the hard inclusion, the flow stress of which was infinite.

Clausing (1967) performed the tensile test of mild steel using notched round tensile specimens and smooth round tensile specimens, observed the longitudinal section of the tensile specimens using an optical microscope, and studied the developments of microcracks and macrocracks. The microcrack was defined as a crack the size of which was approximately equal to the size of the microstructure of the mild steel, whereas the macrocrack was defined as a crack the size of which was much larger than the size of the microstructure of the mild steel. Microcracks, which were mainly either intrapearlitic or interpearlitic, developed prior to the appearance of macrocracks. With increasing the deformation of the specimen, the number of microcracks increased, whereas the size of a microcrack did not increase. Although voids, the size of which were approximately tens of micrometer, were observed experimentally in copper by Puttick (1959) and Rogers (1960), voids were not observed prior to the appearance of the macrocrack, and were created during the development of the macrocrack.

Palmer and Smith (1968) performed the tensile test of a sheet using copper alloys that contained silica particles, beryllia particles, or alumina particles, prepared a thin foil which was machined from a part adjacent to the fracture surface, observed the surface of the thin foil using an electron microscope, and studied the fracture behavior of the copper alloys. Voids nucleated due to the separation of the silica particles, beryllia particles, or alumina particles from the matrix. The tensile test was performed at 293K and 77K. When the tensile test of a sheet using a copper alloy that contained silica particles was performed, the fraction of the silica particles that yielded voids increased with increasing the diameter of the silica particles, whereas the fraction of the silica particles that yielded voids decreased with increasing the temperature.

Liu and Gurland (1968) performed the tensile test of a bar using various kinds of spheroidized steels, the carbon content of which ranged between 0.065% and 1.46%, observed the section of the specimen using an optical microscope and an electron microscope, and studied the fracture behavior of the spheroidized steels. Voids nucleated due to either the separation of the carbide particles from the matrix or the cracking of the carbide particles. In low carbon spheroidized steels, since the carbide particles were principally located on the grain boundaries, voids nucleated and grew on the grain boundaries, whereas in high carbon spheroidized steels, since the carbide particles were not necessarily located on the grain boundaries, voids did not necessarily nucleate and grow on the grain boundaries.

Miller and Smith (1970) performed the tensile test of a bar using various kinds of plain carbon steels, the carbon content of which ranged between 0.1% and 1.2%. Miller and Smith also observed the section of the specimen and the fracture surface of the specimen using an optical microscope and an electron microscope, and suggested a mechanism for the shear cracking in pearlite. The inclusion cracking was less important

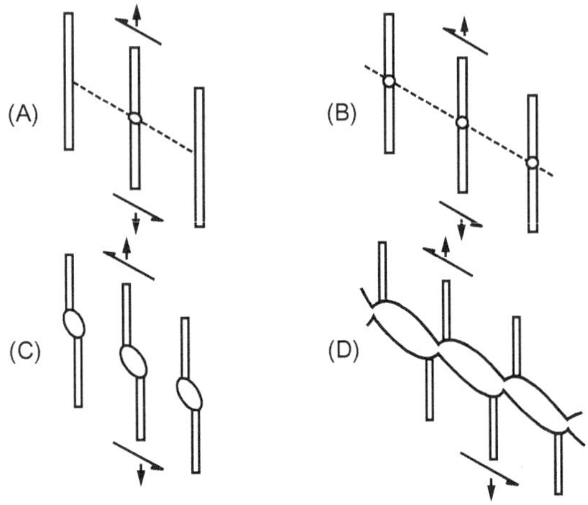

Figure 3.1 Mechanism for shear cracking in pearlite: (A) cracking of a cementite plate, (B) shear zone developing in the ferrite causing cracking of adjacent plates, (C) void growth, (D) void coalescence. *From Miller, L.E., Smith, G.C., 1970. Tensile fracture in carbon steels. J. Iron Steel Inst. 208 (11), 998–1005.*

than the pearlite cracking when inclusions were not large or the carbon content was not low. The orientation of the crack and the orientation of the cementite lamellae in the longitudinal section of the specimen were measured using a large number of cracked pearlite colonies. The angle between the crack and the tensile axis was approximately equal to 50 degrees, whereas the angle between the cementite lamellae and the tensile axis was approximately equal to 5 degrees. Fig. 3.1 shows the mechanism for the shear cracking in pearlite (Miller and Smith, 1970). First, a crack was formed in a cementite lamella. Next, cracks were formed in the adjacent cementite lamellae due to the shear deformation of the ferrite lamellae adjacent to the cementite lamella. Finally, voids were formed and coalesced in the ferrite lamellae and the cementite lamellae.

Gurland (1972) performed the tensile test, the compression test, and the torsion test of a bar using a hypereutectoid steel having a spheroidized structure, and observed the longitudinal section of the specimen using an optical microscope, and studied the initiation of fracture in the microstructure. A crack was observed in a cementite particle in the tensile test, the compression test, and the torsion test. In the tensile test, the length direction of the crack in the cementite particle was perpendicular to the tensile direction, whereas in the compression test, the length direction of the crack in the cementite particle was parallel to the compression

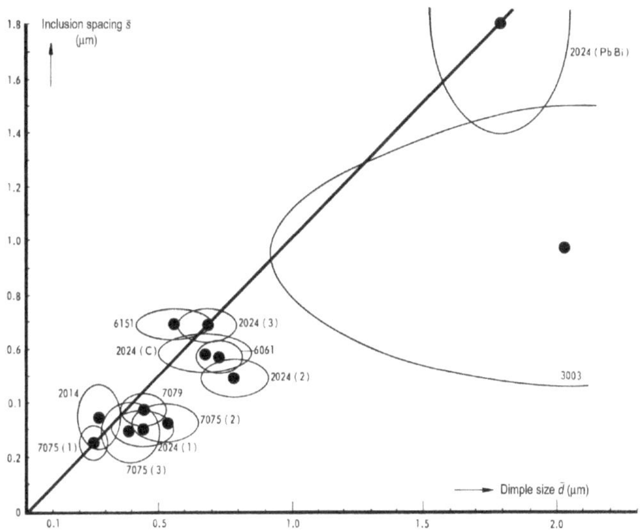

Figure 3.2 Relationship between average dimple spacing and average inclusion distance. *From Broek, D., 1973. The role of inclusions in ductile fracture and fracture toughness. Eng. Fract. Mech. 5 (1), 55–66.*

direction. Hence, the length direction of the crack in the cementite particle was perpendicular to the direction of the maximum principal strain.

Broek (1973) performed the tensile test of a sheet using thirteen kinds of aluminum alloys, observed the fracture surface of the specimen and the section of the specimen using an optical microscope and an electron microscope, and measured the average dimple spacing and the average inclusion distance. There were large inclusions at which the nucleation of voids was observed using an optical microscope and small inclusions at which the nucleation of voids was observed using an electron microscope. If large inclusions nucleated voids, the dimple spacing of the voids should be equal to the distance of the large inclusions. If small inclusions nucleated voids, the dimple spacing of the voids should be equal to the distance of the small inclusions. Fig. 3.2 shows the relationship between the average dimple spacing and the average inclusion distance (Broek, 1973). The ellipses, which surrounded the measurements, indicated the standard deviation of the measurements. The average dimple spacing was almost equal to the average distance of inclusions, which were small inclusions. The experimental result that the average dimple spacing was almost equal to the average inclusion distance indicated that the small inclusions nucleated voids which grew and coalesced. A dislocation model for the void nucleation due to the separation of the inclusions from the matrix, which was based on the solution proposed by Smith (1968), was derived. The relationship between the inclusion volume fraction and the fracture strain calculated using the dislocation model agreed qualitatively with the relationship between the inclusion volume fraction and the fracture strain obtained from the experiment.

Argon and Im (1975) performed the tensile test of spheroidized 1045 steel, Cu–0.6% Cr alloy, and maraging steel using notched round tensile specimens and smooth round tensile specimens; Argon and Im observed the longitudinal section of the specimen using an electron microscope, and estimated the interfacial stress between the inclusions and the matrix for the separation of the inclusions from the matrix. Fig. 3.3 shows the relationship between the coordinate in the axial direction and the fraction of the separated inclusions for the spheroidized 1045 steel (Argon and Im, 1975). The coordinate in the axial direction at which the fraction of the separated inclusions was equal to zero, was obtained by extrapolation using the relationship between the coordinate in the axial direction and the fraction of the separated inclusions. The interfacial stress between the inclusions and the matrix at the coordinate in the axial direction at which the fraction of the separated inclusions was equal to zero, was estimated

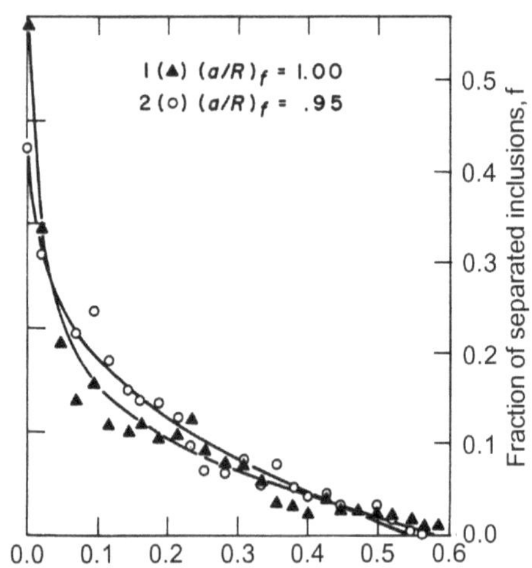

Figure 3.3 Relationship between coordinate in axial direction and fraction of separated inclusions for the spheroidized 1045 steel. *From Argon, A.S., Im, J., 1975. Separation of second phase particles in spheroidized 1045 steel, Cu–0.6pct Cr alloy, and maraging steel in plastic straining. Metall. Trans. A 6 (4), 839–851.*

using the simulation results by Argon et al. (1975a,b). The estimated interfacial stress between the inclusions and the matrix was assumed to be the interfacial stress between the inclusions and the matrix for the separation of the inclusions from the matrix. Because in the spheroidized 1045 steel, the inclusion population was dense and the interaction between two neighboring inclusions was not able to be ignored; the estimated interfacial stress between the inclusions and the matrix was corrected with reference to Argon et al. (1975b).

Inoue and Kinoshita (1977a) performed the tensile test of a bar using various kinds of ferrite—pearlite steels, the carbon content of which ranged between 0.05% and 0.91%. Inoue and Kinoshita observed the longitudinal section of the specimen using an optical microscope and an electron microscope, and investigated the process of the void nucleation in pearlite nodules. It was obvious that the strain received by the pearlite nodule was different from the strain received by the specimen. The strain of the specimen was calculated from the diameter of the specimen, whereas the strain of the pearlite nodule was calculated from the shape of the perlite nodule. Fig. 3.4 shows the relationship between the strain of the specimen and the strain of the perlite nodule (Inoue and Kinoshita, 1977a). The strain of the perlite nodule at which voids nucleated in the perlite nodule scarcely depended on the carbon content of the material. Furthermore, the stress on the perlite nodule at which voids nucleated in the perlite nodule hardly depended on the carbon content of the material.

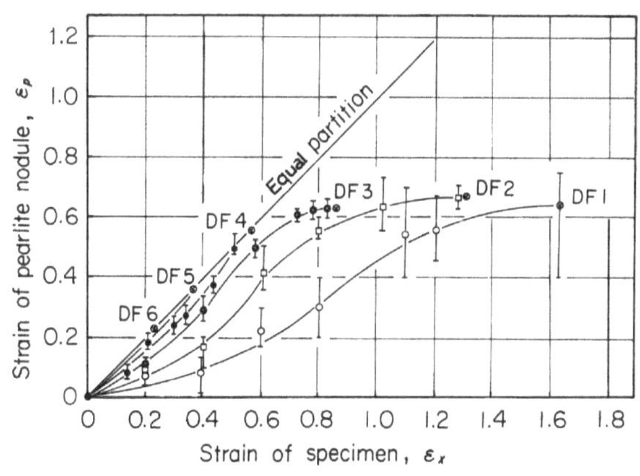

Figure 3.4 Relationship between strain of specimen and strain of perlite nodule. *From Inoue, T., Kinoshita, S., 1977a. Strain partitioning and void formation in ferrite—pearlite steels deformed in tension. Trans. ISIJ 17 (5), 245—251.*

Inoue and Kinoshita (1977b) performed the tensile test of a bar using various kinds of spheroidized steels and ferrite–pearlite steels, the carbon content of which ranged between 0.11% and 0.91%. Inoue and Kinoshita observed the section of the specimen using a scanning electron microscope and a transmission electron microscope, and investigated the process of the void nucleation in spheroidal carbides and pearlite nodules. In both spheroidized steels and ferrite–pearlite steels, with increasing the deformation of the material, the dislocation density around the spheroidal carbides or the pearlite nodules increased, and the spheroidal carbides or the pearlite nodules became surrounded by tangled dislocations. In spheroidized steels, voids nucleated due to either the separation of the spheroidal carbides from the matrix or the cracking of the spheroidal carbides, whereas in ferrite–pearlite steels, voids nucleated due to the cracking of the cementite plates in the pearlite nodules. In low or medium carbon ferrite–pearlite steels, voids grew considerably before the void coalescence because of the large volume fraction of the ferrite matrix, whereas in high carbon ferrite–pearlite steels, voids grew slightly before the void coalescence because of the small volume fraction of the ferrite matrix.

Osakada et al. (1977, 1978) performed the tensile test and the torsion test under various hydrostatic pressure using a pure iron and five kinds of plain carbon steels, and obtained the void nucleation strain and the fracture strain. The void nucleation strain was obtained by changing the hydrostatic pressure during the tensile test or the torsion test from low hydrostatic pressure to high hydrostatic pressure or from high hydrostatic pressure to low hydrostatic pressure. Fig. 3.5 shows the relationships between the mean normal stress and the void nucleation strain and between the mean normal stress and the fracture strain of S25C (Osakada et al., 1977). The mean normal stress was calculated from the approximate stress distribution at the neck of the specimen in the tensile test (Bridgman, 1952). The void nucleation strain was found to correspond to the strain at which cracks began to propagate into the ferrite matrix by the observation using a scanning electron microscope. The ductile fracture criterion, which gave the fracture strain under arbitrary mean normal stress during the tensile test or the torsion test, was proposed with reference to the void nucleation strain and the fracture strain obtained experimentally.

Thomson and Nayak (1980) performed the uniaxial tension and the balanced biaxial tension of a sheet using a cold-rolled aluminum stabilized steel, and measured, in each increment of deformation, the surface roughness and the thickness of the sheet, and the summation of the length of

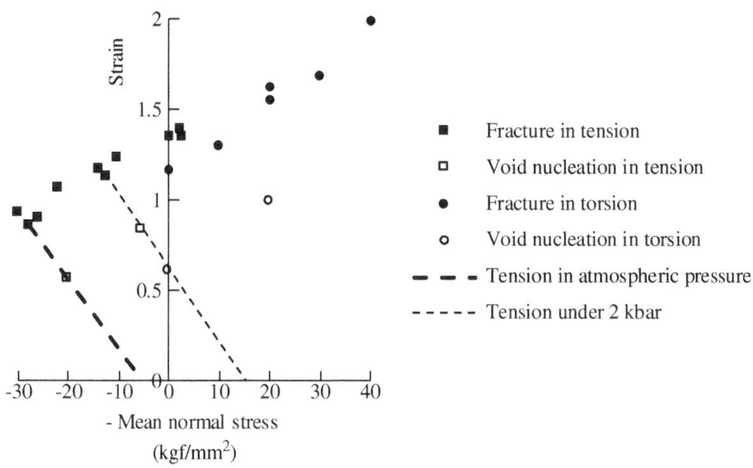

Figure 3.5 Relationships between mean normal stress and void nucleation strain and between mean normal stress and fracture strain of S25C. *From Osakada, K., Watadani, A., Sekiguchi, H., 1977. Ductile fracture of cabin steel under cold metal forming conditions: 1st repor. Tension and torsion tests under pressure. Bull. JSME 20 (150), 1557–1565.*

voids in the thickness direction of the sheet, in the longitudinal section of the sheet. Fig. 3.6 shows the relationship between the equivalent strain and the internal and external thickness reduction factors (Thomson and Nayak, 1980). The internal thickness reduction factor decreased with increasing the summation of the length of voids in the thickness direction of the sheet, whereas the external thickness reduction factor decreased with increasing the surface roughness of the sheet. In uniaxial tension, since the external thickness reduction factor was much smaller than the internal thickness reduction factor when local necking, at which the equivalent strain was equal to 0.26, occurred, voids were not a cause of local necking but a consequence of local necking. In balanced biaxial tension, the external thickness reduction factor was larger than the internal thickness reduction factor when local necking, at which the equivalent strain was equal to 0.36, occurred.

Beremin (1981) performed the tensile test of A508 steel using notched round tensile specimens in various ambient low temperatures, observed the longitudinal section of the tensile specimen using an optical microscope, and obtained the condition of void nucleation from manganese sulfide inclusions. First, the region in the longitudinal section of the tensile specimen, in which either the separation of the inclusions from the matrix

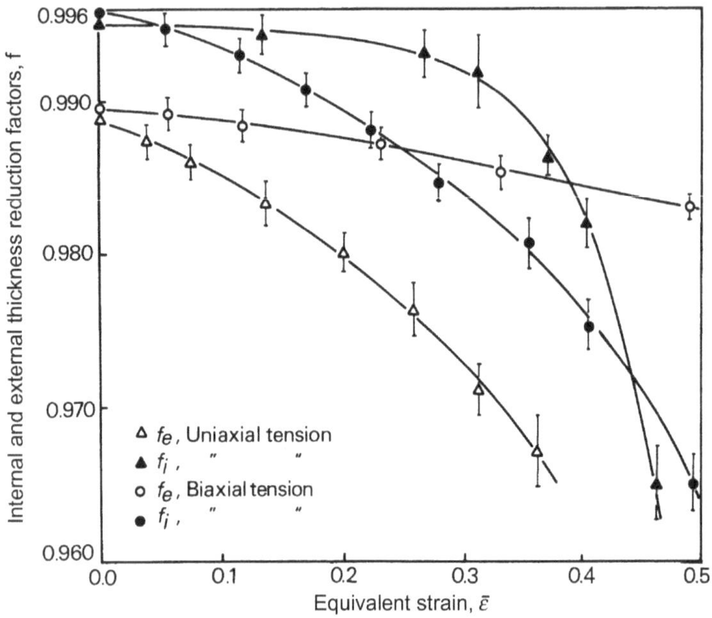

Figure 3.6 Relationship between equivalent strain and internal and external thickness reduction factors. *From Thomson, P.F., Nayak, P.U., 1980. An experimental investigation into the development of thickness non-uniformities leading to failure in sheet steel. J. Mech. Work. Technol. 4 (3), 223–232.*

or the cracking of the inclusions occurred, was determined. Next, the stress induced in the inclusions due to the difference between the deformation of the inclusions and the deformation of the matrix was calculated with reference to Berveiller and Zaoui (1978), whereas the stress distribution in the longitudinal section of the tensile specimen, which had been already calculated using the elastic–plastic finite-element software (Beremin, 1980), was utilized. Finally, the interfacial stress between the matrix and the inclusion at void nucleation was estimated. Fig. 3.7 shows the relationship between the difference between the equivalent stress and the yield stress and the maximum principal stress at void nucleation (Beremin, 1981). When the axis of the tensile specimen was made to coincide with the circumferential direction of a shell, the cracking of the inclusions occurred, whereas when the axis of the tensile specimen was made to coincide with the thickness direction of the shell, the separation of the inclusions from the matrix occurred. The oblique lines indicated the estimated interfacial stress between the matrix and the inclusion at void nucleation.

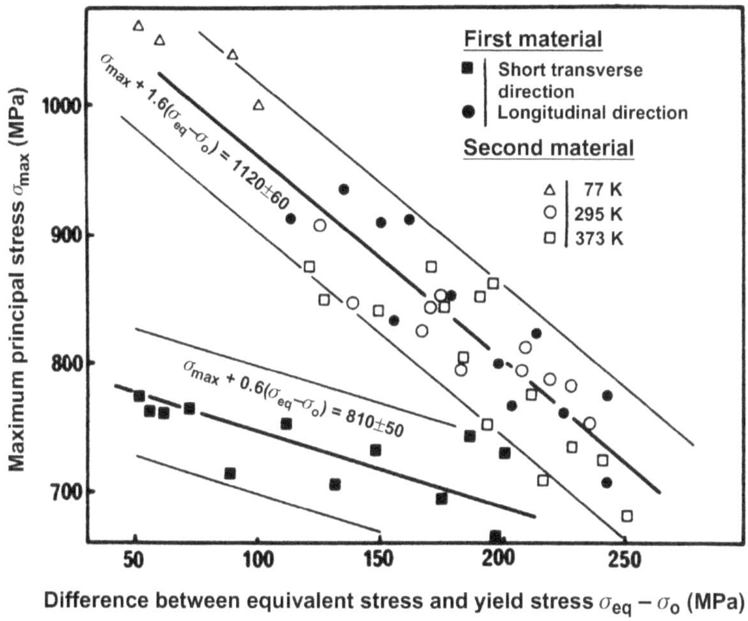

Figure 3.7 Relationship between the difference between equivalent stress and yield stress and maximum principal stress at void nucleation. *From Beremin, F.M., 1981. Cavity formation from inclusions in ductile fracture of A508 steel. Metall. Trans. A 12 (5), 723–731.*

Fisher and Gurland (1981a) performed the tensile test of a bar using low carbon spheroidized steel and medium carbon spheroidized steel, observed the longitudinal section and the cross section of the specimen using an electron microscope, and studied systematically void nucleation, to validate the model proposed by Fisher and Gurland (1981b). Although voids nucleated due to the separation of cementite particles from ferrite matrix, voids nucleated principally on the surface of large cementite particles which were located ferrite grain boundaries. The difference between the rate of void volume fraction and the rate of void volume fraction due to void growth was equal to the rate of void volume fraction due to void nucleation, which was calculated using the rate of void volume fraction obtained experimentally and the void growth model proposed by Rice and Tracey (1969).

Beaver (1983) obtained the fracture forming limit in punch stretching of a sheet using two kinds of medium strength age-hardened Al–Mg–Si alloys AA 6009 and AA 6082, observed the longitudinal section of the

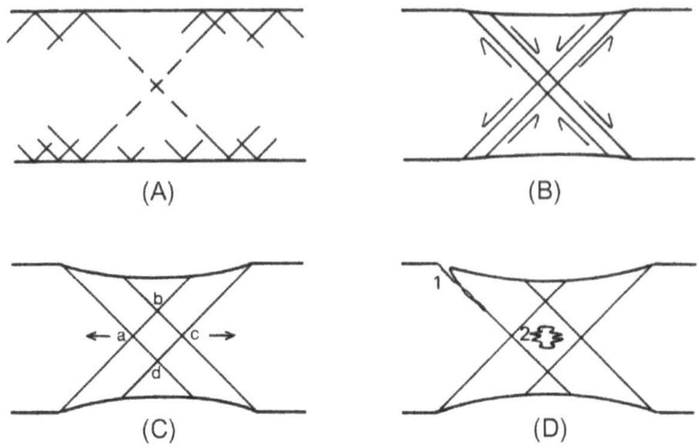

Figure 3.8 Schematic representation of the mechanism of void nucleation for aluminum alloys in biaxial tension: (A) shear zone formation at the sheet surface, (B) shear instability band formation when shear zone has propagated through the sheet thickness, (C) shear instability band growth and enlargement of the plane-strain region "abcd", (D) fracture initiation by void nucleation and growth from either the sheet surface or within the central plane-strain region. *From Beaver, P.W., 1983. Localized thinning, fracture and formability of aluminium sheet alloys in biaxial tension. J. Mech. Work. Technol. 7 (3), 215–231.*

specimen using an optical microscope and the fracture surface of the specimen using an electron microscope, and identified the mechanism of void nucleation. Fig. 3.8 shows the schematic representation of the mechanism of the void nucleation for the aluminum alloys in biaxial tension (Beaver, 1983). First, two shear bands, which were inclined by approximately 45 degrees against the thickness direction of the sheet and which intersected at the center in the thickness direction of the sheet, appeared. Next, necking of the sheet occurred in the thickness direction of the sheet, and the region in which the two shear bands intersected was subjected to plane-strain deformation. Finally, voids nucleated, grew, and coalesced in the region in which the two shear bands intersected. Void nucleation and void growth were not observed prior to the formation of the two intersecting shear bands.

Thomson and Hancock (1984) performed the tensile test of a commercially pure iron using notched round tensile specimens and notched plate tensile specimens. Thomson and Hancock observed the longitudinal section of the tensile specimens using an optical microscope, and studied the condition of void nucleation and the strength of the interface between the matrix and the inclusion. The stress distribution and the strain distribution in the notched round tensile specimen was calculated using a

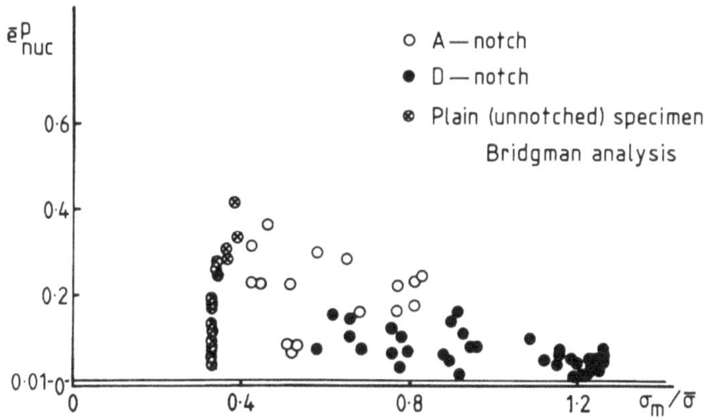

Figure 3.9 Relationship between stress triaxiality and equivalent plastic strain at void nucleation in axisymmetric tensile test. *From Thomson, R.D., Hancock, J.W., 1984. Ductile failure by void nucleation, growth and coalescence. Int. J. Fract. 26 (2), 99–112.*

commercial elastic–plastic finite-element software. Fig. 3.9 shows the relationship between the stress triaxiality and the equivalent plastic strain at void nucleation in the axisymmetric tensile test (Thomson and Hancock, 1984). Although voids nucleated exclusively due to the separation of the iron oxide particles from the matrix, no criterion was obtained on the nucleation of voids. The radial stress on the interface between the matrix and the inclusion was within the range of 0 to 24 times the yield stress of the matrix.

Babout et al. (2004a) performed using a X-ray tomograph (Maire et al., 2001) the in situ tensile test of commercially pure aluminum and aluminum alloy 2124 reinforced by spherical ceramic particles whose volume fractions were 4% or 20%. Babout et al. studied the effect of the flow stress of the matrix and the volume fraction of the inclusions on void nucleation. When the matrix was the pure aluminum, the separation of the inclusions from the matrix principally appeared irrelevantly to the volume fraction of the ceramic particles, whereas when the matrix was the aluminum alloy, the cracking of the inclusions principally appeared irrelevantly to the volume fraction of the ceramic particles. With increasing the flow stress of the matrix, the numerical fraction of the separation of the inclusions from the matrix or the cracking of the inclusions significantly increased when the elongation of the specimen was specified.

Benzerga et al. (2004a) performed the tensile test of X52 steel plate using notched round tensile specimens in various notch-root radii, observed the fracture surface, the longitudinal section, and the cross section of the specimen using both an optical microscope and an electron microscope, and studied on the anisotropies of void nucleation, void growth, and void coalescence. The tensile direction was made to coincide with either the rolling direction, the width direction, or the thickness direction of the steel plate. Manganese sulfide inclusions were elongated during rolling in the rolling direction of the steel plate. When the tensile direction coincided with the rolling direction of the steel plate, voids nucleated principally due to the cracking of the inclusion, followed by the separation of the inclusions from the matrix along the interface parallel to the tensile direction. When the tensile direction coincided with the width direction of the steel plate, voids grew and coalesced principally in the rolling direction of the steel plate.

Shabrov et al. (2004) performed the tensile test of SAE 4330 steel containing 0.04% Ti using notched round tensile specimens in various notch-root radii, observed the longitudinal section of the specimen using an electron microscope, and obtained the void nucleation criterion using the stress and strain distributions calculated by the elastic–plastic finite-element method. The material fractured due to the cracking of inclusions which were cuboidal titanium nitrides. The void nucleation criterion that voids nucleated when the summation of the equivalent stress and the mean normal stress to which a prescribed coefficient was multiplied became a certain value, which was proposed by Needleman (1987) as the void nucleation criterion due to the separation of the inclusions from the matrix, was shown to be effective.

Maire et al. (2008) performed the in situ tensile test of dual-phase steel using an X-ray tomograph and an electron microscope, obtained the void volume fraction, the number of voids, the size of voids, and the height/width ratio of voids during tensile test, and modified the void growth model proposed by Rice and Tracey (1969) to consider void nucleation. Voids nucleated both the cracking of the inclusions which were martensite and the separation of the inclusions from the matrix which were ferrite. Fig. 3.10 shows the relationship between the distance to center in the length direction of the material and the void volume fraction in various deformation steps of the material (Maire et al., 2008). The relationship between the equivalent plastic strain of the matrix and the size of large voids which had already nucleated obtained experimentally, agreed with the relationship between the equivalent plastic strain of the matrix and the size of a void calculated using the Rice and Tracey void growth model.

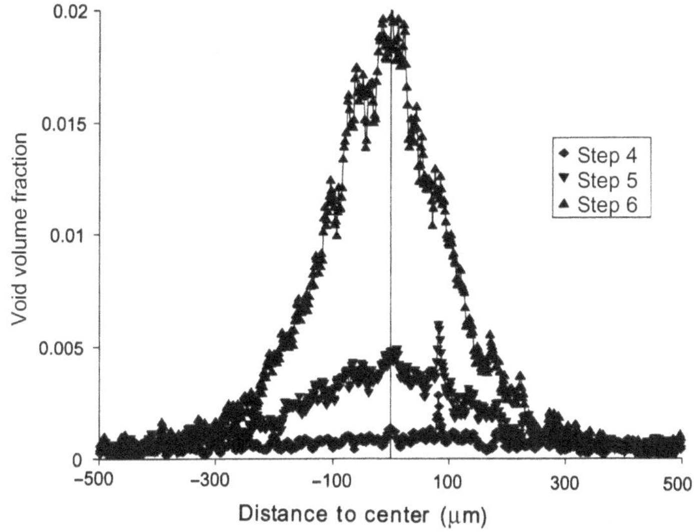

Figure 3.10 Relationship between distance to the center in length direction of material and void volume fraction in various deformation steps of material. *From Maire, E., Bouaziz, O., Di Michiel, M., Verdu, C., 2008. Initiation and growth of damage in a dual-phase steel observed by X-ray microtomography. Acta Mater. 56 (18), 4954–4964.*

3.3 Microscopic phenomena of void growth

Coffin and Rogers (1967) performed the drawing of a strip using tough pitch copper, oxygen-free high thermal conductivity copper, 6061-T6 aluminum alloy, and 60–40 brass in various hydrostatic pressure, and studied the influence of the hydrostatic pressure on the material damage during drawing. The stress distribution during strip drawing was calculated using the slip-line field solution proposed by Hill and Tupper (1948). The material damage was principally evaluated by the change in the material density after multipass drawing, which was the average in the thickness direction of a strip. Fig. 3.11 shows the relationship between the strip thickness and the density of the 6061-T6 aluminum alloy in various hydrostatic pressure (Coffin and Rogers, 1967). Because the material fractured at the center in the thickness direction of the strip, the experiment of multipass strip drawing was performed under two kinds of drawing conditions, in which, according to the slip-line field solution, the mean normal stresses at the center in the thickness direction of the strip were approximately equal to each other.

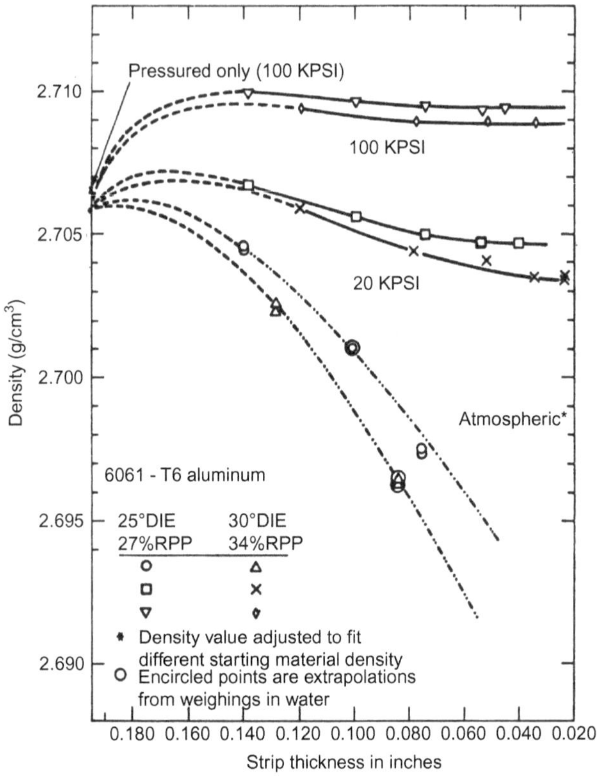

Figure 3.11 Relationship between strip thickness and density of 6061-T6 aluminum alloy in various hydrostatic pressure. *From Coffin, L.F., Rogers, H.C., 1967. Influence of pressure on the structural damage in metal forming processes. Trans. ASM 60, 672–686.*

French et al. (1973) performed the tensile test of a bar using β-brass in various hydrostatic pressure, and obtained the effect of the hydrostatic pressure on the material fracture. For hydrostatic pressures below 300 MPa, the fracture strain increased linearly and rapidly with the increase of the hydrostatic pressure because of the suppression of the void nucleation and the void growth. For hydrostatic pressures above 300 MPa, the fracture strain increased slowly with the increase of the hydrostatic pressure because of the localization of the void nucleation to the region close to the fracture surface. French and Weinrich (1973) performed the tensile test of a bar using α-brass in various hydrostatic pressure, and obtained the effect of the hydrostatic pressure on the material fracture. For hydrostatic pressures below 350 MPa, the fracture strain increased linearly with the increase of the hydrostatic pressure because of the suppression of the void nucleation and the void growth. For hydrostatic pressures above 450 MPa, the fracture strain scarcely changed despite the increase of the hydrostatic pressure because of the fracture due to shear deformation.

French and Weinrich (1975a) performed the tensile test of a bar using tough pitch copper in various hydrostatic pressure, and obtained the effect of the hydrostatic pressure on the material fracture. For hydrostatic pressures below 300 MPa, the fracture strain increased linearly with the increase of the hydrostatic pressure. For hydrostatic pressures above 300 MPa, localized shear deformation occurred at the neck of the specimen and the shape of the fracture surface was a chisel point. French and Weinrich (1975b) performed the tensile test of a bar using a pure aluminum and an aluminum–copper alloy in various hydrostatic pressure, and obtained the effect of the hydrostatic pressure on the material fracture. In the pure aluminum, the shape of the fracture surface was a double cup for hydrostatic pressures below 125 MPa, whereas the shape of the fracture surface was a chisel point for hydrostatic pressures above 150 MPa. In the aluminum–copper alloy, the fracture strain increased linearly and rapidly with the increase of the hydrostatic pressure for hydrostatic pressures below 300 MPa, whereas the fracture strain increased slowly with the increase of the hydrostatic pressure because of the fracture due to shear deformation for hydrostatic pressures above 300 MPa.

Schmitt and Jalinier (1982) performed the uniaxial tension and the balanced biaxial tension of a sheet using aluminum killed steel, electrolytic copper, and 3003 aluminum alloy; Schmitt and Jalinier measured the material density using an electronic thermocontrolled balance and an electron microscope, and modeled the growth of voids. When voids nucleated due to the separation of the inclusions from the matrix, the change in the material density in uniaxial tension was smaller than the change in the material density in balanced biaxial tension. When voids nucleated due to the cracking of the inclusions, the change in the material density in uniaxial tension was almost the same as the change in the material density in balanced biaxial tension.

Marini et al. (1985b) performed the tensile test of low carbon steels containing dispersed spherical alumina particles using notched round tensile specimens, measured the radius of voids at the neck of the notched round tensile specimen using an electron microscope, and evaluated the void growth model proposed by Rice and Tracey (1969). Void growth was defined as the natural logarithm of the current radius of voids divided

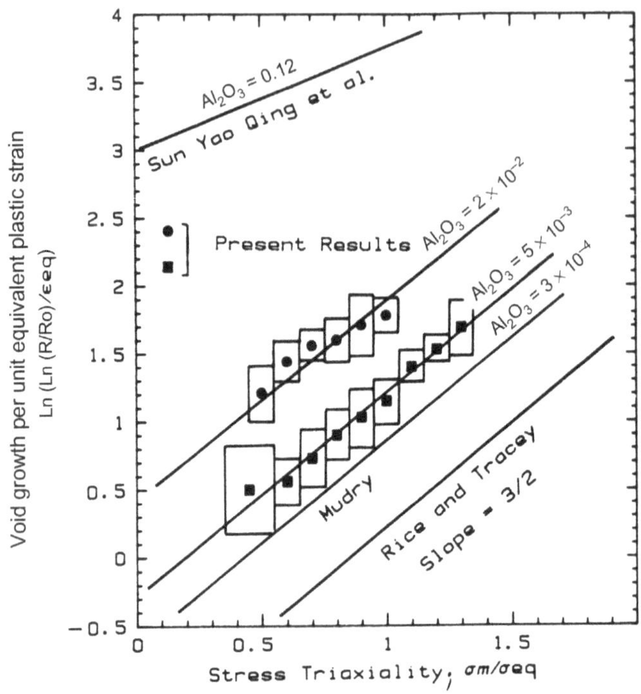

Figure 3.12 Relationship between stress triaxiality and void growth per unit equivalent plastic strain. *From Marini, B., Mudry, F., Pineau, A., 1985b. Experimental study of cavity growth in ductile rupture. Eng. Fract. Mech. 22 (6), 989–996.*

by the initial radius of voids. Fig. 3.12 shows the relationship between the stress triaxiality and the void growth per unit equivalent plastic strain (Marini et al., 1985b). The effect of the stress triaxiality on the void growth per unit equivalent plastic strain obtained experimentally by the authors was equal to the effect of the stress triaxiality on the void growth per unit equivalent plastic strain obtained analytically using the Rice and Tracey void growth model. However, the void growth per unit equivalent plastic strain obtained experimentally was much larger than the void growth per unit equivalent plastic strain obtained analytically. The reason for the difference between the void growth per unit equivalent plastic strain obtained experimentally and the void growth per unit equivalent plastic strain obtained analytically was presumed to be the interaction between neighboring voids.

Spitzig et al. (1988) performed the tensile test and the compression test of a bar using various kinds of iron compacts, the initial void volume fraction of which ranged between 0.3% and 11.1%. Spitzig et al. measured

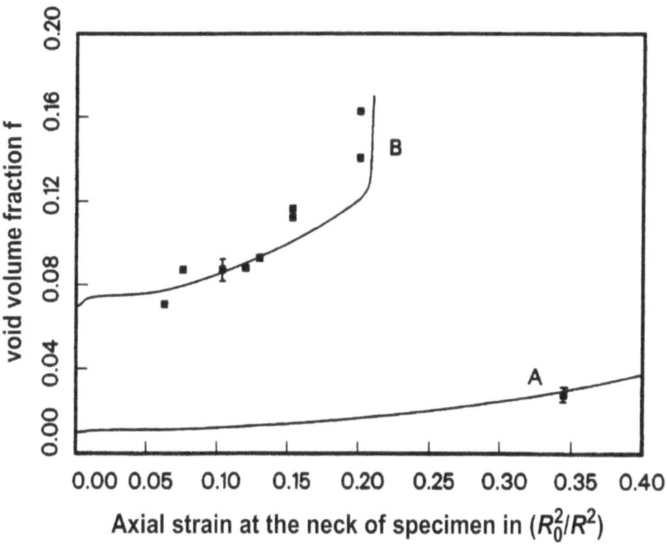

Figure 3.13 Relationship between axial strain at the neck of specimen and void volume fraction. *From Becker, R., Needleman, A., Richmond, O., Tvergaard, V., 1988. Void growth and failure in notched bars. J. Mech. Phys. Solids 36 (3), 317–351.*

void shapes on the longitudinal section of the specimen using an optical microscope, and evaluated several void growth models. The maximum diameter of voids and the minimum diameter of voids calculated using the void growth model proposed by Budiansky et al. (1982) slightly differed from the average maximum diameter of voids and the average minimum diameter of voids obtained experimentally. The void volume fraction calculated using the Gurson yield function (1977) agreed with the void volume fraction obtained experimentally.

Becker et al. (1988) performed the tensile test of various kinds of iron compacts, the initial void volume fraction of which ranged between 0.4% and 7%, using notched round tensile specimens, measured the void area fraction on the longitudinal section of the specimen using an optical microscope, and evaluated the Gurson yield function (1977) modified by Tvergaard (1981). Void nucleation was not assumed to occur in the simulation using the axisymmetric elastic–viscoplastic finite-element method. Material constants in the modified Gurson yield function were determined from the simulation using the representative volume element in which the axisymmetric elastic–viscoplastic finite-element method was used. Fig. 3.13 shows the relationship between the axial strain at the neck of the specimen and the void volume fraction (Becker et al., 1988).

Kao et al. (1990) performed the tensile test of a bar using spheroidized 1045 steel in various hydrostatic pressure, measured the void volume fraction on the longitudinal section of the specimen and observed the fracture surface of the specimen using an electron microscope, and suggested a mechanism for the ductile fracture in the tensile test of a bar. With increasing the hydrostatic pressure, the strain at void nucleation increased, and the ratio of the increment of void volume fraction to the increment of strain decreased, whereas the void volume fraction at fracture scarcely changed despite the change of the hydrostatic pressure. With increasing the hydrostatic pressure, the void growth in the radial direction of the specimen decreased, the minimum size of the carbide particles on the surface of which voids nucleated increased, and the number of small voids, which assisted the coalescence of large voids, decreased.

Pardoen et al. (1998) and Pardoen and Delannay (1998a) performed the tensile test of as-drawn pure copper and annealed pure copper using notched round tensile specimens in various notch-root radii and smooth round tensile specimens, measured the density of the specimen by the Archimedes method, and evaluated several void growth models and several void calescence criteria. The simulation of the tensile test was performed using a commercial elastic–plastic finite-element software. The void volume fraction during deformation calculated using the void growth model proposed by Rice and Tracey (1969) agreed with the void volume fraction during deformation obtained experimentally, when the coefficient on the spherically void growth rate in the Rice and Tracey void growth model, which depended on the strain-hardening exponent of the material, was optimized. The void volume fraction during deformation calculated using the Gurson yield function (1977) modified by Leblond et al. (1995) agreed with the void volume fraction during deformation obtained experimentally. Fig. 3.14 shows the relationship between the equivalent strain and the void volume fraction at the neck of the smooth round tensile specimen of the as-drawn pure copper (Pardoen and Delannay, 1998a). The void coalescence criterion proposed by Brown and Embury (1973) was effective in low stress triaxiality and was not effective in high stress triaxiality, whereas the void coalescence criterion proposed by Thomason (1985a, b) was effective in both low and high stress triaxialities when the void shape was assumed to be spherical.

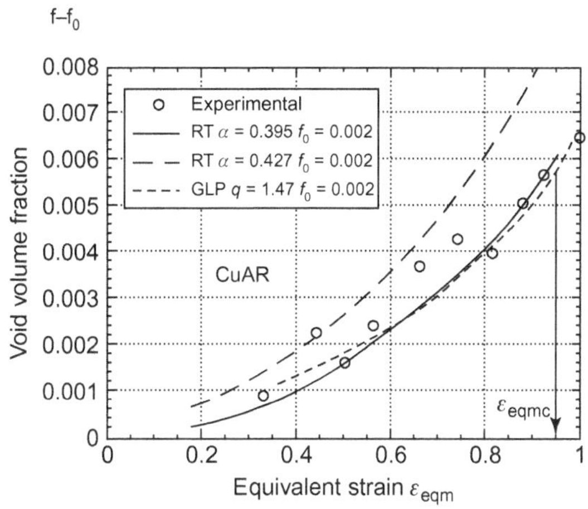

Figure 3.14 Relationship between the equivalent strain and void volume fraction at the neck of smooth round tensile specimen of as-drawn pure copper. *From Pardoen, T., Delannay, F., 1998a. Assessment of void growth models from porosity measurements in cold-drawn copper bars. Metall. Mater. Trans. A 29 (7), 1895–1909.*

3.4 Microscopic phenomena of void coalescence

Edelson and Baldwin (1962) performed the tensile test of bars using a pure copper with various void volume fractions and using copper alloys with various inclusion volume fractions and with various inclusion compositions, and obtained the effect of voids and inclusions on the mechanical properties of the pure copper and the copper alloys. The bars made of the pure copper and the bars made of the copper alloys with various inclusion compositions, such as iron, molybdenum, chromium, alumina, lead, and graphite, were prepared by powder metallurgical methods. Fig. 3.15 shows the relationship between the summation of the void volume fraction and the inclusion volume fraction and the fracture strain (Edelson and Baldwin, 1962). The fracture strain was defined as the natural logarithm of the initial cross-sectional area of the specimen divided by the fractured cross-sectional area of the specimen. The fracture strain depended only on the summation of the void volume fraction and the inclusion volume fraction, and depended neither on the inclusion size nor on the inclusion composition.

Beachem (1963) observed fracture surfaces using an electron microscope, and classified the fracture surfaces into three modes in terms of the modes for void coalescence, which depended on the stress state in the

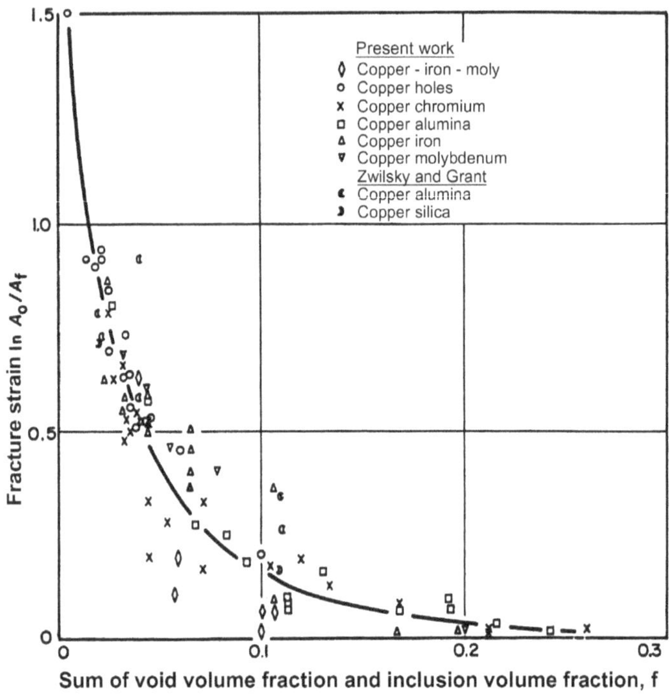

Figure 3.15 Relationship between the sum of void volume fraction and inclusion volume fraction and fracture strain. *From Edelson, B.I., Baldwin, W.M., 1962. The effect of second phases on the mechanical properties of alloys. Trans. ASM 55, 230–250.*

material prior to the material fracture. Fig. 3.16 shows the three modes for the void coalescence (Beachem, 1963). Fig. 3.16A shows the normal rupture, which appeared, for example, at the center of the necked cross section of the bar which was subjected to uniaxial tension. Fig. 3.16B shows the shear rupture, which appeared, for example, when the material was subjected to simple shear. Fig. 3.16C shows the tearing, which appeared, for example, at the notch root of the notched bar which was subjected to uniaxial tension. Equiaxed dimples were formed in the normal rupture, whereas elongated dimples were formed in the shear rupture and the tearing. The usual fracture surface was composed of a mixture of the fracture surface of the normal rupture, the fracture surface of the shear rupture, and the fracture surface of the tearing.

Rosenfield et al. (1972) performed the review on the fracture of steels containing pearlite, and demonstrated the effect of microstructure, such as pearlitic microstructure and spheroidized microstructure, on ductile

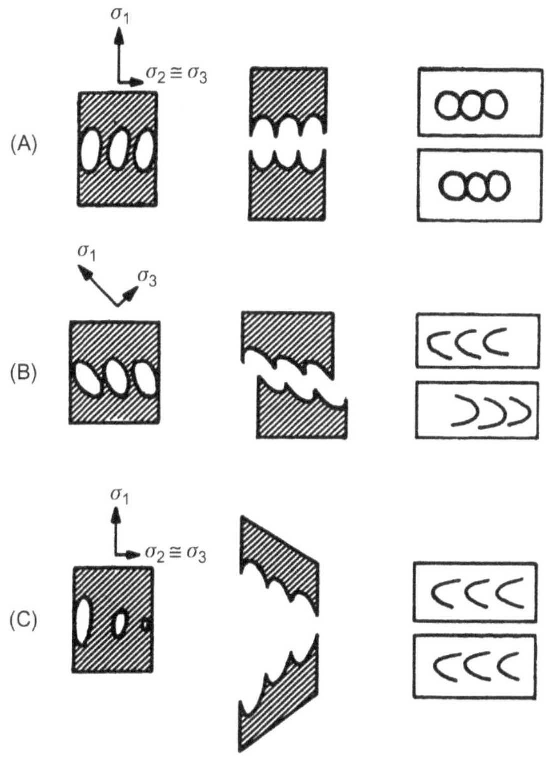

Figure 3.16 Three modes for void coalescence: (A) normal rupture, (B) shear rupture, (C) tearing. *From Beachem, C.D., 1963. An electron fractographic study of the influence of plastic strain conditions upon ductile rupture processes in metals. Trans. ASM 56, 318–326.*

fracture, cleavage fracture, and fatigue fracture. Fig. 3.17 shows the mechanism for the crack growth and the crack coalescence in pearlitic microstructure and spheroidized microstructure (Rosenfield et al., 1972). The mechanism for the crack growth and the crack coalescence in pearlitic microstructure proposed by Miller and Smith (1970) was employed. In pearlitic microstructure, since the deformation of the ferrite lamellae was localized in the shear band, the deformation required to coalesce cracks was small, whereas in spheroidized microstructure, since the deformation of the ferrite matrix was localized after large deformation, the deformation required to coalesce cracks was large. A schematic representation of the crack distribution in hypoeutectoid steel having pearlitic microstructure prior to crack coalescence in the tensile specimen of a bar was demonstrated. In the schematic representation, a pearlite colony was divided into

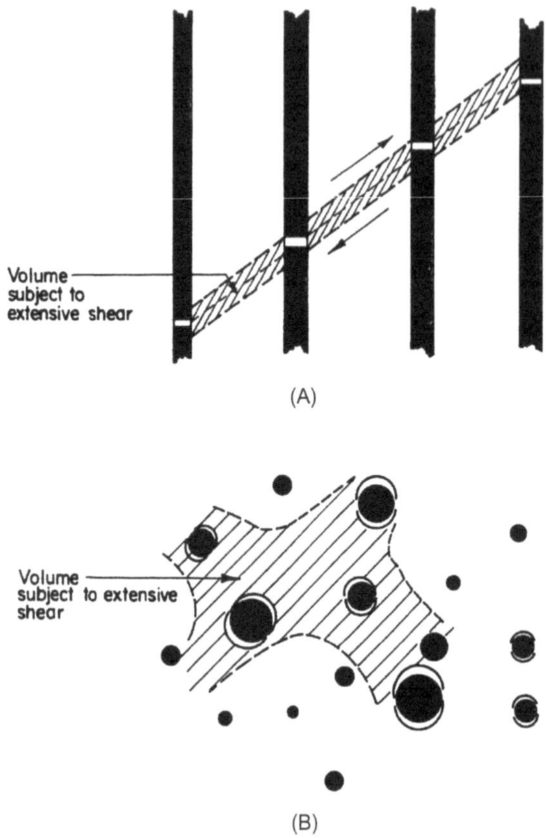

Figure 3.17 Mechanism for crack growth and crack coalescence in pearlitic microstructure and spheroidized microstructure: (A) pearlitic microstructure and (B) spheroidized microstructure. *From Rosenfield, A.R., Hahn, G.T., Embury, J.D., 1972. Fracture of steels containing pearlite. Metall. Trans. 3 (11), 2797–2804.*

two pieces by a crack and the angle between the crack and the tensile axis was approximately equal to 45 degrees.

Cox and Low (1974) performed the tensile test of AISI 4340 steel and 18 Ni, 200 grade maraging steel using notched round tensile specimens and smooth round tensile specimens, observed the fracture surface of the specimen and the section of the specimen using an electron microscope, and studied the mechanism of ductile fracture: void nucleation, void growth, and void coalescence. The cracking of carbo-nitride inclusions was observed in the maraging steel, whereas the separation of manganese sulfide inclusions from the matrix was observed in the AISI 4340 steel. The stress triaxiality had no measurable effect on the void nucleation, whereas the stress triaxiality had considerable effect on the void growth, in both the maraging steel and the AISI 4340 steel. In the maraging steel, the void coalescence occurred when voids contacted with each other. However, the mechanism of the void coalescence for the AISI 4340 steel was different from the mechanism of the void coalescence for the maraging steel. In the AISI 4340 steel, first, large voids nucleated and grew. Next, numerous small voids, which appeared like a crack and were similar to the void sheet described by Rogers (1960), nucleated on the line which connected neighboring two large voids. The angle between the numerous small voids and the tensile axis was approximately equal to 45 degrees. Finally, the neighboring two large voids coalesced.

Weinrich and French (1976) performed the tensile test of a sheet using α-brass, $\alpha-\beta$ brass, and tough pitch copper in various hydrostatic pressure. Weinrich and French observed the fracture surface of the specimen using an electron microscope, and obtained the effect of the hydrostatic pressure on the fracture surface. When the hydrostatic pressure was smaller than a specified hydrostatic pressure, the shape of the fracture surface, on which numerous dimples were observed, was planer. The specified hydrostatic pressure depended on the material. When the hydrostatic pressure was larger than the specified hydrostatic pressure, the shape of the fracture surface, on which no dimples were observed, was chisel point due to localized shear deformation at the neck of the specimen.

Roberts et al. (1976) performed the in situ tensile test of hypoeutectoid steels using a scanning electron microscope, and observed on the surface of the specimen, the void nucleation due to the separation of the inclusions from the matrix, the void growth around the inclusions, and the void coalescence. The void nucleation and the void growth observed on the surface of the specimen was not identical to the void nucleation and the void growth occurred on the interior plane of the specimen, which was parallel to the surface of the specimen and on which the void nucleation and the void growth were observed by sectioning the prestrained specimen. Fig. 3.18 shows the relationship between the distance between the two neighboring voids and the void length in the tensile direction of the two neighboring voids (Roberts et al., 1976). When the void length in the tensile direction of the two neighboring voids became equal to the distance between the two neighboring voids, the distance between the two neighboring voids began to decrease rapidly due to void coalescence, which was consistent with the void coalescence criterion proposed by Brown and Embury (1973).

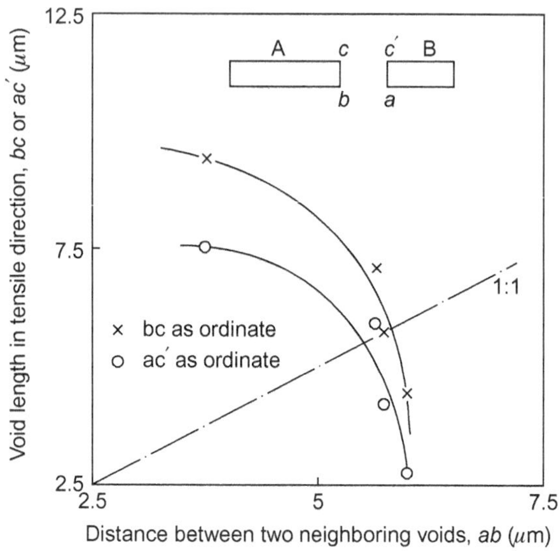

Figure 3.18 Relationship between the distance between two neighboring voids and void length in tensile direction of two neighboring voids. *From Roberts, W., Lehtinen, B., Easterling, K.E., 1976. An in situ SEM study of void development around inclusions in steel during plastic deformation. Acta Metall. 24 (8), 745–758.*

French and Weinrich (1976) performed the tensile test of a sheet using α-brass containing few inclusions, observed the fracture surface and the necked region of the specimen using an optical microscope, a scanning electron microscope, and a transmission electron microscope, and investigated the mechanism of the shear mode of ductile fracture. Subgrain structures observed using a transmission electron microscope corresponded to shear bands observed using an optical microscope. Because neither inclusions nor voids were found in the subgrain structures, dimples observed on the fracture surface were not a precursor to the fracture but a result of the fracture. French and Weinrich (1979) performed the tensile test of a sheet using a spheroidized plain carbon steel, observed the fracture surface and the necked region of the specimen using an optical microscope, a scanning electron microscope, and a transmission electron microscope, and investigated the mechanism of the shear mode of ductile fracture. First, local necking begun and the deformation of the material became localized into shear bands. Next, in the process of local necking, voids which nucleated due to the separation of the inclusions from the matrix, grew and coalesced in the shear bands. Finally, the material fractured, and numerous dimples which originated from the voids were observed on the fracture surface. The measured shear strains to fracture were smaller than the shear strain to fracture calculated using the void coalescence criterion proposed by McClintock et al. (1966).

Brownrigg et al. (1983) performed the tensile test of a bar using a spheroidized 1045 steel in various hydrostatic pressures, observed the longitudinal section and the fracture surface of the specimen using an electron microscope, and obtained the effect of the hydrostatic pressure on the material fracture. With increasing the hydrostatic pressure, the strain at fracture increased linearly, whereas the dimple size on the fracture surface decreased linearly. The number of voids per unit area on the fracture surface was approximately constant despite the change in the hydrostatic pressure, and the void area fraction on the fracture surface scarcely depended on the hydrostatic pressure. Hence, the size of a void on the fracture surface was approximately constant despite the change in the hydrostatic pressure. Furthermore, when inclusions were assumed to be points in a Voronoi diagram, the deformation of a Voronoi cell on the fracture surface was approximately constant despite the change in the hydrostatic pressure.

Marini et al. (1985a) performed the tensile test of A508 steel using notched round tensile specimens, obtained the effect of prestrain on the material fracture, and evaluated the material fracture using the void growth model proposed by Rice and Tracey (1969). The stress distribution and the strain distribution in the notched round tensile specimen were calculated using the in-house elastic–plastic finite-element software (Beremin, 1980). The axis of the prestrained tensile specimen was made to coincide with the axis of the virgin tensile specimen. Because the initial notch-root radius of the prestrained tensile specimen was made to differ from the initial notch-root radius of the virgin tensile specimen, the stress triaxiality at the neck of the prestrained tensile specimen was different from the stress triaxiality at the neck of the virgin tensile specimen. The criterion that the material fractured when the ratio of the current radius of voids to the initial radius of voids, which was calculated using the Rice and Tracey void growth model, became a certain value, was shown to be effective.

Bourcier et al. (1986) performed the tensile test of a bar using pure titanium and Ti–6Al–4V alloy in various void volume fractions, observed the section and the fracture surface of the specimen using an electron microscope, and obtained the void growth and the void coalescence both experimentally and numerically. The relationship between the strain and the void volume fraction obtained experimentally reasonably agreed with the relationship between the strain and the void volume fraction calculated by the finite-element method using the representative volume element in which a void was assumed. The relationship between the void volume fraction and the strain to fracture obtained experimentally agreed with the relationship between the void volume fraction and the strain to flow localization calculated by the elastic–plastic finite-element method in which an initial imperfection was assumed (Saje et al., 1982), when the void volume fraction was transformed into the void area fraction appropriately.

Dubensky and Koss (1987) and Magnusen et al. (1988) performed the tensile test of a sheet containing numerous holes which simulated voids, using 1100-0 Al, 7075-T6 Al, 70–30 α-brass, and low carbon aluminum killed steel; they observed the surface of the specimen using an optical microscope, and obtained the elongation to fracture in various hole shapes and hole arrays. With increasing the minimum space between two neighboring holes, the elongation to fracture drastically increased. With increasing the hole diameter, the elongation to fracture decreased. With increasing the hole area fraction, the elongation to fracture slightly decreased. When the hole diameter and the hole area fraction were specified, the elongation to fracture for the specimen in which the regular array of holes was prepared was larger than the elongation to fracture for the specimen in which the random array of holes was prepared. Fig. 3.19 shows the relationship between the array of holes and the elongation to fracture (Dubensky and Koss, 1987). With increasing the strain-hardening exponent of the material, the ratio of the elongation to fracture for the specimen containing holes in random array to the elongation to fracture for the specimen containing holes in regular array increased. The region of strain localization between two neighboring holes in random array was inclined by approximately 45 degrees against the tensile direction for the specimen of the plate, in which plane-strain deformation was approximately obtained, and was inclined by approximately 90 degrees against the tensile direction for the specimen of the sheet, in which plane-stress state was approximately obtained.

Sun (1991) performed the tensile test of 37Mn5 steel using notched round tensile specimens in various notch-root radii and smooth round tensile specimens. Sun observed the longitudinal section and the fracture surface of the specimen using an electron microscope, and suggested a

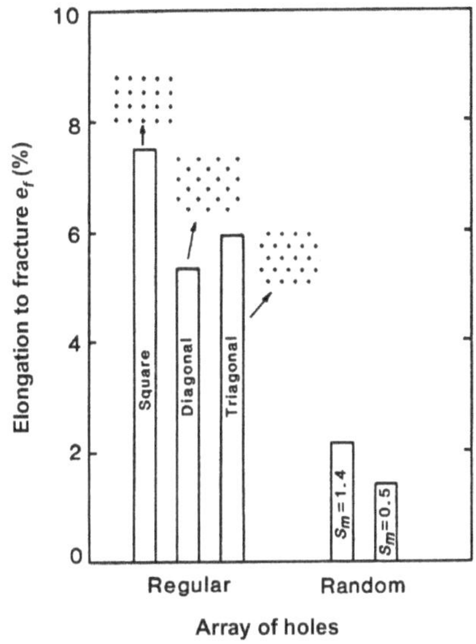

Figure 3.19 Relationship between array of holes and elongation to fracture. *From Dubensky, E.M., Koss, D.A., 1987. Void/pore distributions and ductile fracture. Metall. Trans. A 18 (11), 1887–1895.*

mechanism of the void coalescence in various stress triaxialities. The distance between two neighboring voids was assumed to be the distance between the center of a void and the center of a neighboring void. Internal necking occurred principally due to the decrease in the distance between two neighboring voids at low stress triaxiality, whereas internal necking occurred principally due to the void growth in the direction perpendicular to the tensile direction at high stress triaxiality.

Geltmacher et al. (1996) performed the tensile test and the punch stretching of a sheet in which holes of equal size were made regularly or randomly using 3003 aluminum alloy, observed the surface of the specimen using an optical microscope, and obtained the imposed strain to flow localization in uniaxial tension and balanced biaxial tension. First, the specimen in which two holes, which were in line in the direction perpendicular to the tensile direction, were made in various spaces between the two holes, was used. Fig. 3.20 shows the relationship between the space between the two holes divided by the hole diameter and the imposed equivalent strain to flow localization (Geltmacher et al., 1996). Next, the specimen in which 63 holes were made in a specified hole area fraction and in various minimum spaces between two neighboring holes was used.

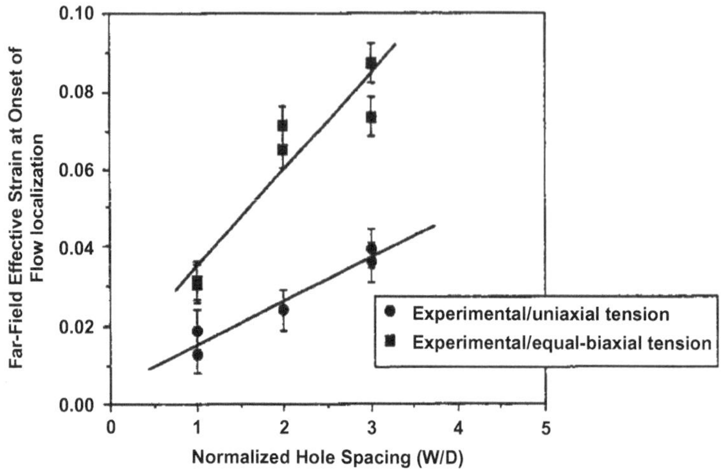

Figure 3.20 Relationship between the space between two holes divided by hole diameter and imposed equivalent strain to flow localization. *From Geltmacher, A.B., Koss, D.A., Matic, P., Stout, M.G., 1996. A modeling study of the effect of stress state on void linking during ductile fracture. Acta Mater. 44 (6), 2201–2210.*

Hole linking depended principally on the space between two neighboring holes in balanced biaxial tension, whereas hole linking depended principally on the direction in which two neighboring holes were in line in uniaxial tension.

Pardoen and Delannay (1998b) performed the tensile test of as-drawn pure copper and annealed pure copper using notched round tensile specimens in various notch-root radii, which were prestrained, that is, unloaded, recrystallized, and reloaded during the tensile test, obtained the equivalent strain at void coalescence, and evaluated the void coalescence criterion proposed by Thomason (1985a,b). The simulation of the tensile test was performed to obtain the void volume fraction rate using the Gurson yield function (1977) modified by Leblond et al. (1995). Fig. 3.21 shows the relationship between the equivalent strain in prestraining and the equivalent strain at void coalescence for the as-drawn pure copper (Pardoen and Delannay, 1998b). During recrystallization in prestraining, the flow stress of the material was assumed to recover, whereas the void shape and the void configuration were not assumed to change. In the Thomason void coalescence criterion, the assumption of ellipsoidal void shape, which was calculated using the void growth model proposed by Rice and Tracey (1969), was shown to be more appropriate than the assumption of spherical void shape.

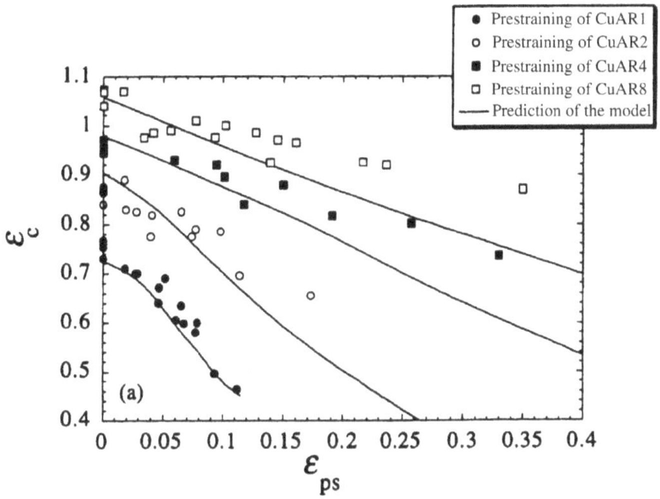

Figure 3.21 Relationship between equivalent strain in prestraining and equivalent strain at void coalescence for as-drawn pure copper. *From Pardoen, T., Delannay, F., 1998b. The coalescence of voids in prestrained notched round copper bars. Fatig. Fract. Eng. Mater. Struct. 21 (12), 1459–1472.*

Goto et al. (1999) performed the tensile test of HY-100 steel plate using notched round tensile specimens in various notch-root radii and smooth round tensile specimens, observed the fracture surface of the specimen using an electron microscope, and obtained the effect of the stress triaxiality on the strain to fracture at the center of the specimen. When the tensile axis was made to be parallel to the width direction of the plate, the following two ductile fracture mechanisms which depended on the imposed stress triaxiality were recognized. When the stress triaxiality was low, the strain to fracture at the center of the specimen was large and equiaxed voids were observed at the fracture surface of the specimen. When the stress triaxiality was high, the strain to fracture at the center of the specimen was small and voids which originated from elongated manganese sulfide inclusions were observed at the fracture surface of the specimen.

Chae et al. (2000) performed the tensile test of HY-100 steel plate using notched round tensile specimens in various notch-root radii, some of which were prestrained, that is, unloaded, machined to enlarge the notch-root radius, and reloaded during the tensile test, and obtained the effect of the prestrain on the strain to fracture at the center of the specimen. The strain to fracture at the center of the specimen which was

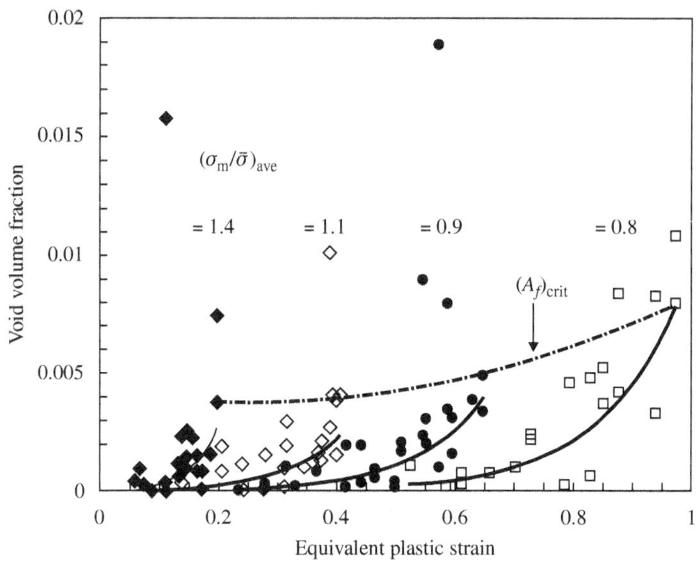

Figure 3.22 Relationship between equivalent plastic strain and void volume fraction in various stress triaxialities. *From Chae, D., Koss, D.A., 2004. Damage accumulation and failure of HSLA-100 steel. Mater. Sci. Eng. A 366 (2), 299–309.*

prestrained, that is, loaded at a high stress triaxiality, and reloaded at a low stress triaxiality, was smaller than the strain to fracture at the center of the specimen which was loaded only at the low stress triaxiality. The simulation result calculated using a commercial elastic–plastic finite-element software indicated qualitatively the effect of the prestrain on the strain to fracture at the center of the specimen obtained experimentally, under the assumption that the material fractured when the equivalent strain between the two neighboring voids became a certain value.

Chae and Koss (2004) performed the tensile test of HSLA-100 steel plate using notched round tensile specimens in various notch-root radii. Chae and Koss observed the longitudinal section of the specimen using an optical microscope, and obtained the relationship between the strain and the void volume fraction in various stress triaxialities using the stress and strain distributions calculated using an elastic–plastic finite-element software. Fig. 3.22 shows the relationship between the equivalent plastic strain and the void volume fraction in various stress triaxialities (Chae and Koss, 2004). Because the strain at void nucleation was comparatively small and assumed to be zero, the material fracture was assumed to occur by void growth and void coalescence. The relationship between the stress

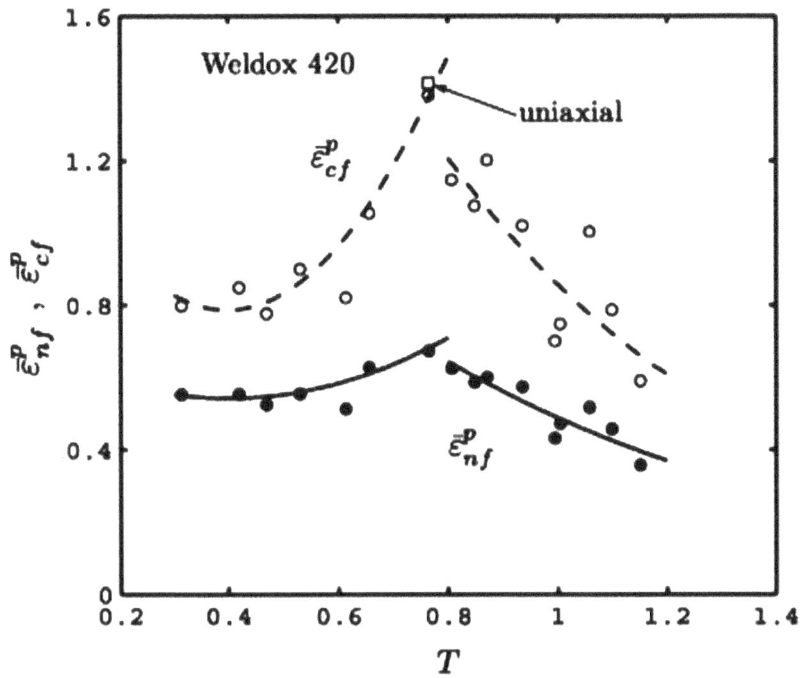

Figure 3.23 Relationship between stress triaxiality and equivalent plastic strain at fracture. *From Barsoum, I., Faleskog, J., 2007a. Rupture mechanisms in combined tension and shear, Exp. Int. J. Solids Struct. 44 (6), 1768–1786.*

triaxiality and the equivalent plastic strain at the start of void coalescence was derived by combining the void volume fraction at the start of void coalescence expressed by the exponential function of the equivalent plastic strain with the void growth model proposed by Rice and Tracey (1969).

Barsoum and Faleskog (2007a) performed the combined test of the tensile test and the torsion test of Weldox 420 steel plate and Weldox 960 steel plate using notched tube tensile and torsion specimens; they observed the fracture surface of the specimen using an electron microscope, and investigated the effects of the stress triaxiality and the Lode parameter on the fracture mechanism. Fig. 3.23 shows the relationship between the stress triaxiality and the equivalent plastic strain at fracture (Barsoum and Faleskog, 2007a). The black circles denoted the average equivalent plastic strain over the notch, whereas the white circles denoted the equivalent plastic strain at the center of the notch. The stress triaxiality at which the equivalent plastic strain was maximized was rather identical to the stress triaxiality at which the Lode parameter was minimized. The fracture surface was covered with equiaxed large, deep dimples which indicated the normal rupture when the stress triaxiality was large, whereas the fracture surface was covered with elongated small, shallow dimples which indicated the shear rupture when the stress triaxiality was small.

Weck et al. (2008a) used an X-ray tomograph and an X-ray radiograph to perform the in situ tensile test of a model material which contained numerous spherical inclusions of the same size in various inclusion volume fractions, observed void nucleation, void growth, and void coalescence, and evaluated the void coalescence criterion proposed by Brown and Embury (1973). The outer part of the model material was made of pure aluminum, whereas the inner part of the model material was made of pure aluminum matrix and ZrO_2/SiO_2 inclusions. Voids nucleated due to the separation of the inclusions from the matrix. With increasing the inclusion volume fraction, both the strain to void nucleation and the strain to void coalescence decreased. The relationship between the inclusion volume fraction and the strain to void coalescence calculated using the Brown and Embury void coalescence criterion agreed with the relationship between the inclusion volume fraction and the strain to void coalescence obtained experimentally in the case of smooth round tensile specimens.

Weck and Wilkinson (2008) used an electron microscope to perform the in situ tensile test of an AA5052 aluminum alloy sheet of 100 μm thickness in which holes of 10 μm hole diameter were made in various hole configurations, obtained the strain to hole coalescence, and evaluated several void coalescence criteria. With decreasing the angle between the line connecting two neighboring holes and the tensile axis, the strain to hole coalescence increased. Fig. 3.24 shows the schematic diagram explaining the hole coalescence in the case that the angle between the line connecting two neighboring holes and the tensile axis is equal to 45 degrees (Weck and Wilkinson, 2008). With increasing the distance between two neighboring holes, the strain to hole coalescence increased. The strain to void coalescence calculated using the void coalescence criterion proposed by Brown and Embury (1973) agreed with the strain to hole coalescence obtained experimentally when the line connecting two neighboring holes was perpendicular to the tensile axis. The strain to void coalescence calculated using the void coalescence criterion proposed by McClintock et al. (1966) agreed with the strain to hole coalescence obtained experimentally when the angle between the line connecting two neighboring holes and the tensile axis was equal to 45 degrees.

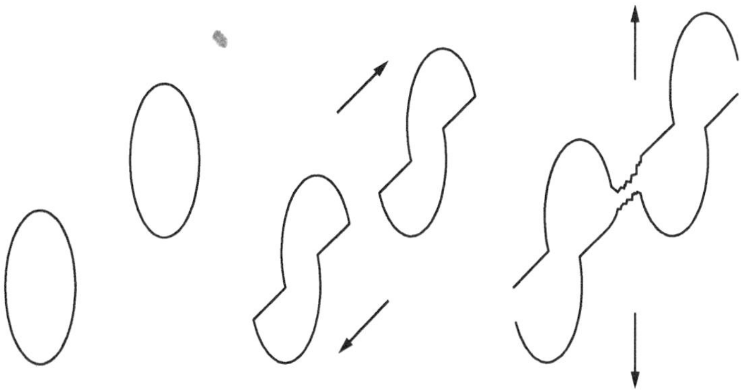

Figure 3.24 Schematic diagram explaining hole coalescence in case that angle between line connecting two neighboring holes and tensile axis is equal to 45 degrees. *From Weck, A., Wilkinson, D.S., 2008. Experimental investigation of void coalescence in metallic sheets containing laser drilled holes. Acta Mater. 56 (8), 1774–1784.*

Weck et al. (2008b) used an X-ray tomograph to perform the in situ tensile test of a pure copper sheet and a copper alloy sheet in which holes were made in various hole configurations, obtained the strain to hole coalescence, and evaluated the void growth model proposed by Rice and Tracey (1969) and several void coalescence criteria. Both the thickness and the hole diameter of the pure copper sheet were equal to 10 μm, whereas both the thickness and the hole diameter of the copper alloy sheet were equal to 40 μm. The holed sheet was held between two sheets without holes so that a hole approximately became a spherical void. At the neck of the specimen, the relationship between the equivalent plastic strain and the void length in the tensile direction, calculated using the Rice and Tracey void growth model, agreed with the relationship between the equivalent plastic strain and the void length in the tensile direction obtained experimentally. The strain to void coalescence calculated using the void coalescence criterion proposed by Thomason (1985a,b) agreed with the strain to void coalescence obtained experimentally when the line connecting two neighboring voids was perpendicular to the tensile axis.

CHAPTER 4

Microscopic ductile fracture criteria

4.1 Introduction

Because microscopic ductile fracture phenomena are divided into void nucleation, void growth, and void coalescence (Dodd and Bai, 1987), it is natural to divide microscopic ductile fracture criteria into the following three categories: void nucleation criteria, void growth models, and void coalescence criteria. There are a number of review papers that deal with microscopic ductile fracture criteria (Tvergaard, 1990; Thomason, 1998; Besson, 2010; Benzerga and Leblond, 2010; Benzerga et al., 2016; Pineau et al., 2016).

A criterion is required to perform the analysis or the simulation of void nucleation or void coalescence, whereas no criterion is required to perform the analysis or the simulation of void growth. Hence, a large number of researches on the analysis or the simulation of void growth have been performed. McClintock (1968a) analyzed the growth of a cylindrical void in a rigid perfectly plastic material, whereas Rice and Tracey (1969) analyzed the growth of a spherical void in a rigid perfectly plastic material. Both the McClintock void growth model and the Rice and Tracey void growth model are still fundamental void growth models. Researches on the analysis or the simulation of void nucleation or void coalescence have not been sufficiently performed and should be intensively performed to clarify microscopic ductile fracture criteria.

Although numerous voids exist in a material, the analysis or the simulation of void nucleation, void growth, or void coalescence is generally performed using the representative volume element, in which there is only one void and on the boundary of which periodic boundary conditions are assumed, to simplify the analysis or the simulation. The representative volume element is also called the unit cell.

Ductile Fracture in Metal Forming
DOI: https://doi.org/10.1016/B978-0-12-814772-6.00004-7

There are two types of void nucleation criteria for both the separation of the inclusions from the matrix and the cracking of the inclusions in the analysis of void nucleation. In one type of void nucleation criterion, void nucleation is determined due to energy. That is, a void is assumed to nucleate when the elastic energy which is released from the separation of the inclusions from the matrix or the cracking of the inclusions is smaller than the surface energy which is created on the interface between the matrix and the inclusion or the interface of the cracked inclusion.

In the other type of void nucleation criterion, void nucleation is determined due to stress; a void is assumed to nucleate when the interfacial stress between the matrix and the inclusion or the maximum principal stress in the inclusion is larger than a certain value. The void nucleation criterion due to energy is not the sufficient condition but the necessary condition for void nucleation. Moreover, when the inclusion diameter is not extremely small, the void nucleation criterion due to energy is unconditionally satisfied if the void nucleation criterion due to stress is satisfied. Hence, the void nucleation criterion due to stress is mainly dealt with.

There are two types of void coalescence criteria in the analysis of void coalescence. In one type of void coalescence criterion, void coalescence is determined geometrically; a void is assumed to coalesce when the void contacts the boundary of the representative volume element. Because flow localization, which occurs prior to the void coalescence, is not considered in the analysis of void coalescence, the strain to fracture calculated using this type of void coalescence criterion is generally larger than the strain to fracture obtained experimentally. This type of void coalescence criterion is used, for instance, by McClintock (1968a).

In the other type of void coalescence criterion, void coalescence is determined energetically; a void is assumed to coalesce when the energy required for the deformation mode of void coalescence, that is, internal necking deformation mode is smaller than the energy required for the deformation mode of void noncoalescence, that is, homogeneous deformation mode. Because flow localization is considered in the analysis of void coalescence, the strain to fracture calculated using this type of void coalescence criterion roughly agrees with the strain to fracture obtained experimentally. This type of void coalescence criterion is used, for instance, by Thomason (1968).

In the following sections, classical papers on microscopic ductile fracture criteria are introduced for each of the following processes: void nucleation, void growth, and void coalescence.

4.2 Microscopic criteria of void nucleation

Gurland and Plateau (1963) proposed a void nucleation criterion due to the cracking of the inclusions in a simple form, in which a void was assumed to nucleate when the elastic energy which was accumulated in the inclusion became larger than the surface energy of the surfaces that newly appeared on the interface of the cracked inclusion. The stress at which a void was assumed to nucleate calculated using the proposed void nucleation criterion was shown to agree approximately with the stress at which a crack was initially observed obtained experimentally in the uniaxial tension of an aluminum casting alloy.

Ashby (1966) proposed, with reference to the dislocation theory (Hirth and Lothe, 1982), a theory of strain hardening of a single crystal in which inclusions were dispersed. In the proposed theory, the inclusions were assumed to deform only elastically, whereas the single crystal was assumed to deform plastically, and the separation of the inclusions from the matrix was assumed to occur when the interfacial stress between the matrix and the inclusion became a certain value. Because the interfacial stress increased roughly linearly with the increase of strain or inclusion diameter, the separation of large inclusions from the matrix appeared prior to the separation of small inclusions from the matrix when the specimen was specified. Once the separation of the inclusions from the matrix occurred, the void volume increased linearly with the increase of strain in the assumption of plane-strain deformation.

Tanaka et al. (1970) calculated the energies of the material in uniaxial tension before the separation of the inclusion from the matrix and after the separation of the inclusion from the matrix, proposed a void nucleation criterion, and showed the relationship between the inclusion diameter and the plastic strain to separation. The inclusion was assumed to deform elastically, whereas the matrix was assumed to deform plastically. When the energy of the material after the separation of the inclusion from the matrix became smaller than the energy of the material before the separation of the inclusion from the matrix, a void was assumed to nucleate. With increasing the inclusion diameter, the plastic strain to separation decreased, which was consistent with the observation due to Palmer and Smith (1968). The void nucleation criterion due to energy described above and also a void nucleation criterion due to stress, in which the plastic strain to separation did not depend on the inclusion diameter, were proposed with reference to Ashby (1966). Fig. 4.1 shows

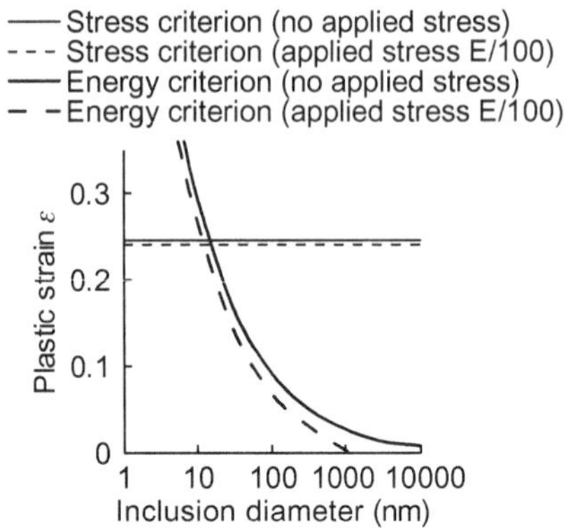

Figure 4.1 Relationship between inclusion diameter and plastic strain to separation. *From Tanaka, K., Mori, T., Nakamura, T., 1970. Cavity formation at the interface of a spherical inclusion in a plastically deformed matrix. Philos. Mag. 21 (170), 267–279.*

the relationship between the inclusion diameter and the plastic strain to separation (Tanaka et al., 1970). When the inclusion diameter was large, since the void nucleation criterion due to energy was always satisfied, void nucleation was dominated by the void nucleation criterion due to stress.

Brown and Stobbs (1971) used an electron microscope to observe the plastic deformation of a matrix of copper in which inclusions of spherical silica particles were dispersed, and proposed a void nucleation criterion due to the separation of the inclusions from the matrix in a simple form. The spherical silica particles were assumed to deform elastically, whereas the copper was assumed to deform plastically. In the proposed void nucleation criterion, a void was assumed to nucleate when the elastic energy accumulated in the inclusion due to Eshelby (1957) became larger than the surface energy created on the interface between the matrix and the inclusion. With increasing the inclusion diameter, the strain to void nucleation increased, which was contrary to the observation due to Palmer and Smith (1968).

Orr and Brown (1974) obtained the stress distribution and the strain distribution in an infinite matrix around a cylindrical inclusion by the finite-difference method. The inclusion was assumed to be rigid, and the matrix was assumed to be either an elastic–plastic linear strain-hardening material or an elastic perfectly plastic material, whereas plane-strain deformation of the matrix was assumed. Principal stresses were applied in the horizontal direction and in the vertical direction on the outer boundary.

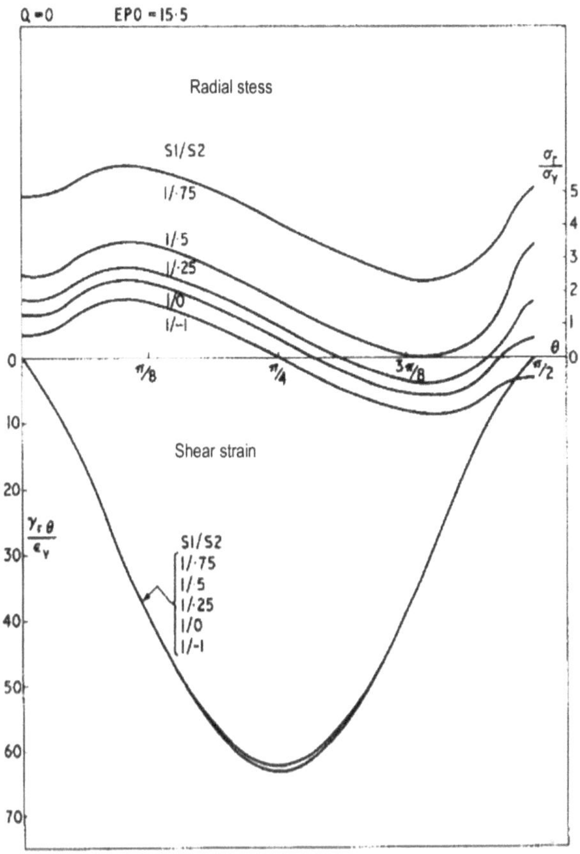

Figure 4.2 Relationship between the angle between direction of maximum principal stress and radial direction and radial stress and shear strain on interface between matrix and inclusion for various ratios of principal stresses in the case of elastic perfectly plastic material. *From Orr, J., Brown, D.K., 1974. Elasto-plastic solution for a cylindrical inclusion in plane strain. Eng. Fract. Mech. 6 (2), 261–274.*

Since the change in the shape of the matrix was not considered in the simulation using the finite-difference method, the maximum effective plastic strain on the outer boundary was restricted to be several percent. Fig. 4.2 shows the relationship between the angle between the direction of the maximum principal stress and the radial direction and the radial stress and the shear strain on the interface between the matrix and the inclusion for various ratios of the principal stresses in the case of the elastic perfectly plastic material (Orr and Brown, 1974).

Argon et al. (1975a) performed the simulation of the axisymmetric tensile test of a smooth bar and naturally necked bars by the elastic–plastic finite-element method, and obtained the equivalent plastic strain distribution and the stress triaxiality distribution in the axial direction in various steps of the simulation in tabular form for spheroidized 1045 steel and Cu–0.6% Cr alloy. The equivalent plastic strain distribution and the stress triaxiality distribution were compared with the equivalent plastic strain distribution and the stress triaxiality distribution calculated with reference to Bridgman (1952).

Argon et al. (1975b) performed the simulation of pure shear of a bar at the center of which a rigid cylinder was placed by the elastic–plastic finite-element method, and obtained the stress distribution and the strain distribution on the rigid cylinder. In the simulation of pure shear of a bar, plane-strain tension in the horizontal direction and plane-strain compression in the vertical direction were performed simultaneously. The matrix was assumed to be an elastic perfectly plastic material, and the radial stress on the rigid cylinder in the direction of plane-strain tension was assumed to be the interfacial stress between the matrix and the inclusion, which was approximated to be the flow stress of the matrix. Since the stress triaxiality of pure shear was equal to zero, the interfacial stress between the matrix and the inclusion when the stress triaxiality was arbitrary was assumed to be the summation of the flow stress of the matrix and the stress triaxiality of the matrix which were calculated in the case that the inclusion was replaced by the matrix. When the inclusion volume fraction was large, the interaction between two neighboring inclusions was unable to be ignored. Hence, the interfacial stress between the matrix and the inclusion, which depended on the inclusion volume fraction, was proposed based on a dislocation model.

Brown and Stobbs (1976) obtained, with reference to a dislocation model, a theoretical stress–strain relationship for the plastic deformation of a matrix of copper in which inclusions of spherical silica particles were dispersed, and proposed a void nucleation criterion due to the separation of the inclusions from the matrix in a simple form. The relationship between the plastic strain and the internal stress obtained in the experiment on the Bauschinger effect (Atkinson et al., 1974) was also used to derive the proposed void nucleation criterion. The spherical silica particles were assumed to deform elastically, whereas the copper was assumed to deform plastically. In the proposed void nucleation criterion, a void was assumed to nucleate when the elastic energy accumulated in the inclusion became larger than the surface energy created on the interface between the matrix and the inclusion. Although the strain at void nucleation calculated did not depend on the inclusion diameter, the strain at void nucleation calculated was not inconsistent with the strain at void nucleation observed.

Chu and Needleman (1980) proposed two kinds of void nucleation criteria for the elastic–plastic finite-element simulation using the Gurson yield function (1977). Chu and Needleman performed the simulation of the punch stretching of a sheet, and obtained the fracture forming limit. In one void nucleation criterion, with reference to Gurland (1972), the increment of void volume fraction was assumed to be proportional to the increment of the equivalent plastic strain of the matrix. Whereas in the other void nucleation criterion, with reference to Argon et al. (1975a,b) and Argon and Im (1975), the increment of void volume fraction was assumed to be proportional to the increment of the maximum normal stress on the interface between the matrix and the inclusion.

Fisher and Gurland (1981b), with reference to the experiment performed by Fisher and Gurland (1981a), proposed a void nucleation criterion due to the separation of the inclusions from the matrix, which depended on both the interfacial stress and the energy, since the void nucleation criterion on the energy was not necessarily satisfied for relatively large inclusions. The void nucleation criterion on the interfacial stress was satisfied when the interfacial stress between the matrix and the inclusion became larger than a critical value, whereas the void nucleation criterion on the energy was satisfied when the elastic energy accumulated in the inclusion became larger than the surface energy created on the interface between the matrix and the inclusion. Fig. 4.3 shows the relationship between the equivalent strain and the critical inclusion radius at which a void nucleates for various surface energies (Fisher and Gurland, 1981b). The void nucleation was likely to occur, with increasing the inclusion diameter, with increasing the flow stress, with increasing the plastic strain, or with increasing the stress triaxiality.

Le Roy et al. (1981) performed the tensile test of a bar using four types of spheroidized plain carbon steels, observed the longitudinal section of the specimen using an electron microscope, obtained the relationship between the equivalent strain and the void volume fraction, and proposed void nucleation models. The strain to void nucleation, which slightly depended on the inclusion volume fraction, was approximately equal to 0.5. The following two types of void nucleation models were proposed:

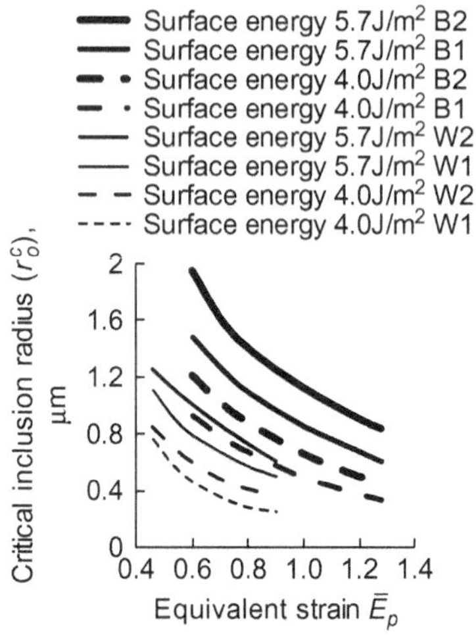

Figure 4.3 Relationship between equivalent strain and critical inclusion radius at which a void nucleates for various surface energies. *From Fisher, J.R., Gurland, J., 1981b. Void nucleation in spheroidized carbon steels. Part 2: Model. Metal Sci. 15 (5), 193–202.*

the continuous void nucleation model in which the number of cracked inclusions increased linearly with increasing the strain and the discontinuous void nucleation model in which all the inclusions cracked at a specified strain. The void growth model proposed by Rice and Tracey (1969) was modified to consider the ratio of the strain rate of the void to the strain rate of the matrix, whereas the void coalescence criterion proposed by Brown and Embury (1973) was used. Fig. 4.4 shows the stress state at void nucleation and the stress state at fracture for axisymmetric deformation (Le Roy et al., 1981). The stress states at void nucleation obtained experimentally were plotted by solid symbols, whereas the stress states at fracture obtained experimentally were plotted by open symbols.

Needleman (1987) performed the simulation of the void nucleation due to the separation of the inclusions from the matrix by the elastic–viscoplastic finite-element method using the cohesive zone model (Barenblatt, 1962). The inclusions were assumed to be rigid, whereas the matrix was assumed to be elastic–viscoplastic, and the representative volume element, which was cylindrical, was subjected to axisymmetric tension. The gap between the inclusion and the matrix, which had a length dimension and beyond which no force between the inclusion and the

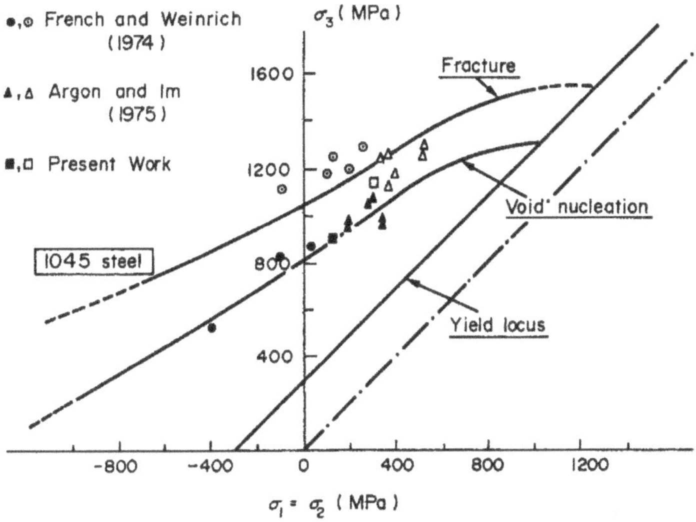

Figure 4.4 From Stress state at void nucleation and stress state at fracture for axisymmetric deformation. *From Le Roy, G., Embury, J.D., Edwards, G., Ashby, M.F., 1981. A model of ductile fracture based on the nucleation and growth of voids. Acta Metall. 29 (8), 1509–1522.*

matrix was exerted, was assumed to be a material constant with reference to the separation stress and the separation work per unit area for iron carbide particles in spheroidized carbon steels (Goods and Brown, 1979). Fig. 4.5 shows the material shapes and the equivalent plastic strain distributions in various steps of the simulation (Needleman, 1987). The separation of the inclusion from the matrix did not occur first at the interface between the matrix and the inclusion on the rotationally symmetric axis. With increasing the inclusion diameter, the strain at void nucleation decreased.

Ghosh and Moorthy (1998) performed the simulation of the void nucleation due to the cracking of the inclusions by the Voronoi Cell finite-element method (Ghosh and Moorthy, 1995; Moorthy and Ghosh, 1996), in which a Voronoi Cell was composed of an inclusion and a matrix that surrounded the inclusion. The inclusion was assumed to be elastic, whereas the matrix was assumed to be elastic–plastic and the deformation of the matrix was assumed to be small. Since the number of finite elements in a Voronoi Cell was only one, the computational time required for the Voronoi Cell finite-element simulation was much smaller than the computational time required for the conventional finite-element

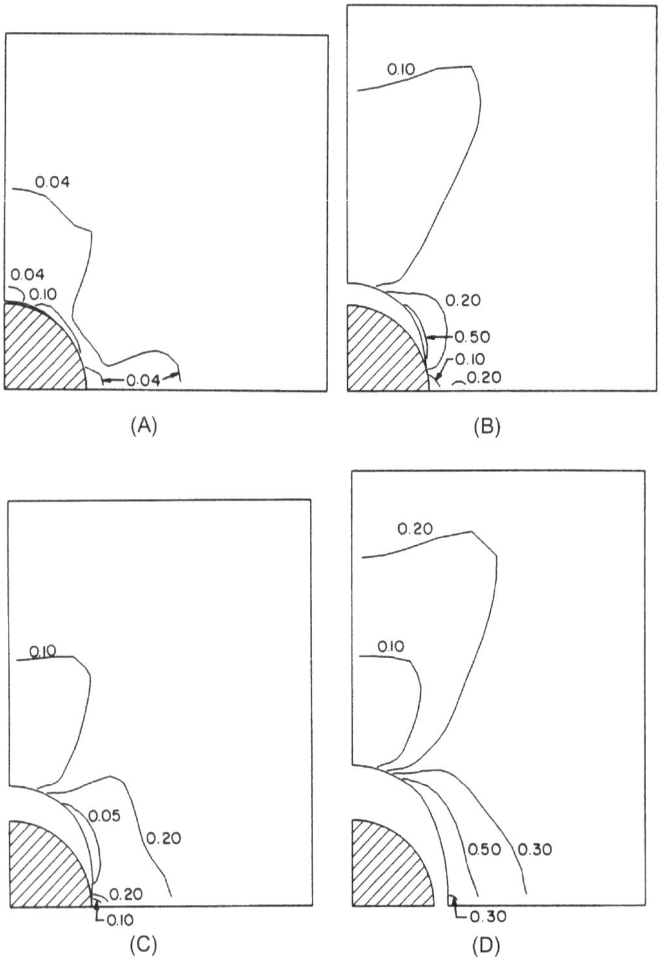

Figure 4.5 Material shapes and equivalent plastic strain distributions in various steps of simulation. *From Needleman, A., 1987. A continuum model for void nucleation by inclusion debonding. Trans. ASME J. Appl. Mech. 54 (3), 525–531.*

simulation. Plane-strain deformation was assumed. When the maximum principal stress in an inclusion became larger than a certain value, the cracking of the inclusion was assumed; the inclusion was divided into two parts and an ellipsoidal void, the flattening of which was large, was assumed. Fig. 4.6 shows the Voronoi Cells in various horizontally applied strains (Ghosh and Moorthy, 1998).

Zhang et al. (2000) completed the simulation method for ductile fracture using the yield function proposed by Gurson (1977) by incorporating

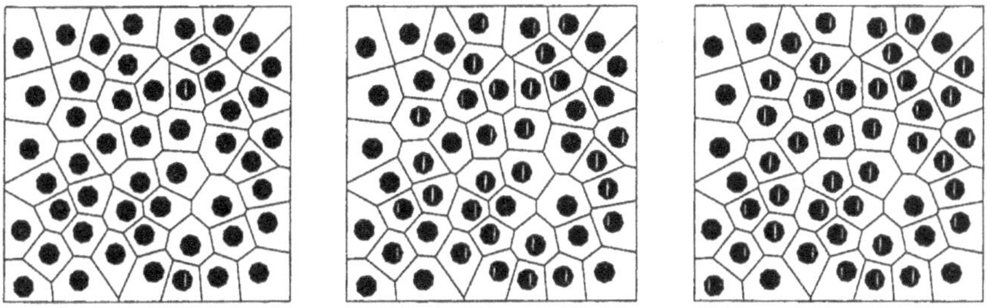

Figure 4.6 Voronoi cells in various horizontally applied strains: (A) $\varepsilon = 0.8\%$, (B) $\varepsilon = 1.4\%$, and (C) $\varepsilon = 2.0\%$. *From Ghosh, S., Moorthy, S., 1998. Particle fracture simulation in non-uniform microstructures of metal-matrix composites. Acta Mater. 46 (3), 965–982.*

into the simulation method the following two kinds of void nucleation criteria: a continuous void nucleation criterion and an impulsive void nucleation criterion. The void coalescence criterion proposed by Thomason (1985a,b) was used in the simulation method for ductile fracture. In the continuous void nucleation criterion, the increment of the void volume fraction due to void nucleation divided by the increment of the equivalent plastic strain was assumed to be constant throughout plastic deformation, whereas in the impulsive void nucleation criterion, the increment of the void volume fraction due to void nucleation occurred exclusively at the beginning of plastic deformation. The experiment and the simulation of the tensile test were performed using notched round tensile specimens and smooth round tensile specimens. Fig. 4.7 shows the relationship between the stress triaxiality and the strain to fracture in various values of the material constant in the void nucleation criteria (Zhang et al., 2000).

Babout et al. (2004b), with reference to Brown and Clarke (1975), proposed a void nucleation criterion due to the following two types of mechanisms: the separation of the inclusions from the matrix and the cracking of the inclusions, and also explained the experimental observations on the two types of mechanisms obtained by Babout et al. (2003). The inclusions were assumed to be elastic, whereas the matrix was assumed to be elastic–plastic. The cracking of the inclusion was assumed to occur when the normal stress in the inclusion became larger than a critical stress, which was the fracture stress of the inclusion. When the maximum of the normal stress in the inclusion was larger than the critical stress, the cracking of the inclusion occurred, whereas when the

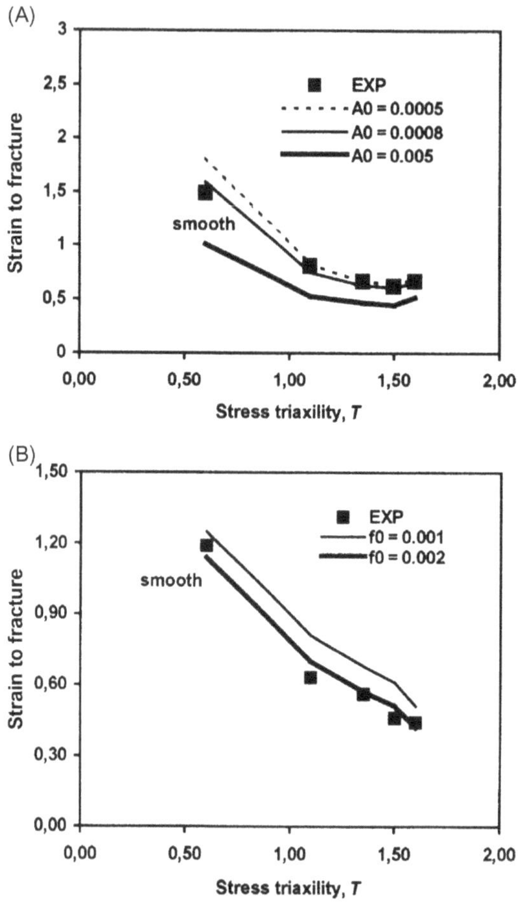

Figure 4.7 Relationship between stress triaxiality and strain to fracture in various values of material constant in void nucleation criteria: (A) X65 pipe steel, (B) Al-4.3% Si alloy. *From Zhang, Z.L., Thaulow, C., Ødegård, J., 2000. A complete Gurson model approach for ductile fracture. Eng. Fract. Mech. 67 (2), 155–168.*

maximum of the normal stress in the inclusion was smaller than the critical stress, the separation of the inclusion from the matrix was assumed to occur. Fig. 4.8 shows the relationship between the yield stress of the matrix, the fracture stress of the inclusion, and the type of the mechanism of void nucleation (Babout et al., 2004b).

4.3 Microscopic model of void growth

McClintock et al. (1966) proposed a void growth model and a void coalescence criterion for a cylindrical void in an infinite body which was

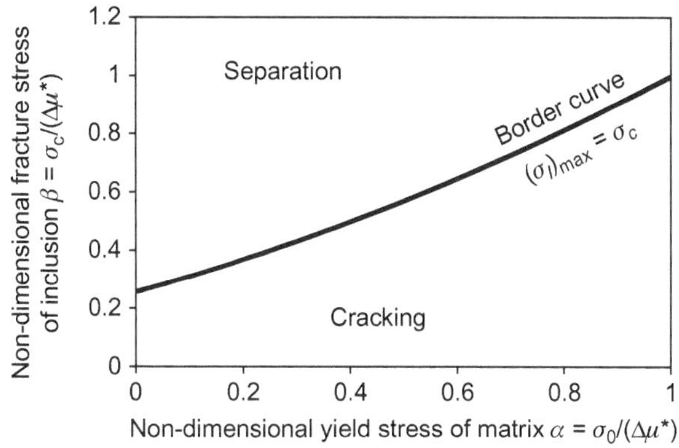

Figure 4.8 Relationship between yield stress of matrix, fracture stress of inclusion, and type of mechanism of void nucleation. *From Babout, L., Brechet, Y., Maire, E., Fougères, R., 2004b. On the competition between particle fracture and particle decohesion in metal matrix composites. Acta Mater. 52 (15), 4517–4525.*

subjected to shear strain by which both the material and the void rotated. Plane-strain deformation was assumed and the representative volume element was used. The void growth model was derived with reference to Berg (1962), whereas the void coalescence criterion was assumed to be satisfied when the void contacted the boundary of the representative volume element. The experiment on the deformation of the material, which contained a cylindrical void, was performed using a plasticine. Fig. 4.9 shows the relationship between the shear strain and the void orientation (McClintock et al., 1966). When the material that contained a cylindrical void was subjected to simple shear deformation, the cylindrical void closed and became a slit. The deformation of the material, which contained initially a cylindrical void, was analyzed before void closure and also after void closure using the void growth model to obtain the elongation and the rotation of the void.

McClintock (1968a) proposed a void growth model and a void coalescence criterion for a cylindrical void in an infinite body which was subjected to normal strain by which both the material and the void did not rotate. Generalized plane-strain deformation was assumed and the representative volume element was used. The void growth model was derived from Berg (1962), whereas the void coalescence was assumed to occur when the void contacted the boundary of the representative volume element. The deformation of the material, which contained a cylindrical void, was experimented using a plasticine. Fig. 4.10 shows the relationship between the equivalent plastic strain and the third flattening of the void

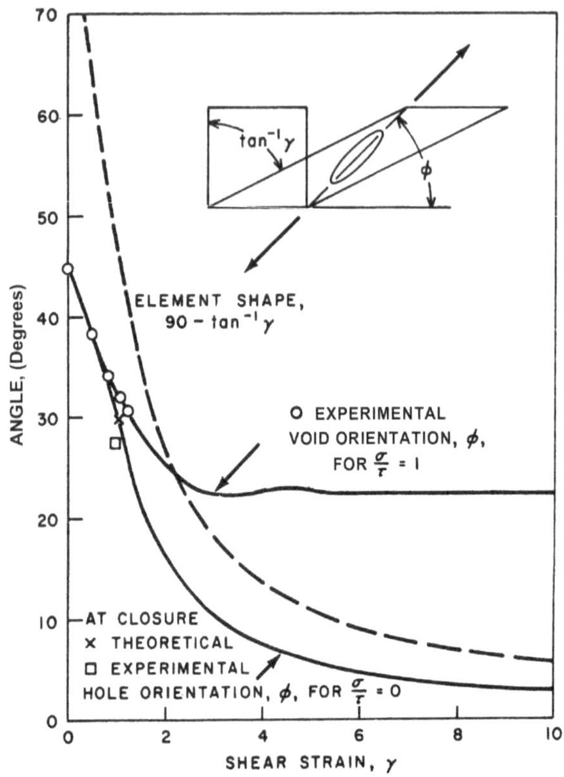

Figure 4.9 Relationship between shear strain and void orientation. *From McClintock, F.A., Kaplan, S.M., Berg, C.A., 1966. Ductile fracture by hole growth in shear bands. Int. J. Fract. Mech. 2 (4), 614–627.*

in balanced biaxial tension, plane-strain bending, and plane-stress bending (McClintock, 1968a). The void volume fraction rate was shown to increase exponentially with the increase in the mean normal stress in the plane perpendicular to the cylindrical void axis.

Rice and Tracey (1969) proposed a void growth model for a spherical void in an infinite body which was subjected to both normal strain and shear strain uniformly. A rigid, perfectly plastic material was assumed and the analysis was based on the principle of maximum plastic work (Hill, 1950) or the upper bound theorem (Prager and Hodge, 1951). The velocity field, which represented the material deformation, was assumed to be a linear combination of the velocity field for the material deformation at an infinite distance, the velocity field for the spherically symmetric volume change of a void, and the velocity field for the spherically

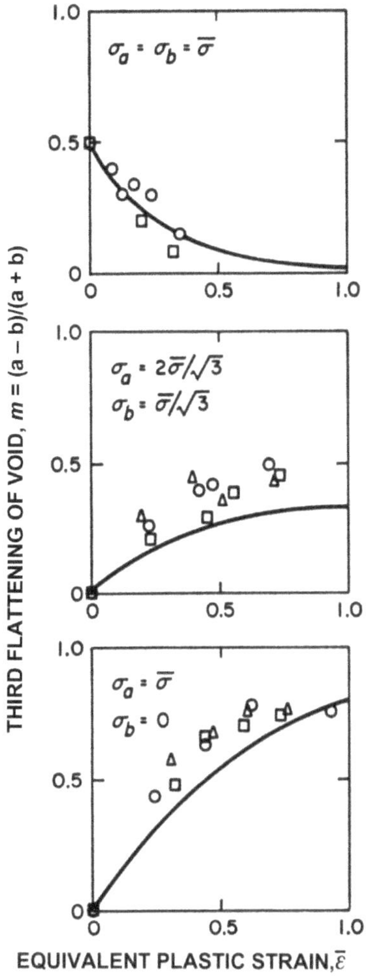

Figure 4.10 Relationship between equivalent plastic strain and third flattening of void in balanced biaxial tension, plane-strain bending, and plane-stress bending. *From McClintock, F.A., 1968. A criterion for ductile fracture by the growth of holes. Trans. ASME J. Appl. Mech. 35 (2), 363–371.*

asymmetric shape change of a void. Because the velocity field for the spherically symmetric volume change of a void was shown to be much more dominant than the velocity field for the spherically asymmetric shape change of a void, the void volume fraction rate was shown to increase exponentially with the increase in the stress triaxiality. Fig. 4.11 shows the relationship between the stress triaxiality and the linear

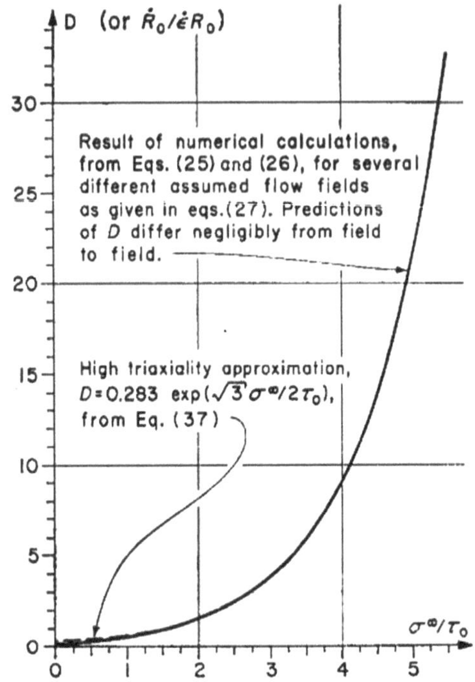

Figure 4.11 Relationship between stress triaxiality and linear combination coefficient for velocity field for spherically symmetric volume change of void. *From Rice, J.R., Tracey, D.M., 1969. On the ductile enlargement of voids in triaxial stress fields. J. Mech. Phys. Solids 17 (3), 201–217.*

combination coefficient for the velocity field for the spherically symmetric volume change of a void (Rice and Tracey, 1969).

Tracey (1971) proposed a void growth model and a void coalescence criterion for a cylindrical void in an infinite body, which is a rigid-plastic power-law strain-hardening material and is a cylinder whose axial length and whose radial length are infinite. Axisymmetric deformation of the material was assumed and axial strain rate was assumed to be constant. When the void growth rate was assumed to be constant, the radial stress required to maintain the imposed void growth rate increased with increasing the void growth rate or with increasing the strain-hardening exponent of the material. When the radial stress was assumed to be constant, the void growth rate required to maintain the imposed radial stress increased with increasing the radial stress or with decreasing the strain-hardening exponent of the material. Void growth at the center of the neck of the specimen was calculated in the uniaxial tensile test, and the void coalescence was assumed to occur when a void contacted a neighboring void. The strain to fracture calculated using the void coalescence criterion was much larger than the strain to fracture obtained experimentally by Edelson and Baldwin (1962).

Needleman (1972) performed the simulation of the plane-strain tensile test of a sheet that contained voids by the elastic–plastic finite-element method using the representative volume element, which was composed of a rectangular prismatic matrix and a cylindrical void. The simulations were performed in various void volume fractions and various strain-hardening exponents of the material. The strain to void coalescence was estimated by linearly extrapolating the load–strain relationship to the point at which the load became zero. The estimated strain to void coalescence was consistent with the strain to void coalescence which was estimated by extrapolating the relationship between the strain and the matrix width at neck to the point at which the matrix width at neck became zero.

Hellan (1975) proposed a void growth model for a spherical void in an infinite body and a void growth model for a cylindrical void in an infinite body in an approximate manner. Radially symmetric deformation of a void and a rigid-plastic power-law strain-hardening material were assumed in the former proposed void growth model, whereas radially asymmetric deformation of a void and a rigid, perfectly plastic material were assumed in the void growth model proposed by Rice and Tracey (1969). Although axisymmetric deformation of a void and a rigid-plastic power-law strain-hardening material were assumed both in the latter proposed void growth model and in the void growth model proposed by Tracey (1971), no axial strain rate was assumed in the latter proposed void growth model, whereas a uniform axial strain rate was assumed in the Tracey void growth model. The relationship between the equivalent plastic strain of the matrix and the void growth calculated using the latter proposed void growth model almost agreed with the relationship between the equivalent plastic strain of the matrix and the void growth calculated using the Tracey void growth model.

Gurson (1977), with reference to the void growth model proposed by Rice and Tracey (1969) and the averaging method proposed by Bishop and Hill (1951) which related microscopic variables and macroscopic variables, proposed a yield function for a rigid, perfectly plastic material which contained spherical voids. Although a yield function for a rigid, perfectly plastic material that contained long cylindrical voids was also proposed, the yield function for spherical voids was principally used, because the assumption of long cylindrical voids was not necessarily realistic. Because the Rice and Tracey void growth model was used, the microscopic velocity field for the spherically symmetric volume change of a void, and the microscopic velocity field for the spherically asymmetric shape change of a void were assumed. The void volume fraction of the material, which was a scalar, was introduced in the yield functions for spherical voids and for long cylindrical voids. The yield function for spherical voids has been extensively used in the field of fracture mechanics.

Tvergaard (1981) performed the simulation of the plane-strain tension of a sheet that contained voids by the elastic–plastic finite-element method using the representative volume element, which was composed of a rectangular prismatic matrix and a cylindrical void. Tvergaard obtained the bifurcation point, and performed the simulation along the bifurcation mode which represented the deformation in a shear band. The bifurcation theory for an elastic–plastic material proposed by Hill (1958) was used. The simulation of the plane-strain tension of a sheet, which contained voids, was performed by the elastic–plastic finite-element method using the yield function proposed by Gurson (1977), and the modification of the Gurson yield function was proposed by the comparison of the simulation results calculated using the representative volume element and the simulation results calculated using the Gurson yield function.

Tvergaard (1982a) performed the simulation of the axisymmetric deformation of a bar that contained voids by the elastic–plastic finite-element method using the representative volume element, which was composed of a cylindrical matrix and an initially spherical void. Tvergaard obtained the bifurcation point, and performed the simulation along the bifurcation mode which represented the deformation in a shear band. The simulation of the spherically symmetric tension of a cylinder which contained a spherical void was performed by the elastic–plastic finite-element method using the representative volume element, and the simulation of the spherically symmetric tension of a sphere was performed by the elastic–plastic finite-element method using the yield function proposed by Gurson (1977). The simulation result calculated using the Gurson yield function substantially coincided with the simulation result calculated using the representative volume element.

Budiansky et al. (1982) proposed a void growth model for a spherical void or a cylindrical void in an infinite body, which is a viscous power-law strain-rate-hardening material and is subjected to axisymmetric tension. A rigid, perfectly plastic material was included in the material, and a void was assumed to be axisymmetric. The deformation of a cylindrical void the axial length of which was infinite was analyzed, and the analytical result was shown to be identical to the analytical result obtained by McClintock (1968a) when a rigid, perfectly plastic material was assumed. With increasing the ratio of the axial stress to the radial stress, a spherical void became a cylindrical void the axial length of which was infinite, and the cylindrical void became a needle the radial length of which was zero. The deformation of a spherical void was analyzed, and the analytical result was shown to be identical to the analytical result obtained by Rice and Tracey (1969) when the velocity field around a void was approximated appropriately for high stress triaxiality and when a rigid, perfectly plastic material was assumed.

Hom and McMeeking (1989) performed the simulation of the deformation of a block that contained voids by the three-dimensional elastic—plastic finite-element method using the representative volume element, which was composed of a cubic matrix and an initially spherical void. When the material was subjected to uniaxial tension, the void shape agreed with the void shape calculated using the void growth model proposed by Rice and Tracey (1969), which indicated that the void coalescence between two neighboring voids did not occur in the applied tensile strain. When the material was subjected to multiaxial tension, the flow of the matrix was localized into the region between the void and the side face of the cube, which indicated that the void coalescence between two neighboring voids began to occur.

Worswick and Pick (1990) performed the simulation of the deformation of a block that contained voids by the three-dimensional elastic—plastic finite-element method using the representative volume element, which was composed of a cubic matrix and an initially spherical void. The material was subjected to uniaxial tension in various imposed hydrostatic stresses, in various initial void volume fractions, and in various strain-hardening exponents of the matrix. The initial spherically symmetric volume changing void growth rate agreed with the initial spherically symmetric volume changing void growth rate calculated using the void growth model proposed by Budiansky et al. (1982), which indicated that the void coalescence between two neighboring voids did not occur in the applied tensile strain.

Gologanu et al. (1993), with reference to the yield function proposed by Gurson (1977), proposed a yield function for a rigid, perfectly plastic material that contained axisymmetric prolate ellipsoidal voids and was subjected to axisymmetric stresses. The averaging method proposed by Bishop and Hill (1951), which related microscopic variables and macroscopic variables, was used. The following two types of microscopic velocity fields were assumed: the microscopic velocity field which mainly represented the axisymmetric volume change of a void, and the microscopic velocity field which represented the axisymmetric shape change of a void. When the void was assumed to be cylindrical, the proposed yield function was identical to the yield function for cylindrical voids proposed by Gurson (1977), whereas when the void was assumed to be spherical, the proposed yield function was identical to the yield function for spherical voids proposed by Gurson (1977). The void volume fraction of the material and also the eccentricity of the void were introduced in the yield function.

Gologanu et al. (1994), with reference to the yield function proposed by Gurson (1977), proposed a yield function for a rigid, perfectly plastic material that contained axisymmetric oblate ellipsoidal voids and was subjected to axisymmetric stresses. The following two types of microscopic velocity fields were assumed: the microscopic velocity field which mainly represented the axisymmetric volume change of a void, and the microscopic velocity field which represented the axisymmetric shape change of a void. When the void was assumed to be spherical, the proposed yield function was identical to the yield function for spherical voids proposed by Gurson (1977). The void volume fraction of the material and also the eccentricity of the void were introduced in the yield function. Gologanu et al. (1997) reviewed both Gologanu et al. (1993) and Gologanu et al. (1994) in a unified manner.

Kuna and Sun (1996) performed the simulation of the deformation of three types of blocks by the three-dimensional elastic–plastic finite-element method, and obtained the effect of the following three types of lattices: cubic primitive lattice, body centered cubic lattice, and hexagonal lattice. The simulation of the deformation of a cylinder was also performed by the axisymmetric elastic–plastic finite-element method for comparison. The initial void volume fraction of the representative volume element was assumed to be constant, and the representative volume element was subjected to quasiaxisymmetric tension in various stress triaxialities. When the length of the block in the direction perpendicular to the tensile direction stopped changing, the void coalescence criterion was assumed to be satisfied. Under each stress triaxiality, both the maximum

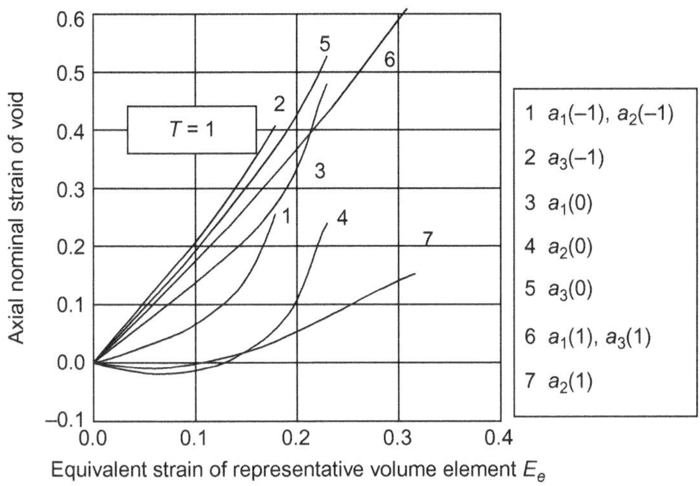

Figure 4.12 Relationship between equivalent strain of representative volume element and axial nominal strain of void. *From Zhang, K.S., Bai, J.B., François, D., 2001. Numerical analysis of the influence of the Lode parameter on void growth. Int. J. Solids Struct. 38 (32–33), 5847–5856.*

loads per unit area and the strains at which the void coalescence criterion was satisfied in the cases of cubic primitive lattice, body centered cubic lattice, hexagonal lattice, and cylindrical lattice were the second largest, the largest, the second smallest, and the smallest of the lattices, respectively.

Zhang et al. (2001) performed the simulation of the deformation of the representative volume element which was composed of a cubic matrix and an initially spherical void by the three-dimensional elastic–plastic finite-element method, and they obtained the effect of the Lode parameter in various stress triaxialities. The initial void volume fraction of the representative volume element was assumed to be 10%. With increasing the stress triaxiality, the effect of the Lode parameter decreased. Fig. 4.12 shows the relationship between the equivalent strain of the representative volume element and the axial nominal strain of the void (Zhang et al., 2001). In the legend, the subscript indicated the axial number, whereas the number in parentheses indicated the value of the Lode parameter.

Ragab (2004a), with reference to the void growth model proposed by Lee and Mear (1992), modified values of parameters introduced by Tvergaard (1981) in the Gurson yield function (1977) modified by Tvergaard (1981), to obtain the effects of the strain hardening of the material and the void shape on the void growth. A spheroidal void and axisymmetric deformation of the material were assumed. Void growth was calculated using the flow rule associated with the Gurson–Tvergaard yield function. The values of the parameters were assumed to be the functions of the stress triaxiality, the strain-hardening exponent of the material, and the ratio of the polar radius of the void to the equatorial radius of the void. Void coalescence was assumed to occur with reference to the void coalescence criterion proposed by Brown and Embury (1973). The fracture strain calculated was compared with the fracture strain obtained experimentally in various literatures, and the fracture strains calculated using the void coalescence criteria proposed by Pardoen and Hutchinson (2000), Benzerga (2002), and McClintock (1968b).

4.4 Microscopic criteria of void coalescence

Thomason (1968) proposed a void coalescence criterion for rectangular prismatic voids that were distributed uniformly in an infinite body which was subjected to normal stress. A rigid, perfectly plastic material and plane-strain deformation were assumed, whereas the representative volume element which consisted of two neighboring voids and the matrix between the two neighboring voids was assumed. The analysis was based on the upper bound theorem (Prager and Hodge, 1951). The void coalescence criterion was assumed to be satisfied when the energy required for the deformation mode of void coalescence, that is, internal necking deformation mode became smaller than the energy required for the deformation mode of void noncoalescence, that is, homogeneous deformation mode. Square prismatic voids were assumed before deformation, and the void volume fraction was assumed be constant during deformation. Fig. 4.13 shows the relationship between the void volume fraction and the strain to fracture in uniaxial tensile test (Thomason, 1968). The relationship obtained analytically was compared with the relationships obtained from the experiment performed by Edelson and Baldwin (1962).

Brown and Embury (1973) proposed a void coalescence criterion for voids that contained inclusions in a simple form, in which void coalescence was assumed to occur when the distance between two neighboring voids in the horizontal direction became equal to the void length in the vertical direction. The proposed void coalescence criterion was derived intuitively from the slip-line field theory (Johnson et al., 1982), because the slip-line field which represented the deformation mode of void coalescence, that is, internal necking deformation mode, was obtained in the

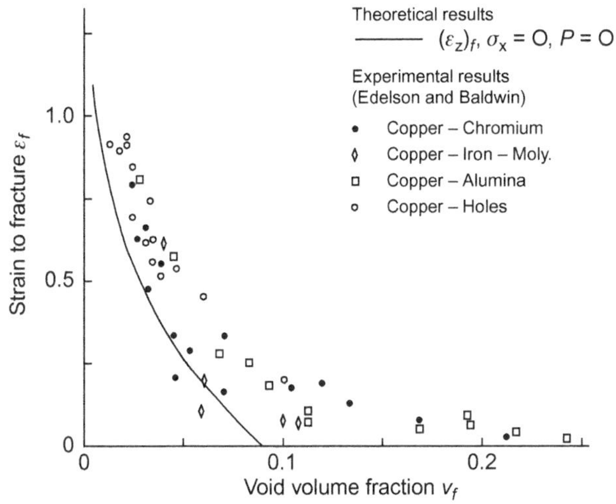

Figure 4.13 Relationship between void volume fraction and strain to fracture in uniaxial tensile test. *From Thomason, P.F., 1968. A theory for ductile fracture by internal necking of cavities. J. Inst. Met. 96, 360–365.*

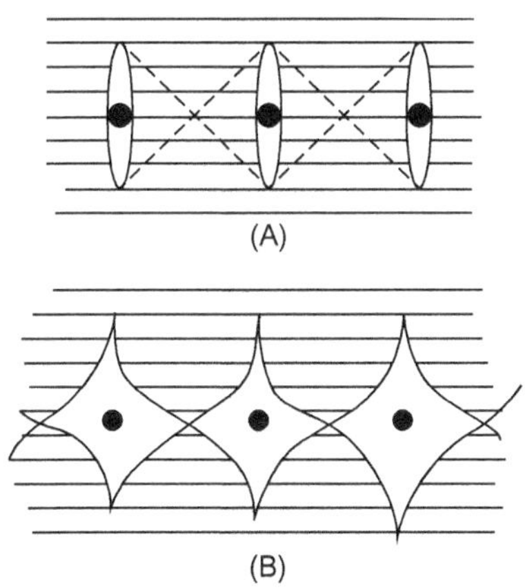

Figure 4.14 Void shapes and void configurations at the beginning of void coalescence and at the ending of void coalescence: (A) at the beginning of void coalescence, (B) at the ending of void coalescence. *From Brown, L.M., Embury, J.D., 1973. The initiation and growth of voids at second phase particles. In: Proceedings of the Third International Conference on the Strength of Metals and Alloys, pp. 164–169.*

region between two neighboring voids when the proposed void coalescence criterion was satisfied. Fig. 4.14 shows the void shapes and the void configurations at the beginning of void coalescence and at the ending of void coalescence (Brown and Embury, 1973). The relationship between

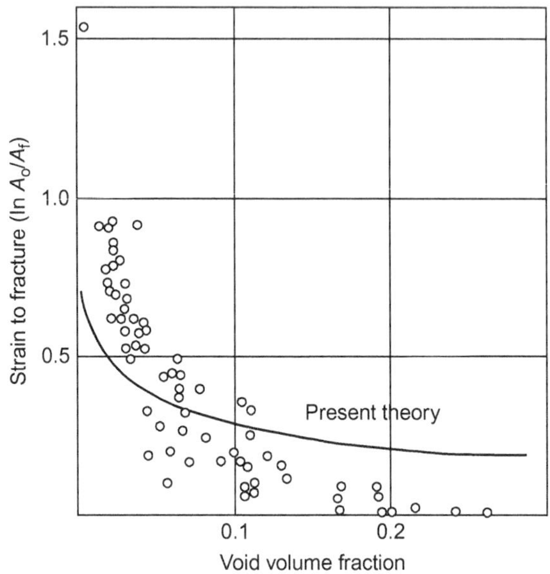

Figure 4.15 Relationships between void volume fraction and strain to fracture calculated from proposed void coalescence criterion and obtained experimentally by Edelson and Baldwin (1962). *From Nagumo, M., 1973. A criterion of ductile fracture on the tensile testing of a perforated mild steel sheet. Acta Metall. 21 (12), 1661–1667.*

the inclusion volume fraction and the strain to void coalescence calculated using the proposed void coalescence criterion was shown to agree with the relationship between the inclusion volume fraction and the strain to void coalescence obtained experimentally when the strain to void nucleation was assumed appropriately.

Nagumo (1973) performed the uniaxial tensile test of a sheet using a low carbon steel in which several holes were pierced regularly, and proposed a void coalescence criterion due to the propagation of cracks which appeared between two neighboring holes. The proposed void coalescence criterion was derived from an energetic point of view; the propagation of cracks was assumed to occur when the change in the energy dissipation due to plastic deformation between two neighboring holes became larger than the change in the energy dissipation due to crack propagation between two neighboring holes. Fig. 4.15 shows the relationships between the void volume fraction and the strain to fracture calculated from the proposed void coalescence criterion and obtained experimentally by Edelson and Baldwin (1962) (Nagumo, 1973).

Melander (1979), using the void coalescence criterion proposed by Melander and Ståhlberg (1980), performed the simulation of the plane-strain tension of a bar in which rectangular prismatic voids, which were the same size, were distributed randomly by the Poisson distribution. A rigid, perfectly plastic material and plane-strain deformation were assumed. The positions of the velocity discontinuity lines were optimized by minimizing the energy dissipated on the velocity discontinuity lines. The randomness was defined as the difference in the coordinates in the tensile direction of two neighboring voids. The void volume fraction was assumed to be 0.25% and 4%, and a hydrostatic stress was imposed on the bar. When the imposed hydrostatic stress was relatively small, although the strain to fracture decreased with decreasing the standard deviation of the randomness, the strain to fracture in the case of randomly distributed voids was larger than the strain to fracture in the case of uniformly distributed voids.

Melander and Ståhlberg (1980) proposed, with reference to Thomason (1968), a void coalescence criterion for rectangular prismatic voids that were distributed in an infinite body which was subjected to normal stress. The analysis was based on the upper bound theorem (Prager and Hodge, 1951), and a rigid, perfectly plastic material and plane-strain deformation were assumed. The representative volume element, which consisted of two neighboring voids and the region between the two neighboring voids, was assumed, and the region between the two neighboring voids was composed of rigid bodies and velocity discontinuity lines. The void coalescence criterion was assumed to be satisfied when the energy required for the deformation mode of void coalescence became smaller than the energy required for the deformation mode of void noncoalescence. The void volume fraction was assumed be constant during deformation. Fig. 4.16 shows the relationship between the void volume fraction and the strain to fracture for various ratios of the size of large voids to the size of small voids (Melander and Ståhlberg, 1980).

Jalinier (1983) proposed void coalescence criteria both due to the separation of the inclusions from the matrix and due to the cracking of the inclusions, and predicted the fracture forming limit. In the void coalescence criterion due to the separation of the inclusions from the matrix, a void was assumed to coalesce when the distance between two neighboring voids became smaller than five times the void radius in the direction connecting the two neighboring voids. The void coalescence criterion due to the cracking of the inclusions was derived with reference to the

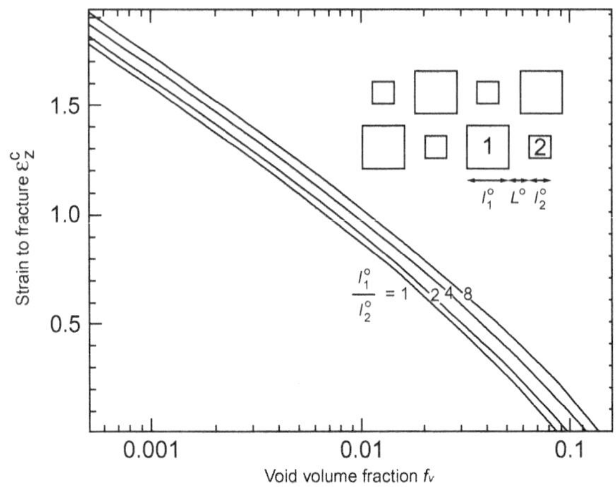

Figure 4.16 Relationship between void volume fraction and strain to fracture for various ratios of the size of large voids to the size of small voids. *From Melander, A., Ståhlberg, U., 1980. The effect of void size and distribution on ductile fracture. Int. J. Fract. 16 (5), 431–439.*

condition of the flow localization into a shear band due to McClintock (1968b). The fracture forming limit was a straight line in the fracture forming limit diagram in the case of the void coalescence criterion due to the separation of the inclusions from the matrix. The fracture forming limit, which agreed with the fracture forming limit obtained from the experiment using a 3003 aluminum alloy, was a curved line in the fracture forming limit diagram in the case of the void coalescence criterion due to the cracking of the inclusions.

Thomason (1985a) proposed a void coalescence criterion for square prismatic voids that were distributed uniformly in an infinite body which was subjected to normal stress. A rigid, perfectly plastic material and three-dimensional deformation were assumed, whereas the representative volume element, which was square prismatic and consisted of a square prismatic void and matrix which surrounded the void, was assumed. The analysis was based on the upper bound theorem (Hill, 1950). The void coalescence criterion was assumed to be satisfied when the energy required for the deformation mode of void coalescence became smaller than the energy required for the deformation mode of void noncoalescence. The void volume fraction was not changed during deformation. The relationship between the void shape and the load at which the void coalescence criterion was satisfied was shown to be remarkably similar to the relationship between the void shape and the

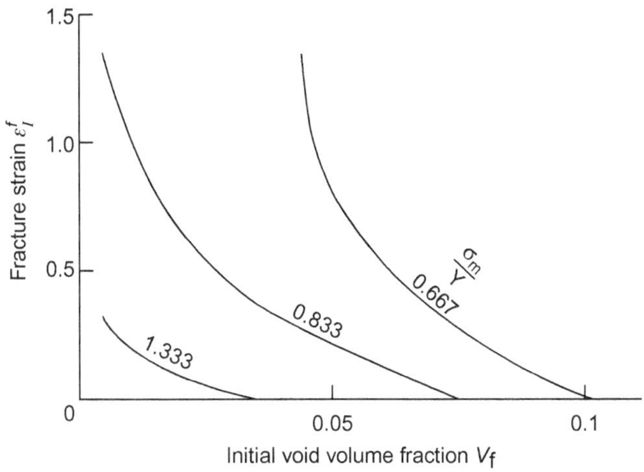

Figure 4.17 Relationship between initial void volume fraction and fracture strain for axisymmetric tension in various mean normal stresses. *From Thomason, P.F., 1985b. A three-dimensional model for ductile fracture by the growth and coalescence of microvoids. Acta Metall. 33 (6), 1087–1095.*

load at which the void coalescence criterion was satisfied for cylindrical voids calculated using the velocity fields proposed by Kudo (1960b). Since the relationship between the void shape and the load at which the void coalescence criterion was satisfied for square prismatic voids had to be obtained numerically, the relationship was expressed as an empirical closed-form solution.

Thomason (1985b), using the void coalescence criterion proposed by Thomason (1985a), proposed a void coalescence criterion for ellipsoidal voids that were distributed uniformly in an infinite body which was subjected to normal stress. The lengths of the principal axes in the ellipsoidal void were made to coincide with the lengths of the principal axes in the square prismatic void used in Thomason (1985a). The representative volume element, which was square prismatic and consisted of an ellipsoidal void and matrix which surrounded the void, was assumed. Because the void shape was calculated using the void growth model proposed by Rice and Tracey (1969), the void volume fraction was changed during deformation. Since the void coalescence criterion was complicated, the void coalescence criterion had to be obtained numerically. Fig. 4.17 shows the relationship between the initial void volume fraction and the fracture strain for axisymmetric tension in various mean normal stresses (Thomason, 1985b).

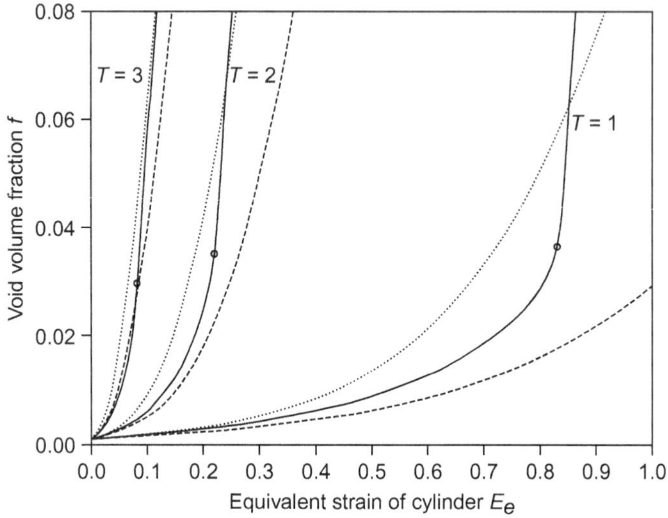

Figure 4.18 Relationship between equivalent strain of cylinder and void volume fraction in various stress triaxialities. *From Koplik, J., Needleman, A., 1988. Void growth and coalescence in porous plastic solids. Int. J. Solids Struct. 24 (8), 835–853.*

Koplik and Needleman (1988) performed the simulation of the void growth in a bar by the elastic–viscoplastic finite-element method and obtained a void coalescence criterion numerically. The representative volume element, which was cylindrical and consisted of an initially spherical void and matrix which surrounded the void, was assumed and subjected to axisymmetric tension. When the radius of the cylinder remained constant and the flow of the matrix was localized into the region between the void and the side face of the cylinder, the void coalescence criterion was assumed to be satisfied. Fig. 4.18 shows the relationship between the equivalent strain of the cylinder and the void volume fraction in various stress triaxialities (Koplik and Needleman, 1988). The void coalescence criterion was satisfied at the white circle. The broken line indicated the relationship obtained using the yield function proposed by Gurson (1977), whereas the dotted line indicated the relationship obtained using the Gurson yield function modified by Tvergaard (1981).

Magnusen et al. (1990) performed the simulation of the void growth and the void coalescence in a sheet in various initial void volume fractions and in various minimum spaces between two neighboring holes using the empirical relationships obtained experimentally, and predicted the strain to fracture. The empirical relationships, which were modified when two

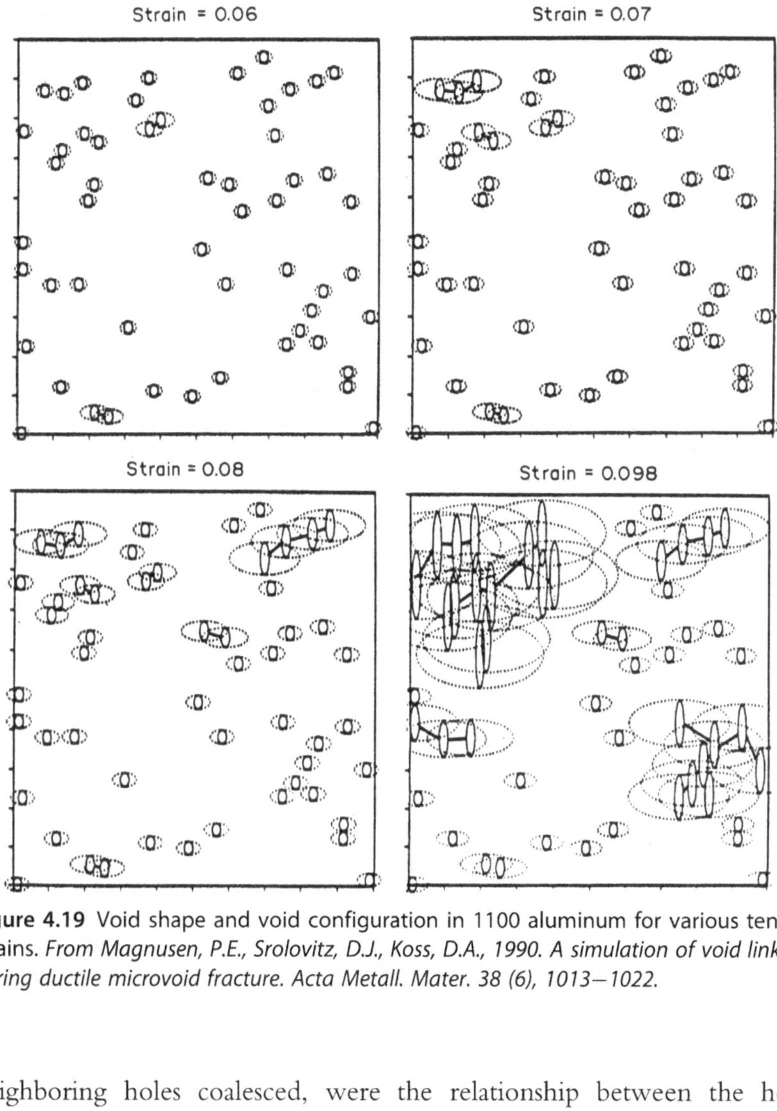

Figure 4.19 Void shape and void configuration in 1100 aluminum for various tensile strains. *From Magnusen, P.E., Srolovitz, D.J., Koss, D.A., 1990. A simulation of void linking during ductile microvoid fracture. Acta Metall. Mater. 38 (6), 1013–1022.*

neighboring holes coalesced, were the relationship between the hole diameter in the tensile direction and the tensile strain calculated from the elongation, and the relationship between the thickness strain around a hole and the thickness strain calculated from the elongation. The void coalescence criterion was assumed to be satisfied when the thickness strain became a critical value. Fig. 4.19 shows the void shape and the void configuration in 1100 aluminum for various tensile strains (Magnusen et al., 1990). The solid ellipse indicated the void shape, whereas the dashed ellipse indicated the location at which the thickness strain was equal to half the critical value. Hence, when two neighboring dashed ellipses overlapped, the void coalescence criterion was satisfied.

Thomason (1993) performed the simulation of the void growth and the void coalescence in a bar in which the diameters of initially cylindrical voids were not uniform and the distances between two neighboring initially cylindrical voids were not uniform, and obtained the strain to fracture. A rigid, perfectly plastic material and plane-strain deformation were assumed. The void growth model proposed by Rice and Tracey (1969) and modified by Huang (1991) was used, whereas the tensile stress for the representative volume element calculated by Thomason (1978) using the slip-line field theory (Johnson et al., 1982) was used to evaluate the void coalescence criterion. The void shape and the void configuration were assumed to be uniform in the tensile direction, whereas the void shape and the void configuration were not assumed to be uniform in the direction perpendicular to the tensile direction. The strain to fracture was minimized irrelevant to the magnitude of the void volume fraction when the initial void volume fraction was uniform in the direction perpendicular to the tensile direction.

Zhang and Niemi (1994a) compared the relationships between the stress triaxiality and the strain to void coalescence calculated using three types of void coalescence criteria. The three types of void coalescence criteria were the critical void growth ratio criterion based on the Rice and Tracey void growth model (1969) used in Rousselier (1987), the critical void volume fraction criterion based on the Gurson yield function (1977) used in Tvergaard and Needleman (1984), and the void coalescence criterion proposed by Thomason (1985a,b). The representative volume element was subjected to axisymmetric tension under a specified stress triaxiality. The critical void growth ratio and the critical void volume fraction under a low stress triaxiality were determined such that the strains to void coalescence calculated using these three types of void coalescence criteria coincided. When the stress triaxiality was not low, the strains to void coalescence calculated using these three types of void coalescence criteria differed.

Zhang and Niemi (1995) proposed a void coalescence criterion derived from the combination of the Gurson yield function (1977) modified by Tvergaard (1981) and the void coalescence criterion proposed by Thomason (1985a,b), and compared the strain to void coalescence with the strain to void coalescence which was calculated using the representative volume element by Koplik and Needleman (1988). Although the

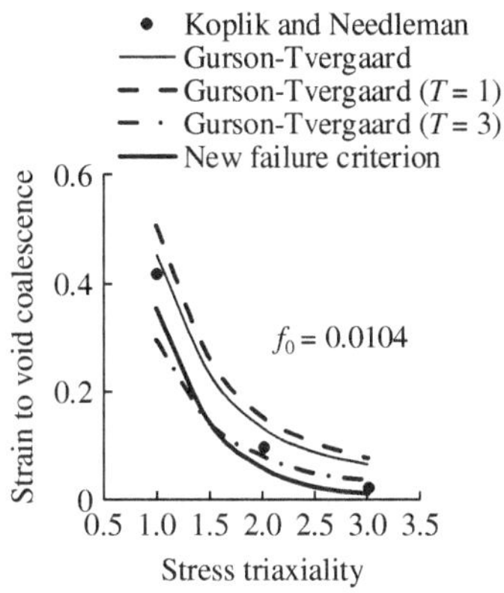

Figure 4.20 Relationship between stress triaxiality and strain to void coalescence. *From Zhang, Z.L., Niemi, E., 1995. A new failure criterion for the Gurson–Tvergaard dilational constitutive model. Int. J. Fract. 70 (4), 321–334.*

void growth model proposed by Rice and Tracey (1969) was used to calculate void growth in the Thomason void coalescence criterion, the Gurson–Tvergaard yield function was used to calculate void growth. Although a void was conventionally assumed to coalesce at a critical void volume fraction in the simulation using the Gurson–Tvergaard yield function, no assumption was made on the critical void volume fraction. Fig. 4.20 shows the relationship between the stress triaxiality and the strain to void coalescence (Zhang and Niemi, 1995).

Bandstra et al. (1998) performed the experiment and the simulation of the tensile test of a bar using HY-100 steel, the axial direction of which was identical to the width direction of a plate, and obtained the relationship between the stress triaxiality and the equivalent strain to fracture. At the fracture surface, elongated manganese sulfide inclusions, the diameter of which was 2–3 μm and the length of which was 30–100 μm, were observed. The simulation of the biaxial tension of a bar was performed using a commercial elastic–plastic finite-element software in various stress triaxialities. Plane-strain deformation was assumed. Several cylindrical voids, the diameter of which was 2.5 μm, were assumed, and the angle between the line that connected two neighboring voids and the tensile direction was assumed to be 45 degrees. Fig. 4.21 shows the relationship between the stress triaxiality and the equivalent strain to fracture (Bandstra et al., 1998). At high stress triaxialities, the effect of the stress triaxiality on the equivalent strain to fracture was small both in the experiment and in the simulation.

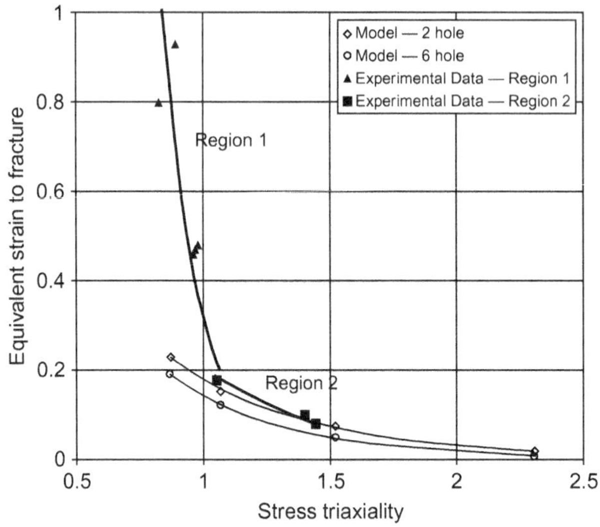

Figure 4.21 Relationship between stress triaxiality and equivalent strain to fracture. *From Bandstra, J.P., Goto, D.M., Koss, D.A., 1998. Ductile failure as a result of a void-sheet instability: experiment and computational modeling. Mater. Sci. Eng. A 249 (1–2), 46–54.*

Benzerga et al. (1999) performed the experiment and the simulation of the tensile test of a bar using X52 steel plate and A508 steel block both of which were anisotropic materials, and evaluated several void coalescence criteria. The yield function proposed by Gologanu et al. (1993, 1994) was used. The following two types of void coalescence criteria were evaluated: the void coalescence criterion derived from the condition for the localization of deformation proposed by Rudnicki and Rice (1975) and the void coalescence criterion proposed by Thomason (1985a, b) and modified by Zhang and Niemi (1994b). Because both X52 steel plate and A508 steel block contained elongated manganese sulfide inclusions, the strain to fracture measured using notched round tensile specimens depended on the relationship between the tensile direction and the three principal directions of the materials. The strain to fracture calculated using notched round tensile specimens depended on the void shape and on the void configuration in the void coalescence criteria.

Komori (1999b) proposed a void coalescence criterion based on the void coalescence criterion presented by Thomason (1968) and the void coalescence criterion presented by Melander and Ståhlberg (1980), and applied the proposed void coalescence criterion to the simulation of the

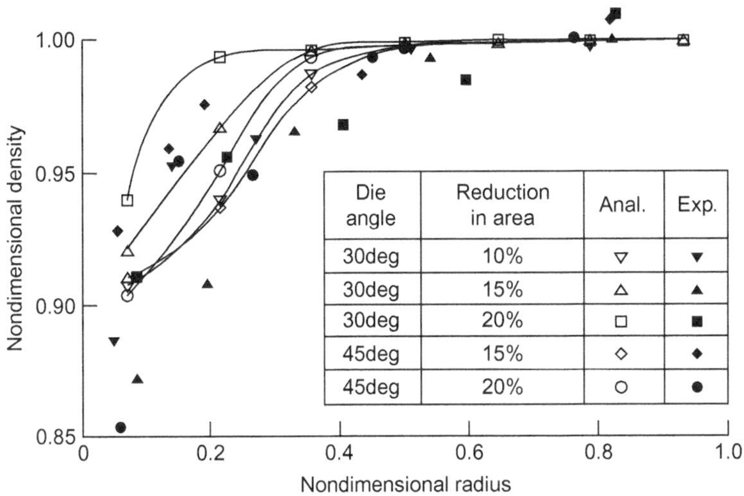

Figure 4.22 Material density distribution in radial direction after drawing through die preceding die at which material fractured. *From Komori, K., 1999b. Proposal and use of a void model for the simulation of ductile fracture behavior. Acta Mater. 47 (10), 3069–3077.*

central burst in the multipass drawing of a bar using tough pitch copper. The void shape was assumed to be a rectangle and the direction of maximum principal strain was assumed to coincide with the direction of maximum principal stress during plastic deformation in both the Thomason, and the Melander and Ståhlberg void coalescence criteria. Whereas, the void shape was assumed to be a parallelogram and the direction of maximum principal strain was not assumed to coincide with the direction of maximum principal stress during plastic deformation in the proposed void coalescence criterion. Fig. 4.22 shows the material density distribution in the radial direction after drawing through the die preceding the die at which the material fractured (Komori, 1999b).

Pardoen and Hutchinson (2000) proposed a void coalescence criterion based on the yield function presented by Gologanu et al. (1993, 1994) and the void coalescence criterion persented by Thomason (1985a,b), and evaluated the proposed void coalescence criterion by the elastic–plastic finite-element simulation using the representative volume element. Axisymmetric deformation of the material was assumed. A rigid, perfectly plastic material was assumed and the void shape was calculated using the void growth model proposed by Rice and Tracey (1969) in the Thomason void coalescence criterion. Whereas, an elastic–plastic

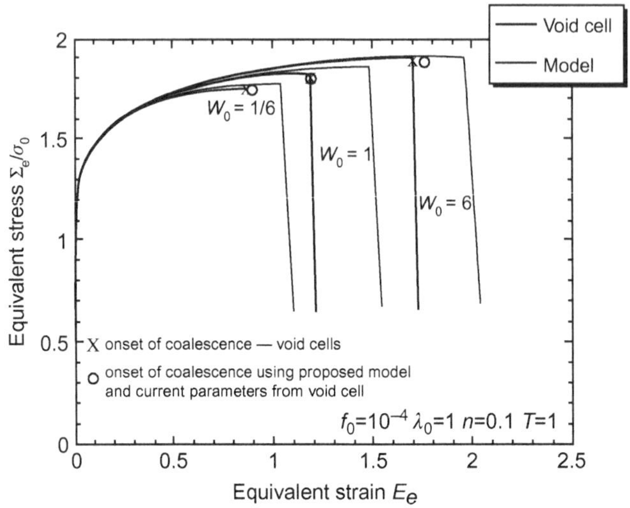

Figure 4.23 Relationship between equivalent strain and equivalent stress in various initial void shapes. *From Pardoen, T., Hutchinson, J.W., 2000. An extended model for void growth and coalescence. J. Mech. Phys. Solids 48 (12), 2467–2512.*

power-law strain-hardening material was assumed and the void shape was calculated using the flow rule associated with the Gologanu et al. yield function in the proposed void coalescence criterion. The analyses were performed in various initial void volume fractions, stress triaxialities, strain-hardening exponents, initial void shapes, and initial void configurations. Fig. 4.23 shows the relationship between the equivalent strain and the equivalent stress in various initial void shapes (Pardoen and Hutchinson, 2000).

Ragab (2000) proposed a ductile fracture criterion in the tensile test of a bar under the condition that the ductile fracture criterion was satisfied when the equivalent stress of the material became zero, and compared the strain to fracture calculated with the strain to fracture obtained experimentally in various literatures. Void growth was calculated using the flow rule associated with the Gurson yield function (1977) modified by Tvergaard (1981), whereas the stress at the center of the neck of the specimen in the tensile test derived analytically by Bridgman (1952) was used. The analysis was performed in various values of a parameter introduced by Tvergaard (1981) in the Gurson–Tvergaard yield function. Void growth was also calculated using the flow rule associated with the yield function proposed by Gologanu et al. (1993), which was obtained

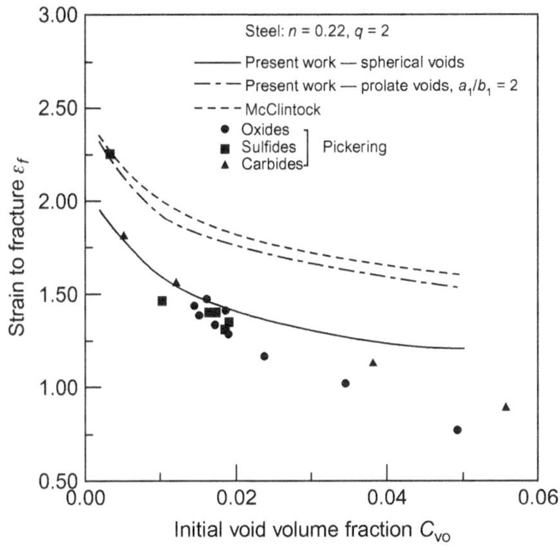

Figure 4.24 Relationship between initial void volume fraction and strain to fracture. *From Ragab, A.R., 2000. Prediction of ductile fracture in axisymmetric tension by void coalescence. Int. J. Fract. 105 (4), 391–409.*

for axisymmetric prolate ellipsoidal voids. Fig. 4.24 shows the relationship between the initial void volume fraction and the strain to fracture (Ragab, 2000). The experimental results for steels in Pickering (1978) were supplemented.

Gologanu et al. (2001a,b) used the yield function proposed by Gologanu et al. (1997) to obtain a void coalescence criterion for the representative volume element which was cylindrical, consisted of an initially spheroidal void and matrix which surrounded the void, and was subjected to axisymmetric tension. The matrix was assumed to be rigid, perfectly plastic. In Gologanu et al. (2001a), because the applied axial stress was larger than the applied radial stress, the upper and lower parts of the representative volume element were assumed to be sound and obey the Mises yield function, whereas the middle part of the representative volume element was assumed to be porous and obey the Gologanu et al. yield function. In Gologanu et al. (2001b), because the applied radial stress was larger than the applied axial stress, the outer part of the representative volume element was assumed to be sound and obey the Mises yield function, whereas the inner part of the representative volume element was assumed to be porous and obey the Gologanu et al. yield function. Both the stress distribution and the strain rate distribution were assumed to be homogeneous in each part. In Gologanu et al. (2001a), the void coalescence criterion was assumed to be satisfied when the upper and lower parts of the representative volume element became rigid, and the analytical results almost agreed with the simulation results obtained by Koplik and Needleman (1988), which were calculated by the elastic–viscoplastic finite-element method using the representative volume element.

Benzerga (2002) proposed a void coalescence criterion based on the void coalescence criterion proposed by Thomason (1985a,b) to improve the void coalescence criterion for an oblate spheroidal void. Benzerga compared the analytical results with the analytical results obtained by Pardoen and Hutchinson (2000) and the simulation results obtained by the elastic–plastic finite-element simulation using the representative volume element. Axisymmetric deformation of the material was assumed, and the applied axial stress was assumed to be larger than the applied radial stress. The representative volume element, which was cylindrical and consisted of an initially spheroidal void and matrix that surrounded the void, was assumed. Analyses were performed in various initial void volume fractions, initial void shapes, initial void configurations, and stress triaxialities.

Ragab (2004b) proposed a void coalescence criterion based on the internal necking of spheroidal voids with reference to the void coalescence criterion proposed by Thomason (1985a,b), and compared the analytical results with the analytical results obtained by Thomason (1985b), Pardoen and Hutchinson (2000), and Benzerga (2002), and the experimental results in various literature. The representative volume element, which was cylindrical and consisted of a spheroidal void and matrix which surrounded the void, was assumed, and a rigid-plastic power-law strain-hardening matrix was assumed. Void growth was calculated using the flow rule associated with the Gurson yield function (1977) modified by Tvergaard (1981), whereas the stress at the center of the neck of the specimen in the tensile test derived analytically by Bridgman (1952) was used. The analysis was performed using optimized values of parameters introduced by Tvergaard (1981) in the Gurson–Tvergaard yield function. Fig. 4.25 shows the relationship between the experimental strain to fracture and the analytical strain to fracture (Ragab, 2004b).

Benzerga et al. (2004b), using the theory of anisotropic ductile fracture proposed by Benzerga et al. (2002), performed the simulation of the tensile test of X52 steel plate using notched round tensile specimens in various notch-root radii, and compared the simulation results using the

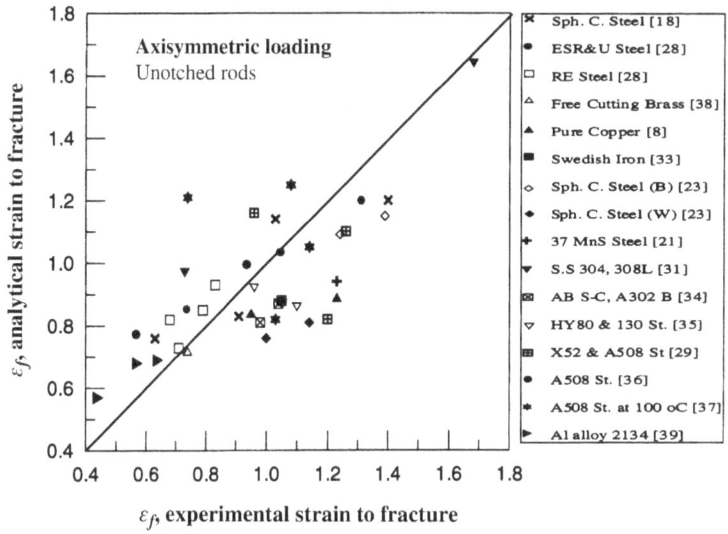

Figure 4.25 Relationship between experimental strain to fracture and analytical strain to fracture. *From Ragab, A.R., 2004b. A model for ductile fracture based on internal necking of spheroidal voids. Acta Mater. 52 (13), 3997–4009.*

three-dimensional elastic–plastic finite-element method with the experimental results obtained by Benzerga et al. (2004a). A spheroidal void was assumed and the void coalescence criterion proposed by Benzerga (2002) was used. The tensile direction was assumed to be identical with either the following three principal directions of the steel plate: the rolling direction, the width direction, or the thickness direction. The effect of the relationship between the tensile direction and the three principal directions of the steel plate on the stress–strain relationship, the strain to fracture, and the void volume fraction at fracture was demonstrated.

Ragab and Saleh (2005) predicted the bendability of sheet metals using the void coalescence criterion proposed by Ragab (2004b), and compared the analytical results with the experimental results in literature. The anisotropic yield function proposed by Hill (1979) was used, whereas the change in the void volume fraction of the material was calculated using the Gurson yield function (1977) modified by Tvergaard (1981), in which values modified by Ragab (2004a), of the parameters introduced by Tvergaard (1981) were used. The relationships between the reduction in the area at fracture and the minimum bending radius predicted from the analysis and calculated using the criterion for the occurrence of surface shear bands proposed by Hutchinson and Tvergaard (1980), agreed with the relationship between the reduction in the area at fracture and the minimum bending radius obtained experimentally.

McVeigh and Liu (2006), based on the damage model proposed by Hao et al. (2000), performed the simulation of central bursts in single pass bar extrusion, and obtained the effects of various forming conditions on the occurrence of central bursts. In the Hao et al. damage model, the damage was incorporated into the plastic potential with reference to Rousselier (1981), whereas the material was assumed to fracture when the ratio of the strain increment due to void to the strain increment due to matrix became larger than a certain value in the representative volume element. The matrix was assumed to obey the Drucker–Prager yield criterion, whereas void growth was calculated using the void growth model proposed by Rice and Tracey (1969). Since voids were assumed to nucleate due to the separation of the inclusions from the matrix, void growth was considered in the plastic potential when the stress triaxiality was positive, whereas void growth was not considered in the plastic potential when the stress triaxiality was negative.

Pardoen (2006) performed the simulation of the tensile test of copper using notched round tensile specimens in various notch-root radii, and obtained the effects of the strain-hardening exponent of the material, the initial void volume fraction, and the initial void shape on the strain to fracture. The yield function proposed by Gologanu et al. (1997) was used to calculate void growth, whereas the void coalescence criterion proposed by Thomason (1985a,b) was used to predict void coalescence. Strain hardening of the material was introduced for both the void growth model and the void coalescence criterion with reference to Pardoen and Hutchinson (2000). Material constants on the initial void volume fraction, the initial void shape, and the initial void configuration were identified prior to the simulation.

Gao and Kim (2006) performed the simulation of the deformation of the representative volume element which was composed of a cubic matrix and an initially spheroidal void by the three-dimensional elastic–plastic finite-element method, and obtained the effects of the stress triaxiality, the Lode parameter, void shape, and void nucleation on the strain to the occurrence of void coalescence. Void coalescence was assumed to occur, with reference to Koplik and Needleman (1988), when the flow of the matrix was localized into the region between two neighboring voids. Void nucleation was assumed to occur, with reference to Tvergaard (1982b), when the deformation of the matrix was simulated using the Gurson yield function (1977) modified by Tvergaard (1981).

Barsoum and Faleskog (2007b) performed the simulation of the combined test of the tensile test and the torsion test of a notched tube using the representative volume element in various stress triaxialities and Lode parameters, and obtained the effects on the strain to the occurrence of void coalescence. Void coalescence was assumed to occur when the deformation gradient rate of the matrix was localized into the region that surrounded the void. When the stress triaxiality was high and the Lode parameter was small, both the tensile deformation of the void due to tension and the shear deformation of the void due to torsion appeared at void coalescence. When the stress triaxiality was low and the Lode parameter was large, the tensile deformation of the void due to tension hardly appeared whereas the shear deformation of the void due to torsion intensely appeared at void coalescence. Fig. 4.26 shows the finite-element mesh at void coalescence in the case that the shear deformation of the void, due to torsion, is predominant (Barsoum and Faleskog, 2007b).

Leblond and Mottet (2008), with reference to Gologanu et al. (2001a), obtained a void coalescence criterion for the representative volume element which was cubic, consisted of an initially spherical void and matrix which surrounded the void, and was subjected to axisymmetric tension and shear. The upper and lower parts of the representative volume element were assumed to be sound and obey the Mises yield function, whereas the middle part of the representative volume element was

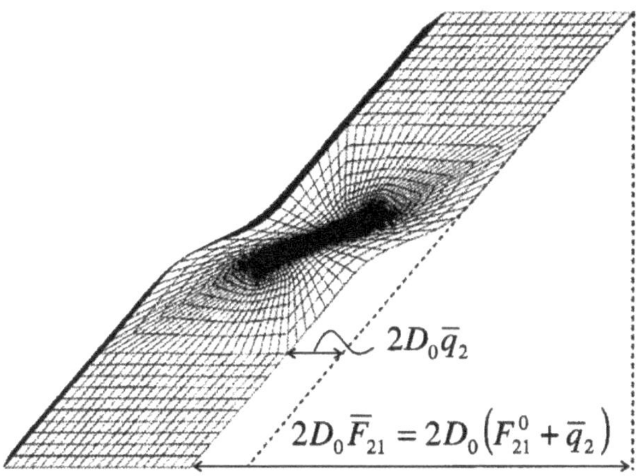

Figure 4.26 Finite-element mesh at void coalescence in case that shear deformation of void due to torsion is predominant. *From Barsoum, I., Faleskog, J., 2007b. Rupture mechanisms in combined tension and shear—micromechanics. Int. J. Solids Struct. 44 (17), 5481–5498.*

assumed to be porous and obey the Gurson yield function (1977) modified by Tvergaard (1981). Void coalescence was assumed to occur when the deformation of the material was localized into the middle part of the representative volume element. The strain to void coalescence analyzed using the yield function and the flow rule agreed with the strain to void coalescence calculated using the elastic–plastic finite-element method.

Ragab (2008), with reference to the void coalescence criterion proposed by McClintock (1968a,b), proposed a void coalescence criterion for biaxially stretched sheet metals, and compared the fracture forming limit calculated with the fracture forming limit obtained experimentally in various literature. A spheroidal void was assumed. The anisotropic yield function proposed by Hill (1979) was used, whereas the change in the void volume fraction of the material was calculated using the Gurson yield function (1977) modified by Tvergaard (1981). The stress distribution at the neck of a sheet derived analytically by Bridgman (1952) was used. The following two kinds of void nucleation criteria were assumed:

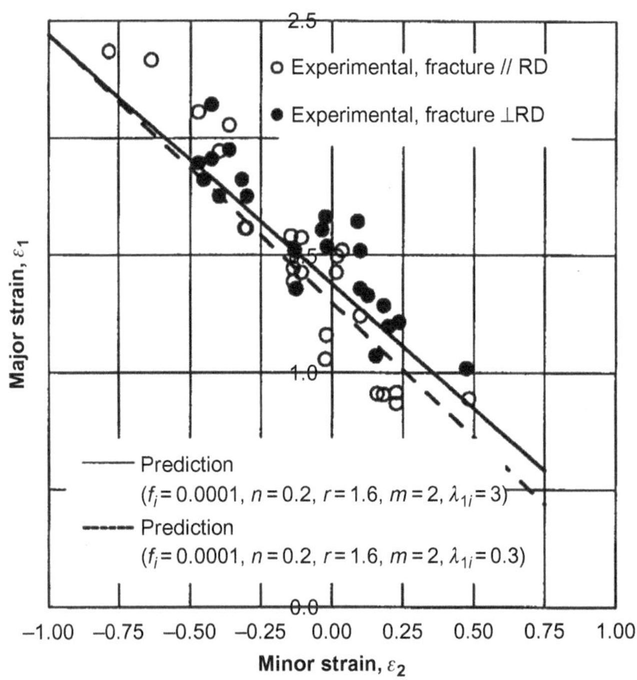

Figure 4.27 Fracture forming limit for low carbon steel. *From Ragab, A.R., 2008. Prediction of fracture limit curves in sheet metals using a void growth and coalescence model. J. Mater. Proc. Technol. 199 (1–3), 206–213.*

step-like nucleation, in which a void nucleated at a specified strain, and continuous nucleation, in which void nucleation rate was proportional to the equivalent strain rate. Fig. 4.27 shows the fracture forming limit for low carbon steel (Ragab, 2008). The fracture forming limit obtained experimentally by Grumbach and Sanz (1972) was supplemented.

CHAPTER 5

Micromechanical modeling using an ellipsoidal void model (author's model)

5.1 Thomason model and Melander and Ståhlberg model

The author's model is derived with reference to the void coalescence criterion proposed by Thomason (1968), which is referred to as the Thomason model (1968) hereafter, and the void coalescence criterion proposed by Melander and Ståhlberg (1980), which is referred to as the Melander and Ståhlberg model (1980) hereafter. Hence, these models are explained in this section.

Fig. 5.1 shows the void shape and the void configuration used in the Thomason model (1968). Fig. 5.1A shows the void shape and the void configuration before deformation. A rectangular void whose width is equal to w_0 in the x-direction, and whose height is equal to h_0 in the y-direction is considered. The rectangular voids are assumed to be distributed homogeneously throughout the material. W_0 indicates the matrix width between two neighboring voids in the x-direction, and H_0 indicates the matrix height between two neighboring voids in the y-direction. Square voids are assumed, and the void configuration in the x-direction is assumed to be identical to the void configuration in the y-direction. Hence, w_0 equals to h_0, and W_0 equals to H_0.

Fig. 5.1B shows the void shape and the void configuration after deformation. The material is elongated in the y-direction; which coincides with the tensile direction. w denotes the void width in the x-direction, h denotes the void height in the y-direction, W indicates the matrix width between two neighboring voids in the x-direction, and H indicates the matrix height between two neighboring voids in the y-direction.

In the deformation, the following assumptions are made:

- Plane-strain deformation occurs.
- The void volume fraction does not change during plastic deformation.
- The deformation gradient of the void coincides with the deformation gradient of the matrix.

Ductile Fracture in Metal Forming
DOI: https://doi.org/10.1016/B978-0-12-814772-6.00005-9

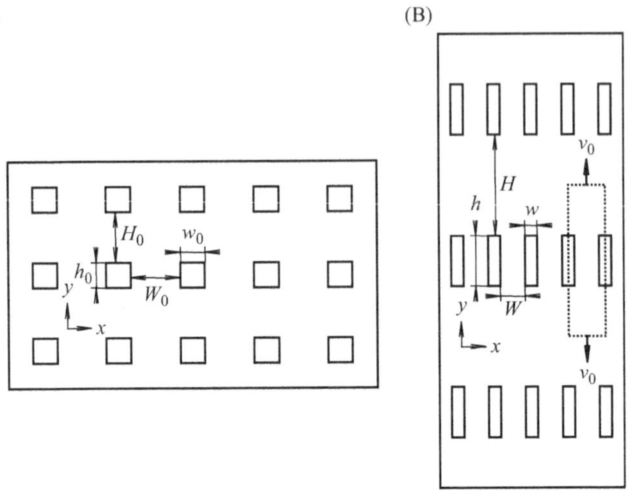

Figure 5.1 Void shape and void configuration used in the Thomason model (1968). (A) Before deformation. (B) After deformation.

In the Thomason model (1968), the following two deformation modes are assumed: homogeneous deformation mode and internal necking deformation mode. When the energy required for the homogeneous deformation mode is smaller than the energy required for the internal necking deformation mode, homogeneous deformation is assumed to occur. However, when the energy required for the homogeneous deformation mode is larger than the energy required for the internal necking deformation mode, internal necking deformation is assumed to occur. In other words, deformation is localized into the region between two neighboring voids, in which the necking of the matrix between the voids occurs and finally the voids coalesce. Hence, ductile fracture is assumed to occur by the coalescence of voids when the energy for the homogeneous deformation coincides with the energy for the internal necking deformation. Therefore, the fracture criterion, that is, the condition that the energy for the homogeneous deformation coincides with the energy for the internal necking deformation, is derived in the following.

A rectangle shown by dotted lines is considered. The length of the rectangle in the x-direction is $W + w$, whereas the length of the rectangle in the y-direction is $H + h$. First, the matrix in the rectangle is assumed to deform homogeneously. When the matrix velocity in the y-direction on the two sides of the rectangle is v_0, the matrix velocity in the x-direction on the remaining two sides of the rectangle is $-(W + w)/(H + h) \cdot v_0$,

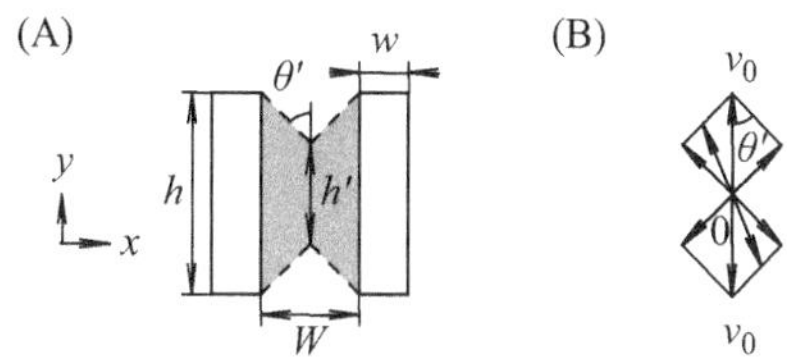

Figure 5.2 Deformation in internal necking region in the Thomason model (1968). (A) Internal necking region. (B) Velocity field on velocity discontinuity line.

because of the condition of volume constancy. Hence, using the yield condition, the energy required for the homogeneous deformation mode is,

$$2v_0 \cdot \sigma_y \cdot (W+w) - 2\frac{W+w}{H+h}v_0 \cdot \sigma_x \cdot (H+h) = 2v_0 \cdot \left(\sigma_y - \sigma_x\right) \cdot (W+w) = 2v_0 \cdot 2k \cdot (W+w) \tag{5.1}$$

where k denotes the shear yield stress of the matrix.

Next, voids are assumed to coalesce by internal necking. Fig. 5.2 shows the deformation in the internal necking region in the Thomason model (1968). Fig. 5.2A shows the internal necking region. In the hatched area, plastic deformation is assumed to occur, whereas the dashed line indicates the velocity discontinuity line. The origin is assumed to locate at the center of the plastic deformation region. Here, h' denotes the minimum height of the plastic deformation region in the y-direction, whereas θ' denotes the angle between the y-direction and the direction of the velocity discontinuity line. Fig. 5.2B shows the velocity field on the velocity discontinuity line denoted as a hodograph.

In the hatched area, the following velocity fields are assumed,

$$u = -\frac{2x}{(h-h')\left(2x/W\right)+h'}v_0 \tag{5.2}$$

$$v = \frac{2h'y}{((h-h')(2x/W)+h')^2}v_0 \tag{5.3}$$

where u and v denote the matrix velocity in the x-direction, and the matrix velocity in the y-direction, respectively. Eqs. (5.2) and (5.3) satisfy the condition of volume constancy and the boundary conditions. Hence, Eqs. (5.2) and (5.3) are a kinematically admissible velocity field.

The amount of velocity discontinuity on the velocity discontinuity line is calculated as follows,

$$\Delta u = \frac{2x}{(h - h')(2x/W) + h'} v_0 \tag{5.4}$$

$$\Delta v = \frac{(h - h')(2x/W)}{(h - h')(2x/W) + h'} v_0 \tag{5.5}$$

where Δu and Δv denote the amount of velocity discontinuity on the velocity discontinuity line in the x-direction, and the amount of velocity discontinuity on the velocity discontinuity line in the y-direction, respectively.

The energy required for the internal necking deformation mode is expressed as the summation of the energy dissipation rate in the hatched area and the energy dissipation rate on the velocity discontinuity line. Hence, the energy for the internal necking deformation in plane-strain deformation is expressed as follows (Kudo, 1960a).

$$\begin{aligned} &\sqrt{2}k\int\sqrt{\dot{\varepsilon}_x^2 + \dot{\varepsilon}_y^2 + \frac{1}{2}\dot{\gamma}_{xy}^2}\,\mathrm{d}V + k\int\sqrt{\Delta u^2 + \Delta v^2}\,\mathrm{d}S \\ &= \sqrt{2}k\int\sqrt{\left(\frac{\partial u}{\partial x}\right)^2 + \left(\frac{\partial v}{\partial y}\right)^2 + \frac{1}{2}\left(\frac{\partial u}{\partial y} + \frac{\partial v}{\partial x}\right)^2}\,\mathrm{d}V + k\int\sqrt{\Delta u^2 + \Delta v^2}\,\mathrm{d}S \end{aligned} \tag{5.6}$$

The integration of the first term in Eq. (5.6) is performed for the plastic deformation region, whereas the integration of the second term in Eq. (5.6) is performed on the velocity discontinuity line.

Eq. (5.6) is calculated by substituting Eqs. (5.2)–(5.5) into Eq. (5.6). Here, the value of h' is optimized so that the energy for the internal necking deformation is minimized.

Since Eq. (5.6) is calculated numerically, numerical simulation is required to obtain the fracture criterion, that is, the condition that the energy for the homogeneous deformation coincides with the energy for the internal necking deformation in the Thomason model (1968).

Fig. 5.3 shows the effect of void volume fraction. Fig. 5.3A shows the relationship between the void volume fraction and the fracture strain in the y-direction, namely, in the tensile direction. With increasing the void volume fraction, the fracture strain decreases. Fig. 5.3B shows the

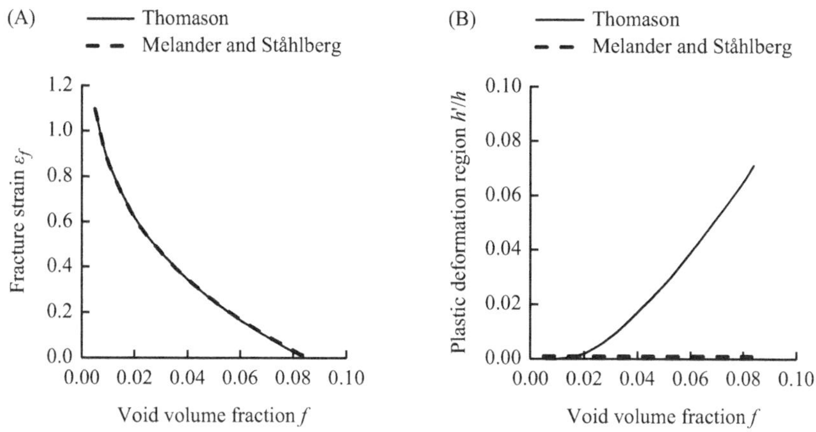

Figure 5.3 Effect of void volume fraction. (A) Fracture strain. (B) Plastic deformation region.

relationship between the void volume fraction and the plastic deformation region. With increasing the void volume fraction, the plastic deformation region represented by h'/h increases.

The fracture criterion, that is, the condition that the energy for the homogeneous deformation coincides with the energy for the internal necking deformation, is obtained numerically in the Thomason model (1968). Hence, the Melander and Ståhlberg model (1980) is proposed, in which h' is assumed to be zero in the Thomason model. First, the energy required for the homogeneous deformation mode in the Melander and Ståhlberg model is the same as the energy required for the homogeneous deformation mode in the Thomason model. Hence, the energy for the homogeneous deformation is already shown in Eq. (5.1). Next, the energy required for the internal necking deformation mode is calculated in the following.

Fig. 5.4 shows the deformation in the internal necking region in the Melander and Ståhlberg model (1980). Fig. 5.4A shows the internal necking region. The dashed line indicates the velocity discontinuity line, whereas there is no plastic deformation region. Here, θ denotes the angle between the y-direction and the direction of the velocity discontinuity line, and $2l$ denotes the length of the velocity discontinuity line, which is equal to $\sqrt{h^2 + W^2}$. Fig. 5.4B shows the velocity field denoted as a hodograph. Here, v_0 denotes the matrix velocity in the y-direction, and Δv denotes the amount of velocity discontinuity, which is equal to $v_0/\cos\theta$. Because the number of the velocity discontinuity line is two, the energy dissipation rate on the velocity discontinuity line is

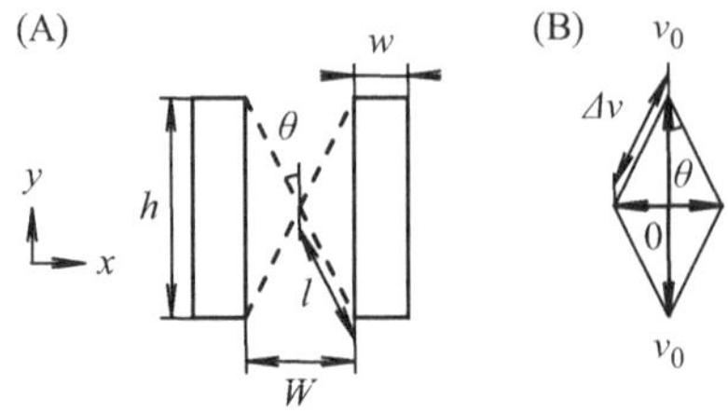

Figure 5.4 Deformation in internal necking region in the Melander and Ståhlberg model (1980). (A) Internal necking region. (B) Velocity field.

$k \cdot \Delta v \cdot 2l \cdot 2 = k \cdot v_0/\cos\theta \cdot \sqrt{h^2 + W^2} \cdot 2$, which is equal to the energy required for the internal necking deformation mode.

Thus, with reference to Eq. (5.1), the fracture criterion, that is, the condition that the energy for the homogeneous deformation coincides with the energy for the internal necking deformation, is expressed as follows.

$$2v_0 \cdot 2k \cdot (W + w) = k \cdot \frac{v_0}{\cos\theta} \cdot \sqrt{h^2 + W^2} \cdot 2 \tag{5.7}$$

Since $\cos\theta$ is equal to $h/\sqrt{h^2 + W^2}$, the following equation is obtained.

$$W = h + \sqrt{2hw} \tag{5.8}$$

Because h_0 is equal to w_0 and H_0 is equal to W_0, the void volume fraction f is defined as follows.

$$f = \frac{w_0^2}{(W_0 + w_0)^2} \tag{5.9}$$

From Eq. (5.9), the following equation is obtained.

$$W_0 = \frac{1 - \sqrt{f}}{\sqrt{f}} w_0 \tag{5.10}$$

The fracture strain in the tensile direction is defined as ε_f. h, w, and W are expressed in the following equation using ε_f.

$$h = w_0 e^{\varepsilon_f}, w = w_0 e^{-\varepsilon_f}, W = W_0 e^{-\varepsilon_f} \tag{5.11}$$

Substituting Eqs. (5.10) and (5.11) into Eq. (5.8), the following equation is obtained, in which the relationship between the void volume fraction and the fracture strain in the tensile direction is expressed analytically.

$$\varepsilon_f = \ln\left(-\frac{1}{\sqrt{2}} + \sqrt{\frac{1}{\sqrt{f}} - \frac{1}{2}}\right) \tag{5.12}$$

The relationship is supplemented in Fig. 5.3A. The relationship calculated using the Melander and Ståhlberg model (1980) agrees well with the relationship calculated using the Thomason model (1968). The relationship is calculated numerically using the Thomason model, whereas the relationship is expressed analytically using Eq. (5.12) in the Melander and Ståhlberg model. Hence, the validity of the simplification of the Thomason model is confirmed.

Since plastic deformation region does not exist in the Melander and Ståhlberg model (1980), plastic deformation region is irrelevant to the magnitude of the void volume fraction in Fig. 5.3B. In the region of high void volume fraction, plastic deformation region calculated using the Thomason model (1968) differs from plastic deformation region assumed in the Melander and Ståhlberg model. However, as shown in Fig. 5.3A, fracture strain calculated using the Melander and Ståhlberg model scarcely differs from fracture strain calculated using the Thomason model.

Melander and Ståhlberg (1980) shows the relationship between the void volume fraction and the fracture strain in the tensile direction for different void size distributions and the relationship between the void volume fraction and the fracture strain in the tensile direction for different inter void distances. However, the relationship calculated using the Melander and Ståhlberg model (1980) is not compared with the relationship calculated using the Thomason model (1968). Hence, Komori (1999b) compares the relationship calculated using the Melander and Ståhlberg model with the relationship calculated using the Thomason model and shows the validity of the Melander and Ståhlberg model.

5.2 Modified Melander and Ståhlberg model

In both the Thomason model (1968) and the Melander and Ståhlberg model (1980), plane-strain tensile test is simulated. Hence, the following assumption is made implicitly in these models:

- The direction of maximum principal strain coincides with the direction of maximum principal stress during plastic deformation.

However, the direction of principal strain changes during plastic deformation in metal-forming processes. Hence, these models cannot be utilized in the simulation of metal-forming processes. Hence, the modified Melander and Ståhlberg model (Komori, 1999b) is proposed and explained in this section.

In the modified Melander and Ståhlberg model (Komori, 1999b), the following assumption is made:

- The direction of maximum principal strain does not coincide with the direction of maximum principal stress during plastic deformation.

In metal-forming processes, a square void before deformation becomes a parallelogrammic void after deformation, because the direction of maximum principal strain does not coincide with the direction of maximum principal stress during plastic deformation. Hence, the void shape is assumed to be a parallelogram in the modified Melander and Ståhlberg model (Komori, 1999b). Parallelogrammic voids are assumed to be distributed uniformly throughout the material.

Fig. 5.5 shows the deformation in the internal necking region in the modified Melander and Ståhlberg model (Komori, 1999b). Two neighboring voids are considered. Fig. 5.5A shows the internal necking region. The dashed line indicates the velocity discontinuity line. Here, v_0 denotes the matrix velocity in the direction of maximum principal stress, θ_1 and θ_2 denote the angles between the direction of maximum principal stress and the directions of the velocity discontinuity lines, and $2l_1$ and $2l_2$ denote the lengths of the velocity discontinuity lines. The distance between two neighboring voids is denoted as L. Fig. 5.5B shows the velocity field denoted as a hodograph. Here, Δv_1 and Δv_2 denote the amount of velocity discontinuity.

First, the amount of velocity discontinuity is calculated. According to a formula for a triangular function, the following equation is obtained.

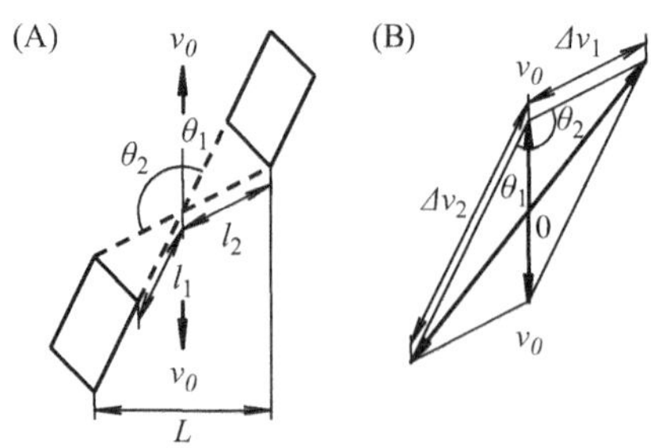

Figure 5.5 Deformation in internal necking region in the modified Melander and Ståhlberg model (Komori, 1999b). (A) Internal necking region. (B) Velocity field.

$$\frac{\Delta v_1}{\sin\theta_1} = \frac{\Delta v_2}{\sin\theta_2} = \frac{2v_0}{\sin(\pi - \theta_1 - \theta_2)} \tag{5.13}$$

From Eq. (5.13), the amount of velocity discontinuity is obtained as follows:

$$\Delta v_1 = \frac{\sin\theta_1}{\sin(\theta_1 + \theta_2)} 2v_0 \tag{5.14}$$

$$\Delta v_2 = \frac{\sin\theta_2}{\sin(\theta_1 + \theta_2)} 2v_0 \tag{5.15}$$

Hence, the energy dissipation rate on the velocity discontinuity line is expressed as follows:

$$k \cdot \Delta v_1 \cdot 2l_2 + k \cdot \Delta v_2 \cdot 2l_1 = 4kv_0 \frac{l_1 \sin\theta_2 + l_2 \sin\theta_1}{\sin(\theta_1 + \theta_2)} \tag{5.16}$$

Eq. (5.16) is also the energy required for the internal necking deformation mode. Next, since the distance between two neighboring voids is equal to L, the energy required for the homogeneous deformation mode is $2v_0 \cdot 2k \cdot L$ with reference to Eq. (5.1).

Hence, the fracture criterion, that is, the condition that the energy for the homogeneous deformation coincides with the energy for the internal necking deformation, is expressed as follows:

$$L = \frac{l_1 \sin\theta_2 + l_2 \sin\theta_1}{\sin(\theta_1 + \theta_2)} \tag{5.17}$$

The relationship between the modified Melander and Ståhlberg model (Komori, 1999b) and the Melander and Ståhlberg model (1980) is explained as follows:

When l_2 is equal to l_1 and θ_2 is equal to θ_1, the following equation is obtained from Eq. (5.17).

$$L = \frac{l_1}{\cos\theta_1} \tag{5.18}$$

The void shape is assumed to be a rectangle for convenience. h_1 is defined as the void height, w_1 is defined as the void width, and W_1 is defined as the matrix width between two neighboring voids. Since L is equal to $W_1 + w_1$, l_1 is equal to $\sqrt{h_1^2 + W_1^2}/2$, and $\cos\theta_1$ is equal to $h_1/\sqrt{h_1^2 + W_1^2}$, Eq. (5.18) becomes the following equation.

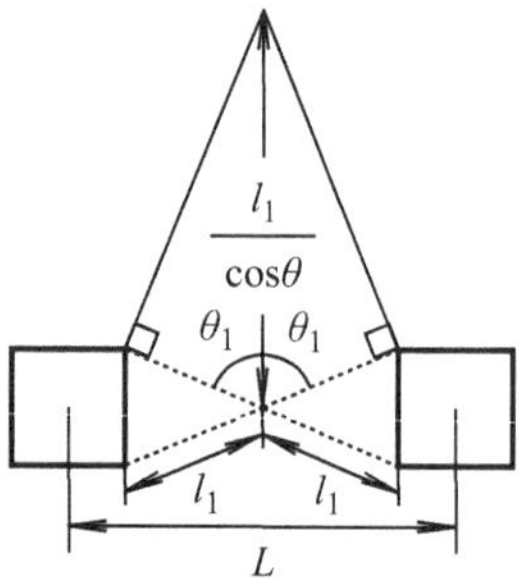

Figure 5.6 Schematic presentation of fracture criterion in the Melander and Ståhlberg model (1980).

$$W_1 = h_1 + \sqrt{2h_1 w_1} \tag{5.19}$$

Hence, Eq. (5.19) is identical to Eq. (5.8). In other words, when the void shape is assumed to be a rectangle, the modified Melander and Ståhlberg model (Komori, 1999b) is identical to the Melander and Ståhlberg model (1980). Therefore, the Melander and Ståhlberg model has proved to be a special model of the modified Melander and Ståhlberg model.

To show a schematic presentation of the fracture criterion is useful (Komori, 2011). In the case of a parallelogrammic void, the fracture criterion is expressed in Eq. (5.18). Fig. 5.6 shows the schematic presentation of the fracture criterion in the Melander and Ståhlberg model (1980). Since the initial shape of a parallelogrammic void is a square, a square void is shown for convenience. When the length of $l_1/\cos\theta_1$ is larger than the length of L, the energy for the internal necking deformation is larger than the energy for the homogeneous deformation. Hence, the fracture criterion is not satisfied. However, when the length of $l_1/\cos\theta_1$ is equal to the length of L, the energy for the internal necking deformation is equal to the energy for the homogeneous deformation. Hence, the fracture criterion is satisfied.

5.3 Ellipsoidal void model from microscopic viewpoint

In all of the Thomason model (1968), the Melander and Ståhlberg model (1980), and the modified Melander and Ståhlberg model (Komori, 1999b), the following assumption is made:

- The void shape is a rectangle.

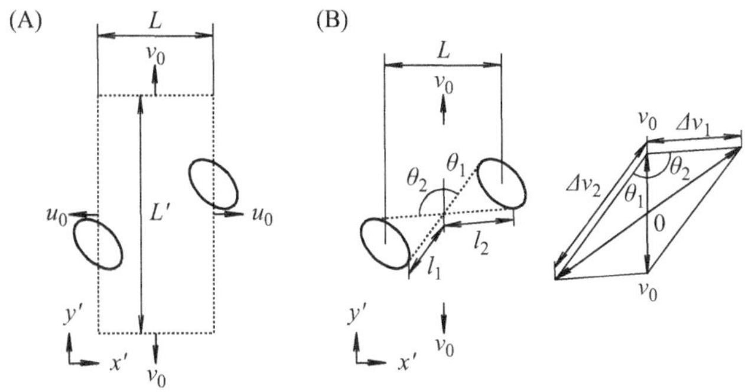

Figure 5.7 Deformation in the ellipsoidal void model. (A) Homogeneous deformation. (B) Internal necking deformation.

However, a rectangular void does not exist in reality. Hence, the ellipsoidal void model (Komori, 2013a) is proposed. In the ellipsoidal void model, the following assumption is made:

- The void shape is an ellipsoid.

Furthermore, in all of these models, the energy required for the homogeneous deformation mode is not evaluated precisely, because the void volume fraction of the material is not considered. Hence, the energy for the homogeneous deformation is modified in the ellipsoidal void model.

Fig. 5.7 shows the deformation in the ellipsoidal void model. First, the energy required for the homogeneous deformation mode is derived. Fig. 5.7A shows the homogeneous deformation. The y'-direction is made to coincide with the direction of the maximum principal stress and the x'-direction is made to coincide with the direction of the minimum principal stress. The rectangle indicated by the dotted lines is considered. The length of the rectangle in the x'-direction is L, which is assumed to be the distance between two neighboring voids, and the length of the rectangle in the y'-direction is L'. The material velocity in the y'-direction is v_0 and the material velocity in the x'-direction is u_0. $u_0 = -\left(L/L'\right)v_0$ because of the condition of volume constancy.

Hence, using the yield condition, the energy for the homogeneous deformation is expressed as follows:

$$\begin{aligned} 2v_0L\sigma_{\max} + 2u_0L'\sigma_{\min} &= 2v_0L\sigma_{\max} - 2\left(\frac{L}{L'}\right)v_0L'\sigma_{\min} \\ &= 2v_0L(\sigma_{\max} - \sigma_{\min}) = 4kv_0L \end{aligned} \tag{5.20}$$

Eq. (5.20) is similar to Eq. (5.1). In calculating the energy for the homogeneous deformation, the void volume fraction of the material is assumed to be zero. However, in reality, the void volume fraction of the material is not zero but f. Since the energy for the homogeneous deformation in the material whose void volume fraction is zero is $4kv_0L$, the energy for the homogeneous deformation in the material whose void volume fraction is f is assumed to be $(1-f)4kv_0L$. This expression is used in the place of $4kv_0L$ hereafter.

Next, the energy required for the internal necking deformation mode is derived. Fig. 5.7B shows the internal necking deformation, in which the velocity field denoted as a hodograph is also shown. The dotted lines indicate the velocity discontinuity line, which is assumed to coincide with the tangent of the two neighboring voids. Here, θ_1 and θ_2 denote the angles between the direction of maximum principal stress and the directions of the velocity discontinuity lines, $2l_1$ and $2l_2$ denote the lengths of the velocity discontinuity lines, and Δv_1 and Δv_2 denote the amount of velocity discontinuity. With reference to Eqs. (5.13)– (5.16), the energy required for the internal necking deformation mode is $4kv_0(l_1\sin\theta_2 + l_2\sin\theta_1)/\sin(\theta_1+\theta_2)$.

Hence, the fracture criterion, that is, the condition that the energy for the homogeneous deformation coincides with the energy for the internal necking deformation, is expressed as follows:

$$(1-f)L = \frac{l_1\sin\theta_2 + l_2\sin\theta_1}{\sin(\theta_1+\theta_2)} \tag{5.21}$$

For convenience, the ratio of the energy for the internal necking deformation to the energy for the homogeneous deformation E is defined as follows:

$$E = \frac{l_1\sin\theta_2 + l_2\sin\theta_1}{(1-f)L\cdot\sin(\theta_1+\theta_2)} \tag{5.22}$$

The fracture criterion is satisfied when E is less than unity.

When l_2 is equal to l_1 and θ_2 is equal to θ_1, the following equation is obtained from Eq. (5.21).

$$(1-f)L = \frac{l_1}{\cos\theta_1} \tag{5.23}$$

Showing a schematic presentation of the fracture criterion is useful (Komori, 2011). In the case of an ellipsoidal void, the fracture criterion is

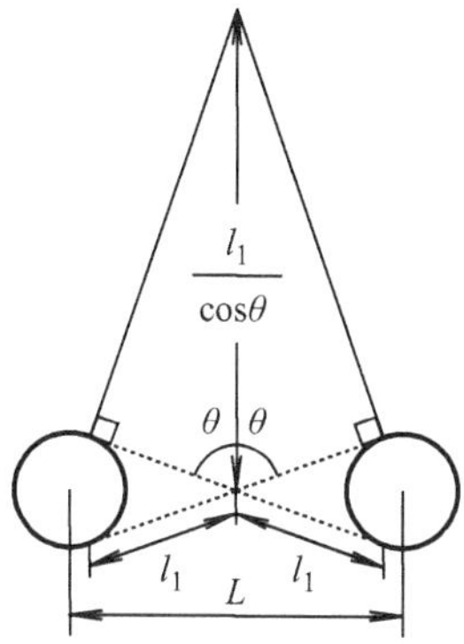

Figure 5.8 Schematic presentation of fracture criterion in the ellipsoidal void model.

expressed in Eq. (5.23). Fig. 5.8 shows the schematic presentation of the fracture criterion in the ellipsoidal model. Since the initial shape of an ellipsoidal void is a circle, a circular void is shown for convenience. When the length of $l_1/\cos\theta_1$ is larger than the length of $(1-f)L$, the energy for the internal necking deformation is larger than the energy for the homogeneous deformation. Hence, the fracture criterion is not satisfied. However, when the length of $l_1/\cos\theta_1$ is equal to the length of $(1-f)L$, the energy for the internal necking deformation is equal to the energy for the homogeneous deformation. Hence, the fracture criterion is satisfied.

In Fig. 5.7, the location of the velocity discontinuity line is determined geometrically. That is, the velocity discontinuity line is assumed to coincide with the tangent of the two neighboring voids. The energy required for the internal necking deformation mode should be minimized when the energy for the internal necking deformation is compared with the energy required for the homogeneous deformation mode. However, the energy for the internal necking deformation is not necessarily minimized when the velocity discontinuity line coincides with the tangent of the two neighboring voids. Hence, the location of the velocity discontinuity line should be optimized so that the energy for the internal necking deformation is minimized.

Fig. 5.9 shows the deformation in the internal necking region in the ellipsoidal void model, in the case that the velocity discontinuity line does not coincide with the tangent of the two neighboring voids. Fig. 5.9A shows the internal necking region. The y'-direction is made to coincide with the direction of the maximum principal stress and the x'-direction is made to coincide with the direction of the minimum principal stress.

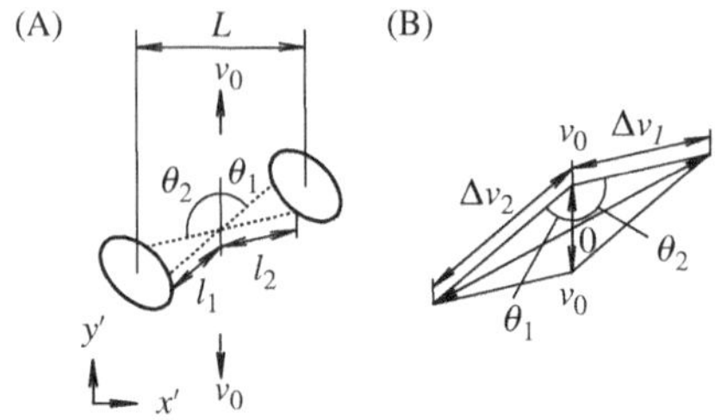

Figure 5.9 Deformation in internal necking region in the ellipsoidal void model. (A) Internal necking region. (B) Velocity field.

L denotes the distance between two neighboring voids and v_0 denotes the material velocity in the y'-direction. θ_1 and θ_2 denote the angles between the direction of maximum principal stress and the directions of the velocity discontinuity lines, and $2l_1$ and $2l_2$ denote the lengths of the velocity discontinuity lines. Fig. 5.9B shows the velocity field denoted as a hodograph. Δv_1 and Δv_2 denote the amount of velocity discontinuity.

Irrelevant to the location of the velocity discontinuity line, with reference to Eqs. (5.13)–(5.16), the energy required for the internal necking deformation mode is $4kv_0(l_1\sin\theta_2 + l_2\sin\theta_1)/\sin(\theta_1 + \theta_2)$. Hence, irrelevant to the location of the velocity discontinuity line, the fracture criterion, that is, the condition that the energy for the homogeneous deformation coincides with the energy for the internal necking deformation, is expressed by Eq. (5.21).

When the location of the velocity discontinuity line is assumed to coincide with the tangent of the two neighboring voids, the energy required for the internal necking deformation mode is calculated immediately. Hence, whether the fracture criterion is satisfied or not is evaluated immediately. However, when the location of the velocity discontinuity line is optimized so that the energy for the internal necking deformation is minimized, the energy for the internal necking deformation is determined through numerical simulation. Hence, whether the fracture criterion is satisfied or not is evaluated through numerical simulation.

5.4 Ellipsoidal void model from macroscopic viewpoint

Fig. 5.10 shows the overview of the entire simulation procedure at each time step. Both macroscopic simulation and microscopic simulation are alternately performed. In the macroscopic simulation, the deformation of the material is simulated by the conventional rigid-plastic finite-element

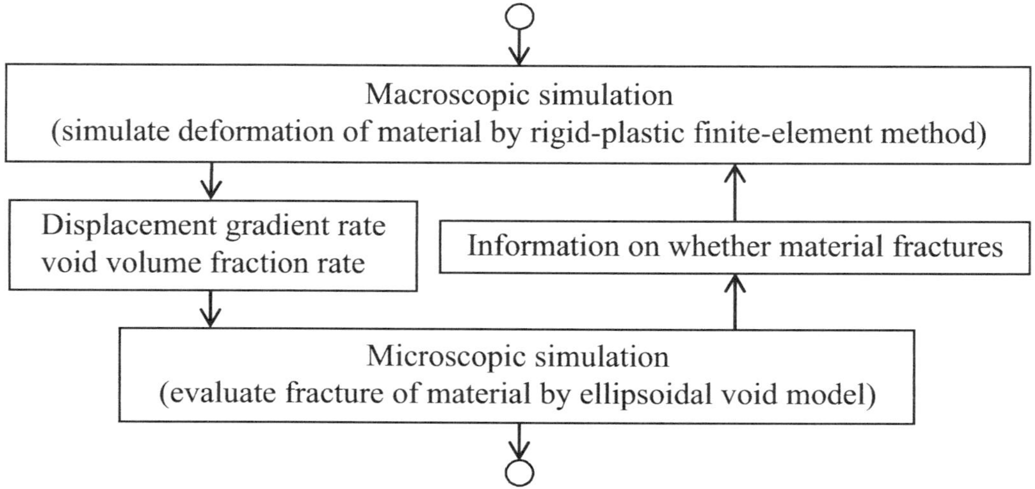

Figure 5.10 Overview of entire simulation procedure at each time step.

method. The displacement gradient rate and the void volume fraction rate of the material, calculated by the macroscopic simulation, are used in the subsequent microscopic simulation. The fracture of the material is evaluated by the ellipsoidal void model in the subsequent microscopic simulation. The information on whether the material fractures is used in the macroscopic simulation of the next time step.

The information transferred from the macroscopic simulation scheme to the microscopic simulation scheme differs from the information transferred from the microscopic simulation scheme to the macroscopic simulation scheme. Hence, the macroscopic simulation and the microscopic simulation are uncoupled. Especially, the void shape and the void configuration in the macroscopic simulation are uncoupled to the void shape and the void configuration in the microscopic simulation.

The deformation of the material is simulated by the conventional rigid-plastic finite-element method (Kobayashi et al., 1989). The following yield function Φ proposed by Gurson (1977) is adopted:

$$\Phi = \frac{3}{2} \cdot \frac{\sigma'_{ij}\sigma'_{ij}}{\sigma_M^2} + 2f\cosh\left(\frac{\sigma_{kk}}{2\sigma_M}\right) - 1 - f^2 = 0 \tag{5.24}$$

where σ_M is the tensile yield stress of the matrix and f is the void volume fraction of the material. Since the yield function Φ is not a function of the second power of stress, it is not easy to perform rigid-plastic simulation using Eq. (5.24). Hence, $\cosh x$ is approximated by $1 + x^2/2$ (Tomita, 1990). The approximated yield function Φ' used in this study is as follows:

$$\Phi' = \frac{3}{2} \cdot \frac{\sigma'_{ij}\sigma'_{ij}}{\sigma_M^2} + \frac{f}{4} \cdot \left(\frac{\sigma_{kk}^2}{\sigma_M^2}\right) - (1-f)^2 = 0 \tag{5.25}$$

According to the plastic potential theory and the flow rule, the relationship between the stress and the strain rate is described by the following equations (Tomita, 1990):

$$\dot{\varepsilon}_{ij} = \frac{3}{2} \cdot \frac{\dot{\overline{\varepsilon}}}{\overline{\sigma}}\left(\sigma'_{ij} + \frac{f}{6}\sigma_{kk}\delta_{ij}\right) \tag{5.26}$$

$$\sigma_{ij} = \frac{2}{3} \cdot \frac{\overline{\sigma}}{\dot{\overline{\varepsilon}}}\left(\dot{\varepsilon}'_{ij} + \frac{2}{3f}\dot{\varepsilon}_{kk}\delta_{ij}\right) \tag{5.27}$$

where $\overline{\sigma}$ is the equivalent stress and $\dot{\overline{\varepsilon}}$ is the equivalent plastic strain rate. $\overline{\sigma}$ and $\dot{\overline{\varepsilon}}$ are defined as follows (Tomita, 1990):

$$\overline{\sigma}^2 = \frac{3}{2}\sigma'_{ij}\sigma'_{ij} + \frac{f}{4}\sigma_{kk}^2 \tag{5.28}$$

$$\dot{\overline{\varepsilon}}^2 = \frac{2}{3}\dot{\varepsilon}_{ij}\dot{\varepsilon}_{ij} + \frac{2}{9}\left(\frac{2}{f} - 1\right)\dot{\varepsilon}_{kk}^2 \tag{5.29}$$

The following evolution equation, which denotes the change in the void volume fraction, is assumed.

$$\dot{f} = (1 - f)\dot{\varepsilon}_{kk} + (\text{void nucleation terms}) \tag{5.30}$$

The first term on the right-hand side of Eq. (5.30) denotes void growth, whereas the second and succeeding terms on the right-hand side of Eq. (5.30) denote void nucleation. Because the terms that denote void nucleation depend on the material, the void nucleation terms are not demonstrated specifically in this section.

5.5 Void shape and void configuration

First, the void shape and the void configuration in two time steps are explained in an idealized material in which no inclusions and no void nucleation are assumed, to explain the fundamental concept in the ellipsoidal void model. Next, the void shape and the void configuration in a large number of infinitesimal time steps are explained in an actual material.

5.5.1 Fundamental concept

Fig. 5.11 shows the void shape and the void configuration. The solid line indicates the contour of voids, whereas the dashed line indicates the line that bisects the region between two neighboring voids. Fig. 5.11A shows the void shape and the void configuration before deformation, and Fig. 5.11B shows the void shape and the void configuration after deformation. Since the fundamental concept of the void shape and the void configuration is explained in this section, no inclusions and no void nucleation are assumed.

The following assumptions are made on the void shape:

- The void shape before deformation is a circle.

- The void shape after deformation is calculated by assuming that the displacement gradient of the void is identical to the displacement gradient of the material.

The following assumptions are made on the void configuration:

- The void configuration before deformation is identical to the centroid configuration of the hexagonal grid.
- The void configuration after deformation is calculated by assuming that the displacement gradient of the void is identical to the displacement gradient of the material.

The assumption that the displacement gradient of the void is identical to the displacement gradient of the material, simplifies the calculations of the void shape and the void configuration, thereby reduces the computational time. The validity of the assumption is examined afterwards.

Fig. 5.12 shows the void shape and the void configuration before deformation. When the area of the hexagonal grid is assumed to be unity, the area of the triangle denoted by the dashed lines becomes one half. If the distance between two neighboring voids is assumed to be p, the area of the triangle becomes $\sqrt{3}p^2/4$. Hence, $p = \sqrt[4]{4/3}$. When the initial void volume fraction is expressed as f_0, the area of a circle that denotes a void becomes f_0.

Fig. 5.13 shows the void shape and the void configuration of two neighboring voids. The solid line indicates the contour of voids. Fig. 5.13A shows the void shape and the void configuration before deformation, and Fig. 5.13B shows the void shape and the void configuration

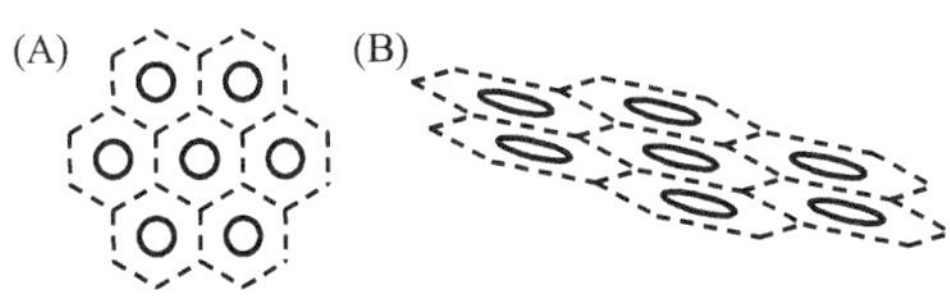

Figure 5.11 Void shape and void configuration. (A) Before deformation. (B) After deformation.

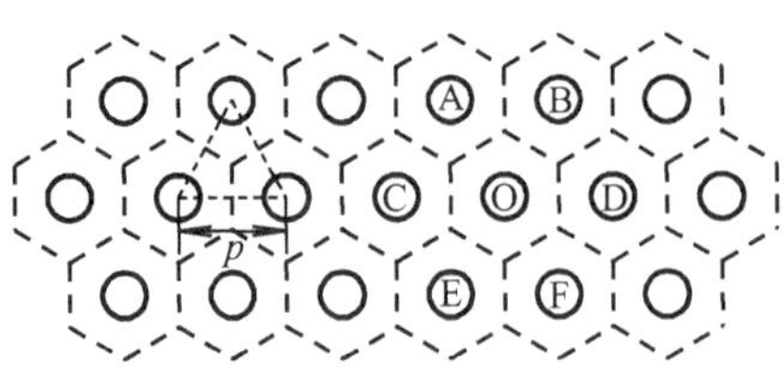

Figure 5.12 Void shape and void configuration before deformation.

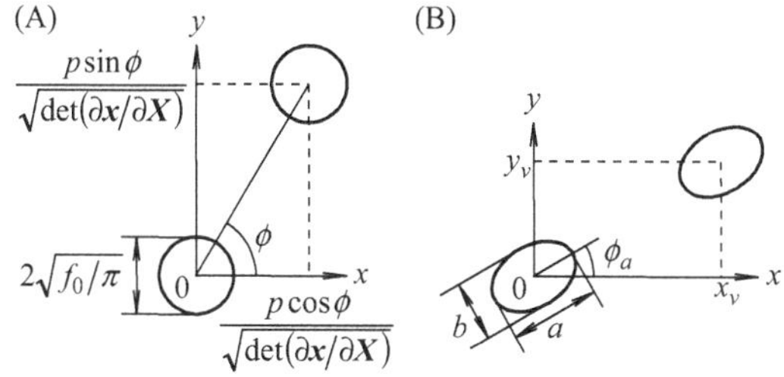

Figure 5.13 Void shape and void configuration of two neighboring voids. (A) Before deformation. (B) After deformation.

after deformation. Two neighboring voids before deformation are still adjacent after deformation. The angle between the line connecting two neighboring voids before deformation and the x-axis is defined as ϕ.

Void O in Fig. 5.12 is considered and referred to as the central void hereafter. Voids A, B, C, D, E, and F are neighboring to the central void. Whether the fracture criterion is satisfied between the central void and the neighboring voids should be determined. Since the coordinate axes are undefined in Fig. 5.12, ϕ has an arbitrary value between 0 and 2π.

The void shape after deformation is calculated by polar decomposition of the deformation gradient $\partial\boldsymbol{x}/\partial\boldsymbol{X}$ (Mase, 1970). The right polar decomposition is written as follows:

$$\frac{\partial \boldsymbol{x}}{\partial \boldsymbol{X}} = \boldsymbol{R} \cdot \boldsymbol{U} \tag{5.31}$$

where $\boldsymbol{R}$ is the orthogonal rotation tensor and $\boldsymbol{U}$ is the right stretch tensor. To perform right polar decomposition, the right Cauchy–Green deformation tensor $\boldsymbol{C}$, which is defined as $(\partial\boldsymbol{x}/\partial\boldsymbol{X})^T \cdot (\partial\boldsymbol{x}/\partial\boldsymbol{X})$, has to be calculated in advance.

The major diameter of the void a and the minor diameter of the void b are calculated as follows:

$$a = \frac{\sqrt{\lambda_{\max}}}{\sqrt[4]{\lambda_{\max}\lambda_{\min}}} \cdot \frac{\sqrt{f}}{\sqrt{\pi}}, \quad b = \frac{\sqrt{\lambda_{\min}}}{\sqrt[4]{\lambda_{\max}\lambda_{\min}}} \cdot \frac{\sqrt{f}}{\sqrt{\pi}} \tag{5.32}$$

where $\lambda_{\max}$ is the maximum principal value of $\boldsymbol{C}$, and $\lambda_{\min}$ is the minimum principal value of $\boldsymbol{C}$.

The angle between the direction of the major diameter of the void a and the x-axis is defined as ϕ_a in Fig. 5.13B and is calculated using the following equation:

$$\phi_a = \phi_{\max} + \phi_{\boldsymbol{R}} \tag{5.33}$$

where $\phi_{\max}$ is the angle between the maximum principal direction of $\boldsymbol{C}$ and the x-axis and $\phi_{\boldsymbol{R}}$ is the angle of rotation due to $\boldsymbol{R}$.

The void configuration after deformation is calculated by the deformation gradient $\partial \boldsymbol{x}/\partial \boldsymbol{X}$. The x-coordinate of the neighboring void x_v and the y-coordinate of the neighboring void y_v are calculated using the following equations:

$$\begin{aligned} x_v &= \frac{p\cos\phi \cdot (\partial x/\partial X) + p\sin\phi \cdot (\partial x/\partial Y)}{\sqrt{\det(\partial \boldsymbol{x}/\partial \boldsymbol{X})}}, \\ y_v &= \frac{p\cos\phi \cdot (\partial y/\partial X) + p\sin\phi \cdot (\partial y/\partial Y)}{\sqrt{\det(\partial \boldsymbol{x}/\partial \boldsymbol{X})}} \end{aligned} \tag{5.34}$$

The ellipsoidal void model is able to be applied to axisymmetric deformation in the macroscopic simulation as well as to plane-strain deformation in the macroscopic simulation. In plane-strain deformation in the macroscopic simulation, the area of the hexagonal grid does not change since the out-of-plane strain is zero. However, in axisymmetric deformation in the macroscopic simulation, the area of the hexagonal grid changes since the out-of-plane strain is not zero. The determinant of the deformation gradient $\det(\partial \boldsymbol{x}/\partial \boldsymbol{X})$ is proportional to the area of the hexagonal grid. Since the area of the hexagonal grid is assumed to be unity, the x-coordinate and y-coordinate of the neighboring void are divided by $\sqrt{\det(\partial \boldsymbol{x}/\partial \boldsymbol{X})}$ to ensure that the out-of-plane strain does not change the area of the hexagonal grid.

5.5.2 Pure metals

A pure metal does not contain inclusions around which voids nucleate. Beevers and Honeycombe (1962) performed the uniaxial tensile test of a bar using pure aluminum and indicated that in the absence of inclusions, cracks were initially observed to form only in some grains and on one slip system in each grain. Beevers and Honeycombe (1962) also postulated that cracks could form by a dislocation mechanism in the heavily distorted region of the neck of inclusion-free metals and alloys. Hence, the void shape and the void configuration are assumed for pure metals in the following (Komori, 2006a, 2016).

In Section 5.5.1, no inclusions and no void nucleation are assumed. Although no inclusions are assumed in pure metals, void nucleation should be assumed in pure metals. Because the initial crack shape is unknown, the following assumption on void nucleation is made.

The following assumptions are made on the void shape:

- The void shape at void nucleation is a circle.
- The void shape after void nucleation is calculated by assuming that the displacement gradient of the void is identical to the displacement gradient of the material.

The following assumptions are made on the void configuration:

- The void configuration at void nucleation is identical to the centroid configuration of the hexagonal grid.
- The void configuration after void nucleation is calculated by assuming that the displacement gradient of the void is identical to the displacement gradient of the material.

In the conventional nonsteady finite-element simulation of metal-forming processes, the whole time step is divided into a large number of infinitesimal time steps and voids nucleate in each time step. Hence, the void shape and the void configuration calculated under the above assumption are not singular. It is not easy to obtain the fracture criterion of the material in which voids of various shapes and voids of various configurations exist. Hence, voids of a single shape and voids of a single configuration are calculated from voids of various shapes and voids of various configurations by taking the weighted average. Then the fracture criterion of the material in which voids of a single shape and voids of a single configuration exist is calculated. Here, calculating the fracture criterion using voids of a single shape and voids of a single configuration is an assumption. The validity of this assumption should be confirmed by further simulation and experiment in the future.

Fig. 5.14 shows the void shape and the void configuration for a pure metal. Voids are assumed to nucleate isotropically in the material at each time step because cracks appear in the absence of inclusions. A circle and a hexagonal grid are embedded in the material after each time step.

After the nth time step, a circle which is embedded in the material after the ith time step, becomes a deformed circle, whereas a hexagonal grid which is embedded in the material after the ith time step becomes a deformed hexagonal grid. $(\partial \boldsymbol{x}/\partial \boldsymbol{X})_{(i)}^{(n)}$ is defined as the deformation gradient which equals $\boldsymbol{I}$, the unit dyadic, after the ith time step and exists after the nth time step. The condition after the zeroth time step implies the condition before deformation.

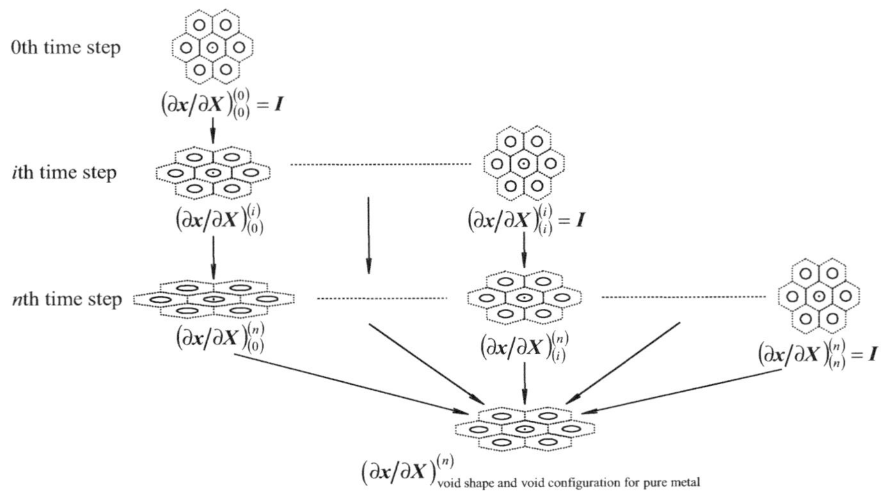

Figure 5.14 Void shape and void configuration for pure metal.

$f_{(i)}^{(j)}$ is defined as the volume fraction of voids that nucleates in the ith time step and which exists after the jth time step. The total void volume fraction after the nth time step is $f_{(0)}^{(n)}+f_{(1)}^{(n)}+\cdots+f_{(n)}^{(n)}$. Here, $f_{(0)}^{(n)}$ is the volume fraction of voids that exists before deformation and which exists after the nth time step.

The void volume fraction after the nth time step $f_{(0)}^{(n)},f_{(1)}^{(n)},\cdots,f_{(n)}^{(n)}$ is assumed to be known beforehand. Using this void volume fraction, the void volume fraction after the $n+1$th time step $f_{(0)}^{(n+1)},f_{(1)}^{(n+1)},\cdots,f_{(n)}^{(n+1)},f_{(n+1)}^{(n+1)}$ is calculated. First, in the simulation of the $n+1$th time step, the total void volume fraction after the $n+1$th time step $f_{(0)}^{(n+1)}+f_{(1)}^{(n+1)}+\cdots+f_{(n)}^{(n+1)}+f_{(n+1)}^{(n+1)}$ and the volume fraction of voids which nucleates in the $n+1$th time step $f_{(n+1)}^{(n+1)}$ are calculated. Next, the growth rate of voids in the $n+1$th time step is assumed to be independent of the number of time steps in which voids nucleate. Hence, the void volume fraction after the $n+1$th time step is calculated as follows:

$$f_{(i)}^{(n+1)}=\frac{f_{(0)}^{(n+1)}+f_{(1)}^{(n+1)}+\cdots+f_{(n)}^{(n+1)}}{f_{(0)}^{(n)}+f_{(1)}^{(n)}+\cdots+f_{(n)}^{(n)}}f_{(i)}^{(n)}(i=0,1,\cdots,n) \tag{5.35}$$

Because the void shape and the void configuration after the nth time step are not singular, a single void shape and a single void configuration after the nth time step are calculated by taking the weighted average to determine whether the fracture criterion of the material is satisfied. Hence, the deformation gradient of a single void shape and the deformation gradient of a single void configuration for a pure metal after the nth time step are assumed to be calculated as follows:

$$\begin{aligned}(\partial\mathbf{x}/\partial\mathbf{X})_{\text{void shape for pure metal}}^{(n)}&=\frac{\sum_{i=0}^{n}f_{(i)}^{(n)}(\partial\mathbf{x}/\partial\mathbf{X})_{(i)}^{(n)}}{\sum_{i=0}^{n}f_{(i)}^{(n)}}\\(\partial\mathbf{x}/\partial\mathbf{X})_{\text{void configuration for pure metal}}^{(n)}&=\frac{\sum_{i=0}^{n}f_{(i)}^{(n)}(\partial\mathbf{x}/\partial\mathbf{X})_{(i)}^{(n)}}{\sum_{i=0}^{n}f_{(i)}^{(n)}}\end{aligned} \tag{5.36}$$

5.5.3 Alloys in which inclusions separate from matrix

When the strength of the inclusion is lower than the strength of the matrix, the inclusion easily deforms during deformation, whereas when the strength of the inclusion is higher than the strength of the matrix, the inclusion deforms only slightly during deformation. Hereafter, the former inclusion is called soft inclusion, whereas the latter inclusion is called hard inclusion.

Fig. 5.15 shows the void shape and the void configuration for the alloy in which the inclusions separate from the matrix. The void configuration for soft inclusions and the void configuration for hard inclusions are assumed to be identical to the void configuration for the pure metal in which voids do not nucleate.

Although the void shape for soft inclusions is assumed to be identical to the void shape for the pure metal in which voids do not nucleate, the void shape for hard inclusions is not assumed to be identical to the void shape for the pure metal in which voids do not nucleate. Hence, the following assumptions are made on the void shape for hard inclusions:

- The void shape before deformation is a circle.
- The void shape after deformation is also a circle.

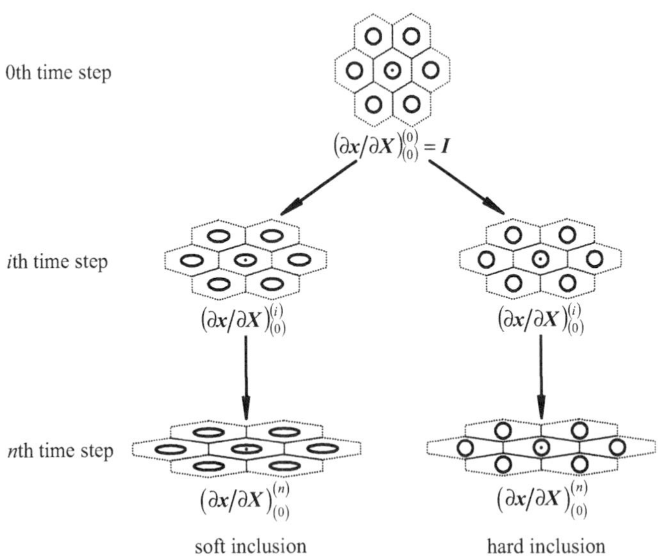

Figure 5.15 Void shape and void configuration for alloy in which inclusions separate from matrix.

Because voids nucleate due to the separation of the inclusions from the matrix, the number of voids is equal to the number of inclusions; the number of voids does not change during deformation.

Hence, the deformation gradient of the void shape and the deformation gradient of the void configuration for soft inclusions after the nth time step are assumed to be calculated as follows:

$$\begin{aligned} (\partial \boldsymbol{x}/\partial \boldsymbol{X})^{(n)}_{\text{void shape for soft inclusion}} &= (\partial \boldsymbol{x}/\partial \boldsymbol{X})^{(n)}_{(0)} \\ (\partial \boldsymbol{x}/\partial \boldsymbol{X})^{(n)}_{\text{void configuration for soft inclusion}} &= (\partial \boldsymbol{x}/\partial \boldsymbol{X})^{(n)}_{(0)} \end{aligned} \tag{5.37}$$

Furthermore, the deformation gradient of the void shape and the deformation gradient of the void configuration for hard inclusions after the nth time step are assumed to be calculated as follows.

$$\begin{aligned} (\partial \boldsymbol{x}/\partial \boldsymbol{X})^{(n)}_{\text{void shape for hard inclusion}} &= \boldsymbol{I} \\ (\partial \boldsymbol{x}/\partial \boldsymbol{X})^{(n)}_{\text{void configuration for hard inclusion}} &= (\partial \boldsymbol{x}/\partial \boldsymbol{X})^{(n)}_{(0)} \end{aligned} \tag{5.38}$$

5.5.4 Alloys in which inclusions crack

Fig. 5.16 shows the void shape and the void configuration for the alloy in which inclusions crack. The black part denotes the inclusion, the white part which is surrounded by an inclusion denotes the void, and the white

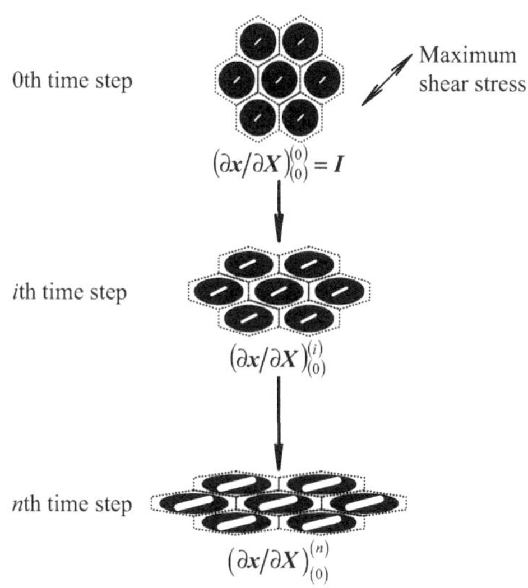

Figure 5.16 Void shape and void configuration for alloy in which inclusions crack.

part which exists between two neighboring inclusions denotes the matrix. The void and the inclusion are assumed to be uniformly distributed in the material.

The void configuration and the inclusion configuration are assumed to be identical to the void configuration for the pure metal in which voids do not nucleate.

The inclusion shape is assumed to be identical to the void shape for the pure metal in which voids do not nucleate. However, the void shape is not assumed to be identical to the void shape for the pure metal in which voids do not nucleate due to the cracking of the inclusion. Hence, the following assumptions are made on the void shape:

- The void shape before deformation is an ellipsoid whose flattening is equal to 0.9.
- The void shape after deformation is also an ellipsoid whose flattening is equal to 0.9.

Because voids nucleate due to the cracking of the inclusion, the number of voids is equal to the number of inclusions; the number of voids does not change during deformation.

Hence, the deformation gradient of the void shape and the deformation gradient of the void configuration for inclusion cracking after the nth time step are assumed to be calculated as follows:

$$\begin{aligned} (\partial\mathbf{x}/\partial\mathbf{X})^{(n)}_{\text{void shape for inclusion cracking}} &= (\partial\mathbf{x}/\partial\mathbf{X})^{(n)}_{(0)} \\ (\partial\mathbf{x}/\partial\mathbf{X})^{(n)}_{\text{void configuration for inclusion cracking}} &= (\partial\mathbf{x}/\partial\mathbf{X})^{(n)}_{(0)} \end{aligned} \tag{5.39}$$

The void shape is basically determined by Eq. (5.39). However, because of the assumption on the void shape, the void shape is modified; the major diameter of the void and the minor diameter of the void are modified so that the flattening of the ellipsoidal void is equal to 0.9, under the condition that the void volume fraction does not change due to the modification.

Furthermore, the deformation gradient of the inclusion shape and the deformation gradient of the inclusion configuration for inclusion cracking after the nth time step are assumed to be calculated as follows.

$$\begin{aligned} (\partial\mathbf{x}/\partial\mathbf{X})^{(n)}_{\text{inclusion shape for inclusion cracking}} &= (\partial\mathbf{x}/\partial\mathbf{X})^{(n)}_{(0)} \\ (\partial\mathbf{x}/\partial\mathbf{X})^{(n)}_{\text{inclusion configuration for inclusion cracking}} &= (\partial\mathbf{x}/\partial\mathbf{X})^{(n)}_{(0)} \end{aligned} \tag{5.40}$$

Figure 5.17 Void shape in hexagonal grid.

In the case of carbon steels, the direction of the major axis of the crack at crack initiation in the pearlite nodules is known to coincide with the direction of the maximum shear stress of the material (Miller and Smith, 1970). Hence, the direction of the major axis of the void before deformation is assumed to be identical to the direction of the maximum shear stress of the material.

In the case of carbon steels, the void remains in the pearlite nodule and does not penetrate into the ferrite matrix until the internal necking between two neighboring voids appears (Inoue and Kinoshita, 1977b). However, in the ellipsoidal void model, because insufficient restrictions are made on the void shape, the void may break the inclusion into two parts and penetrate into the matrix. Hence, the following restriction is made on the void shape.

Fig. 5.17 shows the void shape in the hexagonal grid. When the void penetrates into the matrix, the void shape is modified so that the void does not penetrate into the matrix, under the condition that the void volume fraction does not change due to the modification. Hence, the flattening of the ellipsoidal void becomes smaller than 0.9 after the modification. After the inclusion is broken into two parts, the void is not assumed to nucleate because the inclusion does not fracture furthermore. Hence, the void nucleation terms in Eq. (5.30) are not assumed.

5.6 Void coalescence region

Although the criterion of void coalescence in Section 5.3 is microscopically satisfied for a specified angle between the coordinate axis and the line connecting two neighboring voids, the material does not necessarily develop macroscopic fractures. The criterion of microscopic void coalescence is related to macroscopic material fracture, as shown below.

Fig. 5.18 shows the two neighboring voids and the void coalescence region. The distance between the two neighboring voids before forming is p, which is equal to $\sqrt[4]{4/3}$. Variable ϕ is defined as the angle between an arbitrary axis and the line connecting the two neighboring voids. Because the direction of the arbitrary axis is unknown, ϕ is assumed to be an arbitrary value between 0 and 2π. Furthermore, because the void configuration

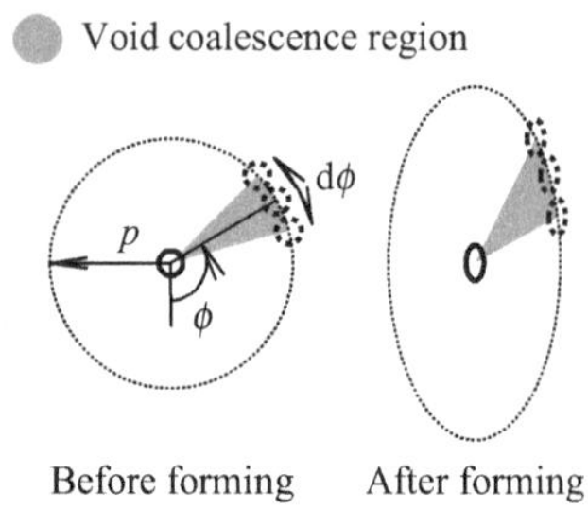

Figure 5.18 Two neighboring voids and void coalescence region.

is assumed to be macroscopically isotropic before forming, the probability density function of ϕ becomes constant regardless of the value of ϕ.

When ϕ is specified, the ratio E in Eq. (5.22) can be calculated for a void after forming. Hence, the ratio E for a void after forming is calculated for all values of ϕ. $d\phi$ is defined as the range of ϕ in which the ratio E for a void after forming is less than unity. The void coalescence region is defined as a sector whose central angle is $d\phi$.

Six other voids surround a given void, as shown in Fig. 5.11. When $d\phi$ is $\pi/3(=2\pi/6)$, the probability that the ratio E is less than unity for a specified ϕ is unity. Since the void configuration is assumed to be macroscopically isotropic before forming, the macroscopic probability of void coalescence is unity. Similarly, when $d\phi$ is $\pi/6(=(2\pi/6)/2)$, the probability that the ratio E is less than unity for a specified ϕ is half. Hence, the macroscopic probability of void coalescence is half, which is assumed to be a reasonable criterion for material fracture. Therefore, when the following equation is satisfied, the material is assumed to be macroscopically fractured.

$$d\phi = \frac{\pi}{6} \tag{5.41}$$

The macroscopic probability of void coalescence of half should be validated by further experiment and simulation in future.

5.7 Prestrain

The void shape in the prestraining process is identical to the void shape in subsequent processes, and the void configuration in the prestraining process is identical to the void configuration in subsequent processes. In other words, no special simulation methods are required to perform the simulation of the prestraining process.

CHAPTER 6

Simulation results using an ellipsoidal void model (author's model)

6.1 Fundamental deformation

The characteristics of the ellipsoidal void model are demonstrated (Komori, 2013a). To obtain the simulation results immediately, the location of the velocity discontinuity line is determined geometrically; the velocity discontinuity line is assumed to coincide with the tangent of the two neighboring voids. Furthermore, no void coalescence region is assumed to simplify the simulation method. To evaluate the ellipsoidal void model, simulations using the representative volume element are also performed. The deformation is assumed to be uniform so that no necking deformation occurs and a simple stress−strain relationship, $\sigma_M = \text{const.}$, is assumed.

6.1.1 Plane-strain tension

Fig. 6.1 shows the relationship between the initial void volume fraction and the nominal fracture strain for plane-strain tension. Fig. 6.1A shows the coordinates and the notation. Since the length of the simulation region in the y-direction is unity, the displacement in the y-direction v^* coincides with the nominal strain. The solid line denotes the material shape and the void shape before deformation, whereas the dotted line denotes the material shape and the void shape after deformation. Fig. 6.1B shows the relationship between the initial void volume fraction and the nominal fracture strain. When no void growth is assumed, only the microscopic simulation is performed because it is not necessary to perform the macroscopic simulation. Fig. 6.1B also shows the relationship for the rectangular void calculated using the ellipsoidal void model. The void shape before deformation is assumed to be square.

When the initial void volume fraction exceeds 12%, the nominal fracture strain of a rectangular void is zero, whereas the nominal fracture strain of an ellipsoidal void is not zero. When the initial void volume

Ductile Fracture in Metal Forming
DOI: https://doi.org/10.1016/B978-0-12-814772-6.00006-0

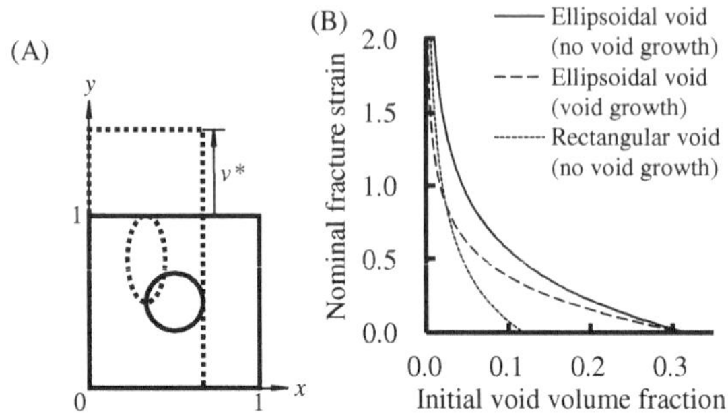

Figure 6.1 Relationship between initial void volume fraction and nominal fracture strain for plane-strain tension: (A) coordinates and notation and (B) relationship between initial void volume fraction and nominal fracture strain.

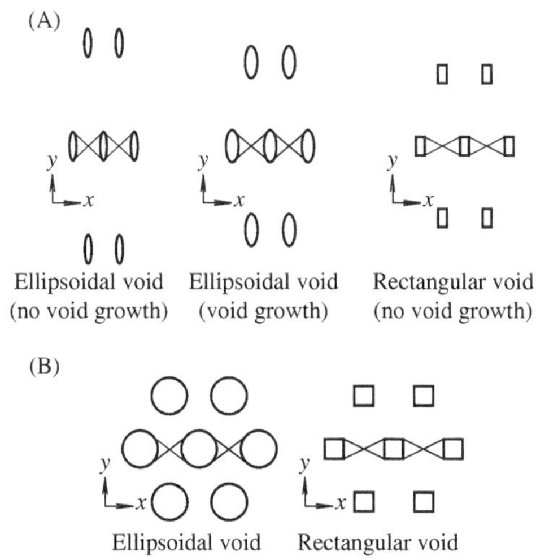

Figure 6.2 Final void shape and final void configuration for plane-strain tension: (A) initial void volume fraction = 0.05 and (B) nominal fracture strain = 0.

fraction is specified, the nominal fracture strain of an ellipsoidal void with void growth is smaller than the nominal fracture strain of an ellipsoidal void without void growth. However, the initial void volume fraction below which the nominal fracture strain is zero for an ellipsoidal void with void growth is the same as the initial void volume fraction below which the nominal fracture strain is zero for an ellipsoidal void without void growth.

Fig. 6.2 shows the final void shape and the final void configuration for plane-strain tension. The final state is the state which satisfies the fracture criterion. The velocity discontinuity lines at the final state are also indicated. Fig. 6.2A shows the final void shape and the final void configuration when the initial void volume fraction is equal to 5%. Since the direction of the minimum principal strain coincides with the x-direction, the angle between the line connecting two neighboring voids before deformation and the x-axis, which is defined as ϕ in Fig. 5.13A, and is equal to zero. The void volume fraction of the final state of an ellipsoidal void when void growth occurs is larger than the void volume fraction of the final state of an ellipsoidal void when no void growth occurs. The angle between the two velocity discontinuity lines when void growth occurs is 81 degrees, whereas the angle between the two velocity discontinuity lines when void growth does not occur is 84 degrees. The angle between the two velocity discontinuity lines of a rectangular void is 57 degrees.

Fig. 6.2B shows the final void shape and the final void configuration when the nominal fracture strain is equal to zero. When the nominal fracture strain is zero, the fracture criterion is satisfied when the nominal strain is zero. The state in which the nominal strain is zero is the initial state. Hence, the final void shape and the final void configuration are identical to the initial void shape and the initial void configuration. The void volume fraction of an ellipsoidal void is 31.5%, whereas the void volume fraction of a rectangular void is 12%. The angle between the two velocity discontinuity lines for an ellipsoidal void is 72 degrees, whereas the angle between the two velocity discontinuity lines for a rectangular void is 51 degrees. The angle between the two velocity discontinuity lines increases with increasing void volume fraction when the void configuration does not change. The difference between the void volume fraction of an ellipsoidal void and the void volume fraction of a rectangular void results in a difference between the angle for an ellipsoidal void and the angle for a rectangular void.

A simulation using the representative volume element is performed to validate the ellipsoidal void model. The deformation of the material is simulated by the conventional rigid-plastic finite-element method (Kobayashi et al., 1989). The condition of volume constancy of the material is satisfied by the penalty method.

Fig. 6.3 shows the representative volume element for plane-strain tension. Fig. 6.3A shows the coordinates and the notation. Although the shape of the representative volume element should be hexagonal, it is not easy to choose suitable boundary conditions for the displacement.

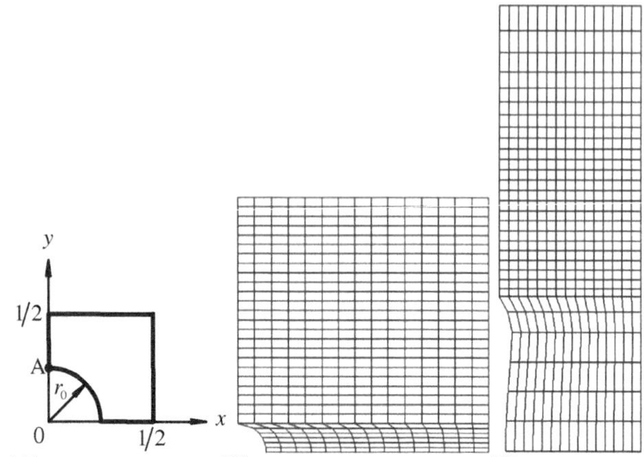

Figure 6.3 Representative volume element for plane-strain tension: (A) coordinates and notation; (B) initial finite-element mesh; and (C) final finite-element mesh.

Furthermore, the rectangular representative volume element is widely used. Hence, the shape of the representative volume element is assumed to be rectangular. r_0 denotes the radius of the cylindrical void, which is equal to $\sqrt{f_0/\pi}$, where f_0 denotes the initial void volume fraction. The following boundary conditions on the displacement are assumed,

$$\begin{aligned} u = 0 \text{ on } x = 0 \text{ and } u = \text{const. on } x = 1/2 \\ v = 0 \text{ on } y = 0 \text{ and } v = \text{const. on } y = 1/2 \end{aligned} \tag{6.1}$$

where u denotes the displacement in the x-direction, whereas v denotes the displacement in the y-direction. Hence, the rectangular representative volume element remains rectangular after the displacement.

Fig. 6.3B shows the initial finite-element mesh when the initial void volume fraction is 1%. The simulation is performed until the displacement in the x-direction u on the outer side of the representative volume element at $y = 0$ is negligible compared with the displacement in the x-direction u on the inner side of the representative volume element at $y = 0$. The flow of the material is then assumed to be localized into the plane $y = 0$. Hence, the material is assumed to fracture.

Fig. 6.3C shows the final finite-element mesh when the initial void volume fraction is 1%. The void elongates mainly in the y-direction. Since the void length in the x-direction is not a maximum on $y = 0$, the void shape is not ellipsoidal. Since the void shape is assumed to be ellipsoidal in the ellipsoidal void model, the void shape in the representative volume element differs from the void shape in the ellipsoidal void model.

Since the deformation of the material near the void is heterogeneous, remeshing is performed at each time step. Remeshing is difficult to perform in the elastic—plastic finite-element method since the nodal force at each node needs to be redistributed after remeshing. However, redistribution of the nodal force at each node is not required after remeshing in the rigid-plastic finite-element method. Hence, remeshing is easy to perform.

Fig. 6.4 shows the simulation results using the representative volume element and the ellipsoidal void model for plane-strain tension. Void growth is assumed in the ellipsoidal void model. Fig. 6.4A shows the relationship between the initial void volume fraction and the nominal fracture strain. With increasing initial void volume fraction, the nominal fracture strain calculated by the representative volume element and the nominal fracture strain calculated by the ellipsoidal void model decrease. The initial void volume fraction above which the nominal fracture strain is zero using the ellipsoidal void model is almost the same as the initial void volume fraction above which the nominal fracture strain is zero using the representative volume element. When the initial void volume fraction is larger than 20%, the nominal fracture strain calculated by the ellipsoidal void model is slightly larger than the nominal fracture strain calculated by the representative volume element. When the initial void volume fraction is smaller than 10%, the nominal fracture strain calculated by the ellipsoidal

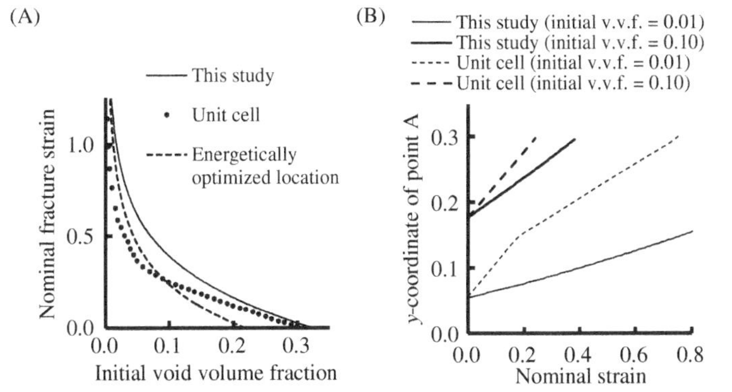

Figure 6.4 Simulation results obtained using representative volume element and ellipsoidal void model for plane-strain tension: (A) relationship between initial void volume fraction and nominal fracture strain and (B) relationship between nominal strain and *y*-coordinate of point A.

void model is larger than the nominal fracture strain calculated by the representative volume element.

In this study, to obtain the simulation results immediately, the location of the velocity discontinuity line is determined geometrically; the velocity discontinuity line is assumed to coincide with the tangent of the two neighboring voids. However, to increase the accuracy of the simulation results, the location of the velocity discontinuity line is optimized energetically such that the energy required for the internal necking deformation mode is minimized. The relationship between the initial void volume fraction and the nominal fracture strain is supplemented in the case that the location of the velocity discontinuity line is optimized energetically. The initial void volume fraction above which the nominal fracture strain is zero using the ellipsoidal void model is smaller than the initial void volume fraction above which the nominal fracture strain is zero using the representative volume element. However, regardless of the initial void volume fraction, the nominal fracture strain calculated by the ellipsoidal void model fairly agrees with the nominal fracture strain calculated by the representative volume element.

Fig. 6.4B shows the relationship between the nominal strain and the y-coordinate of point A. The y-coordinate of point A is equal to half the void length in the y-direction. Since the direction of the maximum principal stress coincides with the y-direction, the void length in the y-direction is essential in the ellipsoidal void model. With increasing nominal strain, the y-coordinate of point A calculated by the ellipsoidal void model and the y-coordinate of point A calculated by the representative volume element increase. However, the y-coordinate of point A calculated by the ellipsoidal void model is smaller than the y-coordinate of point A calculated by the representative volume element.

The following discussion of the ellipsoidal void model is based on Fig. 6.4. When the initial void volume fraction is smaller than 10%, the nominal fracture strain is relatively large. Hence, the difference between the y-coordinate of point A calculated by the ellipsoidal void model and the y-coordinate of point A calculated by the representative volume element is relatively large. Therefore the difference between the nominal fracture strain calculated by the ellipsoidal void model and the nominal fracture strain calculated by the representative volume element is relatively large. When the initial void volume fraction is larger than 20%, the nominal fracture strain is relatively small. Hence, the difference between the y-coordinate of point A calculated by the ellipsoidal void model and the

y-coordinate of point A calculated by the representative volume element is relatively small. Therefore the difference between the nominal fracture strain calculated by the ellipsoidal void model and the nominal fracture strain calculated by the representative volume element is relatively small. The assumption on the void shape in the ellipsoidal void model, namely that the displacement gradient of the void is identical to the displacement gradient of the matrix, should be improved in the future.

6.1.2 Simple shear

Fig. 6.5 shows the relationship between the initial void volume fraction and the nominal fracture strain for simple shear. Fig. 6.5A shows the coordinates and the notation. Since the length of the simulation region in the y-direction is equal to unity, the displacement in the x-direction u^* coincides with the nominal strain. The solid line denotes the material shape and the void shape before deformation, whereas the dotted line denotes the material shape and the void shape after deformation. Fig. 6.5B shows the relationship between the initial void volume fraction and the nominal fracture strain. According to Eq. (5.26), the volumetric strain rate $\dot{\varepsilon}_{kk} = (3/4) \cdot (\dot{\bar{\varepsilon}}/\bar{\sigma}) \cdot f\sigma_{kk}$. Since the stress triaxiality $(\sigma_{kk}/3\bar{\sigma})$ is zero in simple shear, the volumetric strain rate $\dot{\varepsilon}_{kk}$ is zero. Hence, no void growth is assumed to occur. It is thus not necessary to perform the macroscopic simulation, and only the microscopic simulation is performed.

In plane-strain tension, since the direction of the minimum principal strain coincides with the x-direction, the angle between the line connecting two neighboring voids before deformation and the x-axis, which is

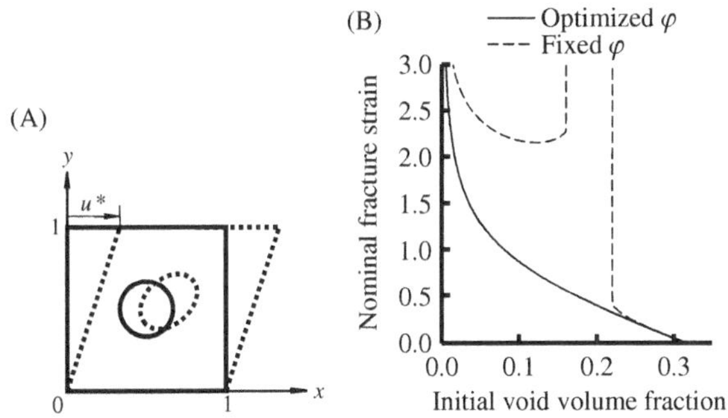

Figure 6.5 Relationship between initial void volume fraction and nominal fracture strain for simple shear: (A) coordinates and notation and (B) relationship between initial void volume fraction and nominal fracture strain.

defined as ϕ in Fig. 5.13A, is zero. Hence, it is not necessary to optimize ϕ. However, it is necessary to optimize ϕ in simple shear since the direction of the minimum principal strain varies during deformation. Furthermore, to demonstrate the effect of optimization, it is beneficial to fix ϕ rather than to optimize ϕ. Hence, ϕ is fixed for comparison. ϕ is set to be $-\pi/4$, since the angle between the direction of the minimum principal strain and the x-axis is equal to $-\pi/4$ when the nominal strain is infinitesimally small. The nominal fracture strain for fixed ϕ is almost the same as the nominal fracture strain for optimized ϕ when the initial void volume fraction exceeds 25%. However, the nominal fracture strain for fixed ϕ differs greatly from the nominal fracture strain for optimized ϕ when the initial void volume fraction is smaller than 22%.

Fig. 6.6 shows the final void shape and the final void configuration for simple shear. The final state is the state that satisfies the fracture criterion. The velocity discontinuity lines at the final state are also indicated. Fig. 6.6A shows the final void shape and the final void configuration when the initial void volume fraction is equal to 22%. Both the final state for fixed ϕ and the final state for optimized ϕ are shown. The central void, which is void O in Fig. 5.12, coalesces with one of the adjacent voids, which are void A, B, C, D, E, or F, for both fixed ϕ and optimized ϕ. $\phi = -45$ degrees for fixed ϕ, whereas $\phi = -38$ degrees for optimized ϕ.

Fig. 6.6B shows the final void shape and the final void configuration when the initial void volume fraction is equal to 16%. Both the final state for fixed ϕ and the final state for optimized ϕ are shown. The central

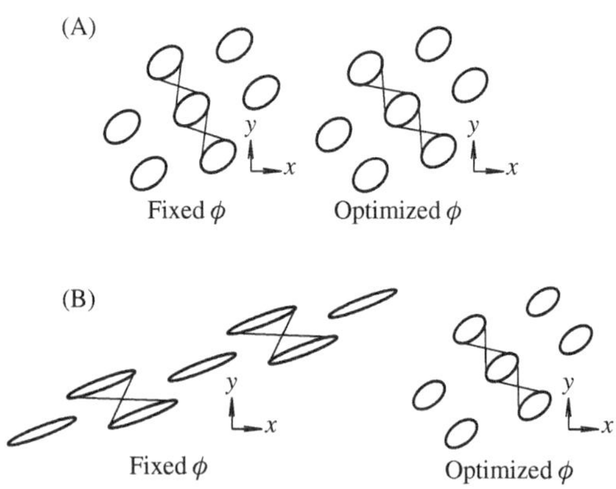

Figure 6.6 Final void shape and final void configuration for simple shear: (A) initial void volume fraction = 0.22 and (B) initial void volume fraction = 0.16.

void coalesces with one of the adjacent voids for optimized ϕ. However, the central void does not coalesce with one of the adjacent voids for fixed ϕ; instead nonadjacent voids, which are voids A and D in Fig. 5.12, coalesce. $\phi = -45$ degrees for fixed ϕ, wheras $\phi = -35$ degrees for optimized ϕ. The optimized value of ϕ increases with decreasing initial void volume fraction.

A simulation using the representative volume element is performed to validate the ellipsoidal void model. Fig. 6.7 shows the representative volume element for simple shear. Fig. 6.7A shows the coordinates and the notation. r_0 denotes the radius of the cylindrical void and is equal to $\sqrt{f_0/\pi}$, where f_0 denotes the initial void volume fraction.

Periodic boundary conditions should be assumed when two neighboring voids coalesce. However, the angle between the line connecting two neighboring voids before deformation and the x-axis, which is ϕ in Fig. 5.13A, should be optimized in simple shear. This angle is not known in advance because the location of the neighboring void is not known in advance. It is thus impossible to perform the simulation until two neighboring voids coalesce, in other words, until the flow of the material is localized. Hence, the simulation is performed until the nominal strain

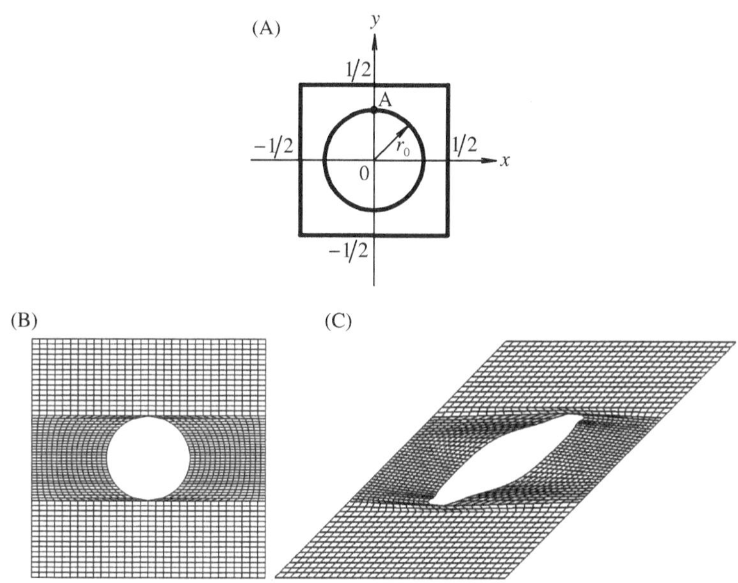

Figure 6.7 Representative volume element for simple shear: (A) coordinates and notation; (B) initial finite-element mesh; and (C) final finite-element mesh.

reaches a certain value. Then, the following boundary conditions for the displacement are assumed,

$$\begin{aligned} u &= y \text{ on } x = \pm 1/2 \text{ and } y = \pm 1/2 \\ v &= 0 \text{ on } x = \pm 1/2 \text{ and } y = \pm 1/2 \end{aligned} \tag{6.2}$$

where u denotes the displacement in the x-direction and v denotes the displacement in the y-direction. Hence, a rectangular representative volume element becomes parallelogrammic after the displacement.

Fig. 6.7B shows the initial finite-element mesh when the initial void volume fraction is equal to 10%. Fig. 6.7C shows the final finite-element mesh when the initial void volume fraction is equal to 10%. The final state is the state in which the nominal strain becomes unity. The void elongates in the direction of the diagonal of the parallelogrammic representative volume element. Since the deformation of the material near the void is heterogeneous, remeshing is performed at each time step.

Fig. 6.8 shows the simulation results obtained using the representative volume element and the ellipsoidal void model for simple shear. Fig. 6.8A shows the relationship between the nominal strain and the x-coordinate of point A. The simulation is performed until the nominal strain becomes two when the initial void volume fraction is 1%, whereas the simulation is performed until the nominal strain becomes unity when the initial void volume fraction is 10%. With increasing nominal strain, the x-coordinate of point A calculated by the representative volume element and the

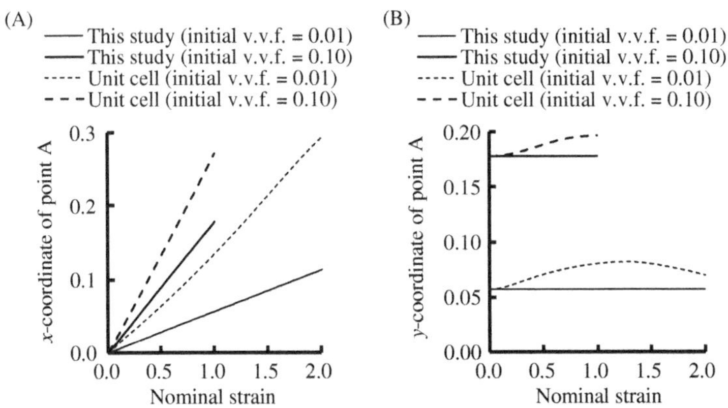

Figure 6.8 Simulation results obtained using representative volume element and ellipsoidal void model for simple shear: (A) relationship between nominal strain and x-coordinate of point A and (B) relationship between nominal strain and y-coordinate of point A.

x-coordinate of point A calculated by the ellipsoidal void model increase. However, the x-coordinate of point A calculated by the ellipsoidal void model is smaller than the x-coordinate of point A calculated by the representative volume element. In other words, the void rotates relative to the material.

Fig. 6.8B shows the relationship between the nominal strain and the y-coordinate of point A. When the nominal strain is increased, the y-coordinate of point A calculated by the ellipsoidal void model remains constant, whereas the y-coordinate of point A calculated by the representative volume element changes slightly. The y-coordinate of point A calculated by the ellipsoidal void model almost coincides with the y-coordinate of point A calculated by the representative volume element.

The square root of the summation of the square of the x-coordinate of point A and the square of the y-coordinate of point A are approximately equal to half the maximum void length. As is shown in Fig. 6.6, the maximum void length is essential in the ellipsoidal void model. Fig. 6.8 shows that the maximum void length calculated by the ellipsoidal void model is smaller than the maximum void length calculated by the representative volume element. Hence, the assumption on the void shape in the ellipsoidal void model, namely that the displacement gradient of the void is identical to the displacement gradient of the matrix, should be improved in the future.

6.1.3 Uniaxial tension

Fig. 6.9 shows the relationship between the initial void volume fraction and the nominal fracture strain for uniaxial tension. Axisymmetric deformation of the material is assumed in the macroscopic simulation of uniaxial tension. Fig. 6.9A shows the coordinates and the notation. Since the length of the simulation region in the z-direction is equal to unity, the displacement in the z-direction w^* coincides with the nominal strain. Fig. 6.9B shows the relationship between the initial void volume fraction and the nominal fracture strain. The relationship when there is no void growth for plane-strain tension is also shown for comparison. When the initial void volume fraction is specified, the nominal fracture strain when there is no void growth in uniaxial tension is larger than the nominal fracture strain when there is no void growth in plane-strain tension. The reason for this difference is that the strain in the out-of-plane direction is not zero in uniaxial tension. When the initial void volume fraction is

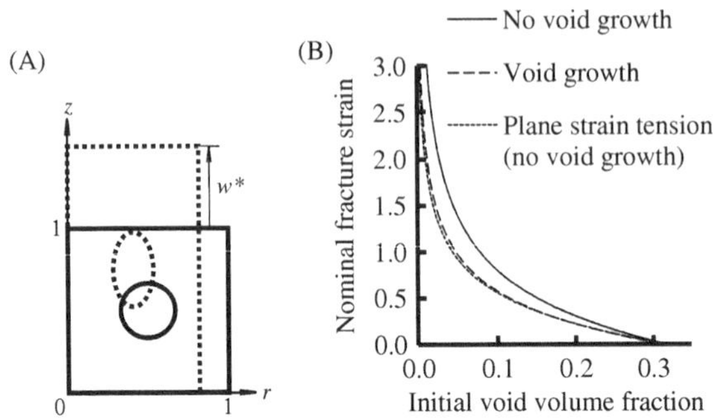

Figure 6.9 Relationship between initial void volume fraction and nominal fracture strain for uniaxial tension: (A) coordinates and notation and (B) relationship between initial void volume fraction and nominal fracture strain.

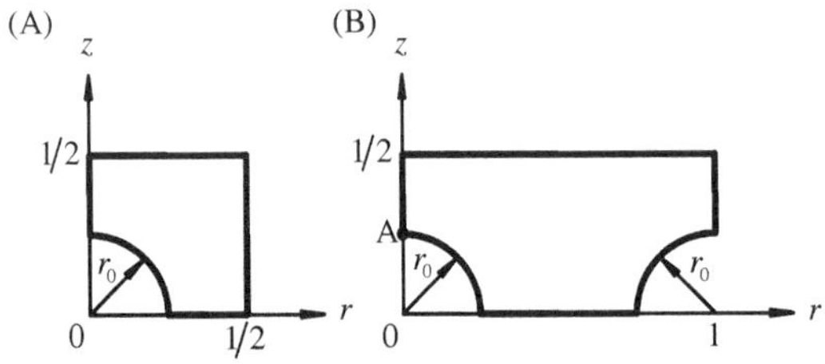

Figure 6.10 Coordinates and notation in representative volume element for uniaxial tension: (A) spherical void and (B) spherical void and toroidal void.

specified, the nominal fracture strain when there is void growth in uniaxial tension is smaller than the nominal fracture strain when there is no void growth in uniaxial tension. Interestingly, the relationship when there is void growth in uniaxial tension almost coincides with the relationship when there is no void growth in plane-strain tension.

The simulation using the representative volume element is performed to validate the ellipsoidal void model. Fig. 6.10 shows the coordinates and notation in the representative volume element for uniaxial tension. Fig. 6.10A shows the representative volume element containing a spherical void. r_0 denotes the radius of the spherical void. Since in the ellipsoidal void model, the void volume fraction is defined in a plane, the void volume fraction is defined in the rz-plane in uniaxial tension. Hence, $r_0 = \sqrt{f_0/\pi}$, where f_0 denotes the initial void volume fraction. The following boundary conditions are assumed for the displacement,

$$u = 0 \text{ on } r = 0 \text{ and } u = \text{const. on } r = 1/2$$
$$w = 0 \text{ on } z = 0 \text{ and } w = \text{const. on } z = 1/2 \tag{6.3}$$

where u denotes the displacement in the r-direction and w denotes the displacement in the z-direction. Hence, the cylindrical cell remains cylindrical after the displacement.

Simulations using a representative volume element containing a spherical void have been performed in previous studies (Koplik and Needleman, 1988; Tvergaard, 1982a). The representative volume element in these studies is the same as the representative volume element in this study, which is shown in Fig. 6.10A. The boundary conditions in these studies are the same as the boundary conditions in this study, which are represented by Eq. (6.3). In these studies, it is assumed that the void configuration in the plane $z = 0$ is identical to the centroid configuration of the hexagonal grid, and that there are neighboring voids to coalesce with the spherical void. However, since the axisymmetric deformation of a cylindrical representative volume element containing a spherical void is assumed, it is clear that there are no neighboring voids to coalesce with the spherical void.

Fig. 6.10B shows the representative volume element containing a spherical void and a toroidal void. r_0 is both the radius of the spherical void and the radius of the toroidal void and $r_0 = \sqrt{f_0/\pi}$. The following boundary conditions are assumed for the displacement,

$$u = 0 \text{ on } r = 0 \text{ and } u = \text{const. on } r = 1$$
$$w = 0 \text{ on } z = 0 \text{ and } w = \text{const. on } z = 1/2 \tag{6.4}$$

such that the cylindrical cell remains cylindrical after the deformation.

When axisymmetry is assumed, an infinite number of representative volume elements can be assumed to exist in the z-direction, whereas only one representative volume element can be assumed to exist in the r-direction. The reason for this assumption is as follows. When axisymmetry is assumed, a spherical void whose center is located on the z-axis can be assumed. However, a spherical void whose center does not lie on the z-axis cannot be assumed. Although there are no toroidal voids in reality, a toroidal void whose center lies on the z-axis must be assumed. When a toroidal void is assumed, it is clear that only one representative volume element can be assumed to exist in the r-direction.

The simulation is performed until the displacement in the r-direction u on the outer side of the representative volume element at $z = 0$ is negligible compared with the displacement in the r-direction u on the inner side of the representative volume element at $z=0$. The flow of the material is then assumed to be localized in the plane $z=0$. Hence, the material is assumed to fracture. In Koplik and Needleman (1988), the flow of the material is assumed to be localized when the radius of the representative volume element remains constant. Hence, it is clear that the condition for assuming the flow localization in this study is virtually the same as the condition for assuming the flow localization in Koplik and Needleman (1988).

Fig. 6.11 shows the final finite-element meshes in the representative volume element for uniaxial tension. The initial finite-element mesh containing a spherical void and a toroidal void is made to coincide with twice the initial finite-element mesh containing a spherical void. Since the deformation of the material near the void is heterogeneous, remeshing is performed at each time step.

Fig. 6.11A shows the final finite-element mesh containing a spherical void and a toroidal void when the initial void volume fraction is equal to 1%. The maximum length of the spherical void in the z-direction differs from the maximum length of the toroidal void in the z-direction.

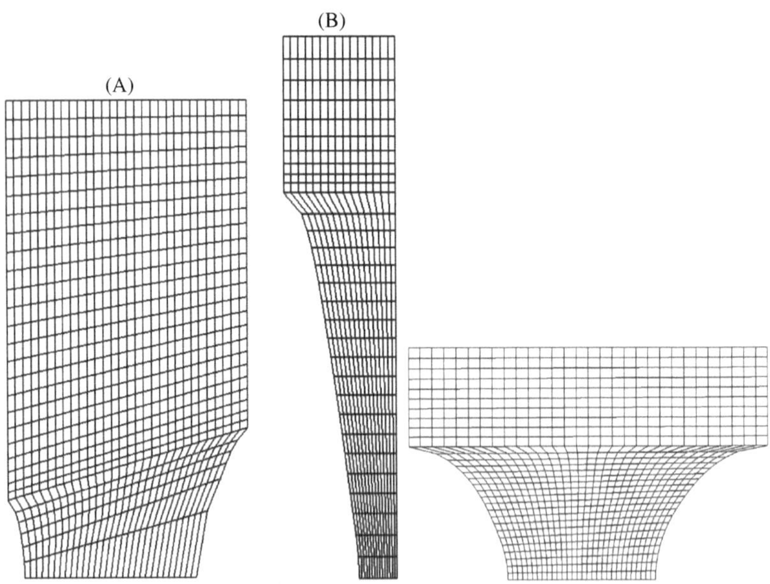

Figure 6.11 Final finite-element meshes in representative volume element for uniaxial tension: (A) initial void volume fraction = 0.01 and (B) initial void volume fraction = 0.2.

Although the maximum length of the spherical void in the r-direction differs from the maximum length of the toroidal void in the r-direction, both the length of the spherical void in the r-direction and the length of the toroidal void in the r-direction are largest on $z = 0$. Fig. 6.11B shows the final finite-element mesh containing a spherical void and the final finite-element mesh containing a spherical void and a toroidal void when the initial void volume fraction is equal to 20%. The representative volume element containing a spherical void has a much larger deformation than the representative volume element containing a spherical void and a toroidal void. In the representative volume element containing a spherical void, the spherical void elongates in the z-direction and becomes needle-like; this needle-like void has been theoretically predicted by other researchers for a viscous material (Budiansky et al., 1982).

Fig. 6.12 shows the simulation results obtained using the representative volume element and the ellipsoidal void model for uniaxial tension. Fig. 6.12A shows the relationship between the initial void volume fraction and the nominal fracture strain. The nominal fracture strain decreases with increasing initial void volume fraction. When the initial void volume fraction is specified, the nominal fracture strain calculated by the representative volume element containing a spherical void is much larger than the nominal fracture strain calculated by the ellipsoidal void model. This is because there are no neighboring voids to coalesce with the spherical void. When the initial void volume fraction is specified, the nominal

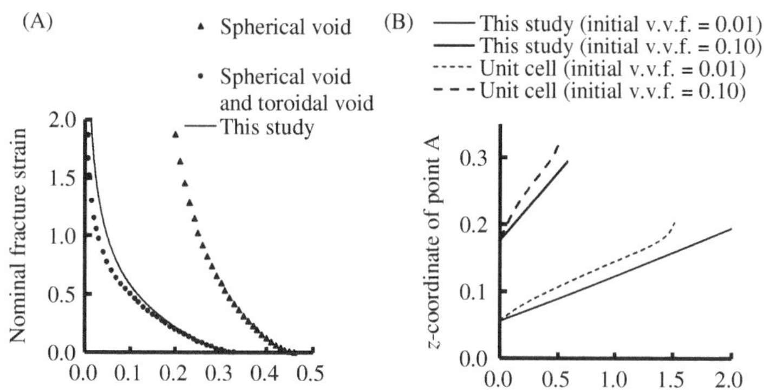

Figure 6.12 Simulation results obtained using representative volume element and ellipsoidal void model for uniaxial tension: (A) relationship between initial void volume fraction and nominal fracture strain and (B) relationship between nominal strain and z-coordinate of point A.

fracture strain calculated by the representative volume element containing a spherical void and a toroidal void is almost equal to the nominal fracture strain calculated by the ellipsoidal void model. This is because the toroidal void coalesces with the spherical void. It is clear that the relationship between the initial void volume fraction and the nominal fracture strain for uniaxial tension calculated by the simulation using the axisymmetric representative volume element containing a spherical void is inappropriate.

Fig. 6.12B shows the relationship between the nominal strain and the z-coordinate of point A. Since the direction of the maximum principal stress coincides with the z-direction, the void length in the z-direction is essential in the ellipsoidal void model. With increasing nominal strain, the z-coordinate of point A is calculated by the ellipsoidal void model and the z-coordinate of point A is calculated by the representative volume element increase. Furthermore, the z-coordinate of point A calculated by the ellipsoidal void model is almost equal to the z-coordinate of point A calculated by the representative volume element. Fig. 6.12B demonstrates that the assumption on the void shape in the ellipsoidal void model, namely that the displacement gradient of the void is identical to the displacement gradient of the matrix, is appropriate for uniaxial tension.

6.1.4 Effect of strain hardening

Fig. 6.13 shows the relationship between the initial void volume fraction and the nominal fracture strain for various strain-hardening exponents for uniaxial tension. The stress–strain relationship of the matrix is assumed to be,

$$\sigma_M = (\varepsilon_M + 0.01)^n \tag{6.5}$$

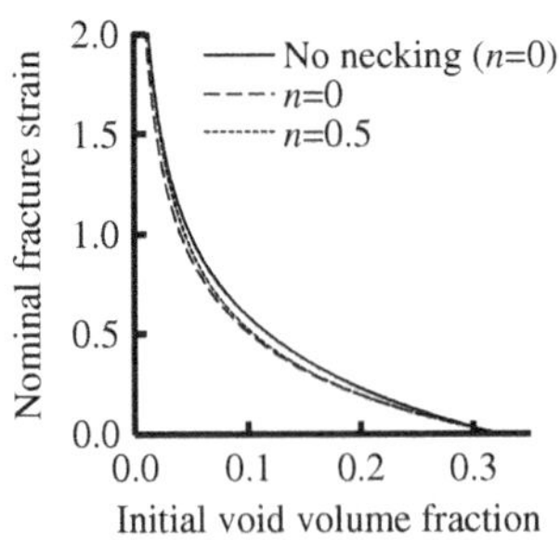

Figure 6.13 Relationship between initial void volume fraction and nominal fracture strain for various strain-hardening exponents for uniaxial tension.

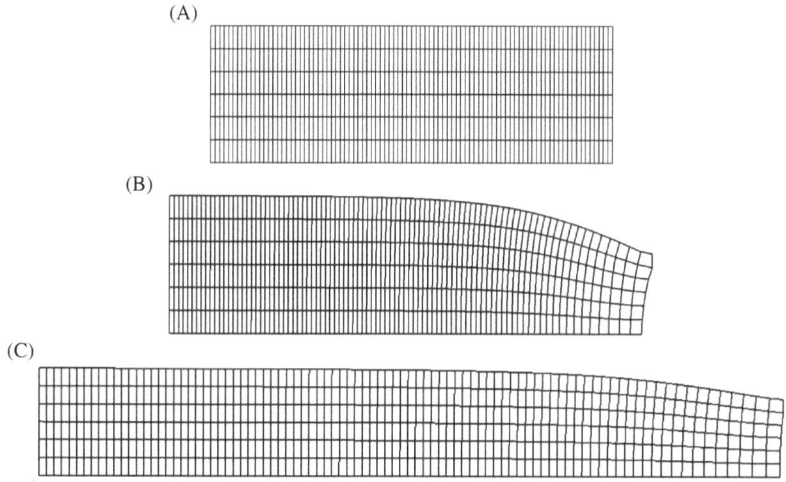

Figure 6.14 Finite-element meshes for various strain-hardening exponents for uniaxial tension. (initial void volume fraction = 0.01): (A) initial state; (B) final state ($n = 0$); and (C) final state ($n = 0.5$).

where ε_M denotes the strain of the matrix and n denotes the strain-hardening exponent. The relationship obtained in Section 6.1.3 is supplemented for the case when necking does not occur since the material is assumed to deform uniformly. The relationship between the initial void volume fraction and the nominal fracture strain depends very little on the strain-hardening exponent.

Fig. 6.14 shows the finite-element meshes for various strain-hardening exponents. The initial void volume fraction is equal to 1%. When $n = 0$, necking occurs if $\varepsilon_M > 0$, whereas when $n = 0.5$, necking occurs if $\varepsilon_M > 0.5$. Hence, the finite-element mesh of the final state when $n = 0$ differs from the finite-element mesh of the final state when $n = 0.5$. However, the cross-sectional area of the final state when $n = 0$ is approximately the same as the cross-sectional area of the final state when $n = 0.5$. The logarithmic fracture strain ε_f is calculated using the formula $\varepsilon_f = \ln(S_0/S_f)$, where S_0 is the initial cross-sectional area and S_f is the final cross-sectional area. Hence, ε_f when $n = 0$, which is 1.09, is approximately the same as ε_f when $n = 0.5$, which is 1.16.

6.1.5 Effect of yield function

The Gurson yield function Φ (1977) is derived with reference to the assumption of spherically symmetric deformation of the material around a spherical void. Hence, the effect of the existence of surrounding voids on the material deformation around a central void is not considered. It is of considerable interest to estimate the effect of the existence of surrounding voids on the material deformation around a central void. Tvergaard (1981) proposes a modified Gurson yield function, such that the simulation result using the modified Gurson yield function approaches the simulation result using the representative volume element. The modified Gurson yield function is referred to as the Gurson–Tvergaard yield function hereafter.

The Gurson–Tvergaard yield function Ψ (1981) is expressed in the following equation.

$$\Psi = \frac{3}{2} \cdot \frac{\sigma'_{ij}\sigma'_{ij}}{\sigma_M^2} + 3f\cosh\left(\frac{\sigma_{kk}}{2\sigma_M}\right) - 1 - (1.5f)^2 = 0 \tag{6.6}$$

Since the yield function Ψ is not a function of the second power of stress, it is not easy to perform rigid-plastic simulation using Eq. (6.6). Hence, cosh$x$ is approximated by $1 + x^2/2$ (Tomita, 1990). The approximated yield function Ψ' used in this study is as follows.

$$\Psi' = \frac{3}{2} \cdot \frac{\sigma'_{ij}\sigma'_{ij}}{\sigma_M^2} + \frac{3f}{8} \cdot \left(\frac{\sigma_{kk}^2}{\sigma_M^2}\right) - (1 - 1.5f)^2 = 0 \tag{6.7}$$

Fig. 6.15 shows the relationship between the initial void volume fraction and the nominal fracture strain for various yield functions for uniaxial tension. The nominal fracture strain calculated by the Gurson–Tvergaard yield function is slightly smaller than the nominal fracture strain calculated by the Gurson yield function for a specified initial void volume fraction. However, the relationship between the initial void volume fraction and

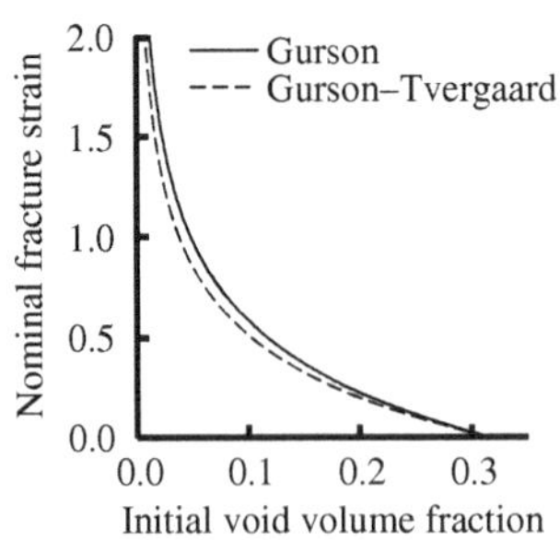

Figure 6.15 Relationship between initial void volume fraction and nominal fracture strain for various yield functions for uniaxial tension.

the nominal fracture strain calculated by the Gurson−Tvergaard yield function agrees reasonably well with the relationship between the initial void volume fraction and the nominal fracture strain calculated by the Gurson yield function.

6.1.6 Effect of stress triaxiality

Fig. 6.16 shows the relationship between the initial void volume fraction and the nominal fracture strain for various stress triaxialities for plane-strain tension. Fig. 6.16A shows the coordinates and the notations. Because the length of the simulation region in the y-direction is unity, the displacement in the y-direction v^* coincides with the nominal strain. σ^* denotes the imposed hydrostatic stress. The solid line denotes the material shape and the void shape before deformation, whereas the dotted line denotes the material shape and the void shape after deformation.

Fig. 6.16B shows the relationship between the initial void volume fraction and the nominal fracture strain. The relationship in the case that the imposed stress triaxiality ($\sigma^*/\overline{\sigma}$) is zero, coincides with the relationship for the ellipsoidal void accompanied by void growth in Fig. 6.1. Because plane-strain tension is assumed, the stress triaxiality ($\sigma_{kk}/3\overline{\sigma}$) is $1/\sqrt{3}$ in the case that the imposed stress triaxiality ($\sigma^*/\overline{\sigma}$) is zero. When the initial void volume fraction is specified, the nominal fracture strain decreases with increasing imposed stress triaxiality ($\sigma^*/\overline{\sigma}$). The fact that

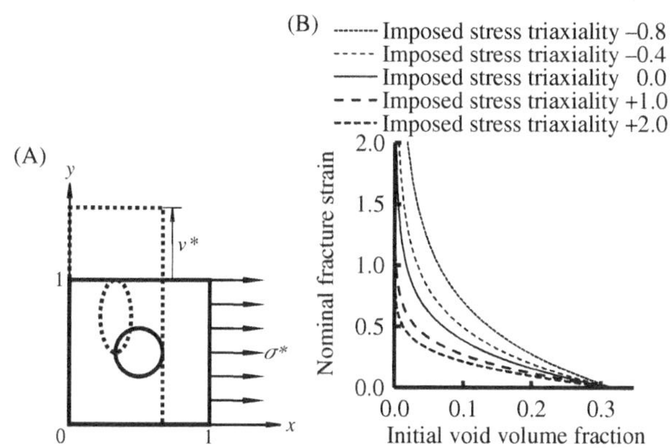

Figure 6.16 Relationship between initial void volume fraction and nominal fracture strain for various stress triaxialities for plane-strain tension: (A) coordinates and notations and (B) relationship between initial void volume fraction and nominal fracture strain.

the stress triaxiality strongly affects ductile fracture is well known by Bridgman (1952).

6.1.7 Effect of plastic deformation mode

Fig. 6.17 shows the relationship between the initial void volume fraction and the nominal fracture strain for the three plastic deformation modes: plane-strain tension, uniaxial tension, and simple shear. Void growth is assumed. The relationship between the initial void volume fraction and the nominal fracture strain for the three plastic deformation modes is already obtained in Sections 6.1.1–6.1.3. When the initial void volume fraction is specified, plane-strain tension has the smallest nominal fracture strain of the three plastic deformation modes, whereas simple shear has the largest nominal fracture strain of the three plastic deformation modes.

The stress triaxiality ($\sigma_{kk}/3\overline{\sigma}$) in plane-strain tension is equal to $1/\sqrt{3}$, the stress triaxiality in uniaxial tension is equal to 1/3, and the stress triaxiality in simple shear is equal to zero. The reason why the nominal fracture strain in plane-strain tension is the smallest of the three plastic deformation mode is conjectured that the stress triaxiality in plane-strain tension is the largest of the three plastic deformation modes. The reason why the nominal fracture strain in simple shear is the largest of the three plastic deformation modes is conjectured that the stress triaxiality in simple shear is the smallest of the three plastic deformation modes.

6.1.8 Comparison with experimental results

Fig. 6.18 shows the experimental results and the simulation results for uniaxial tension. Fig. 6.18A shows the relationship between the initial void volume fraction and the nominal fracture strain for copper. The relationship

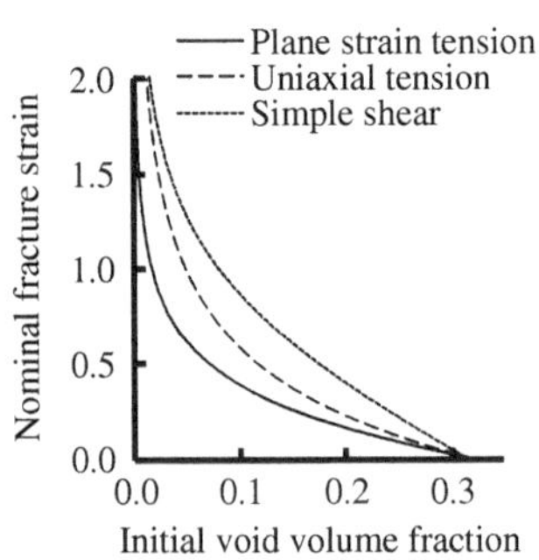

Figure 6.17 Relationship between initial void volume fraction and nominal fracture strain for three plastic deformation modes.

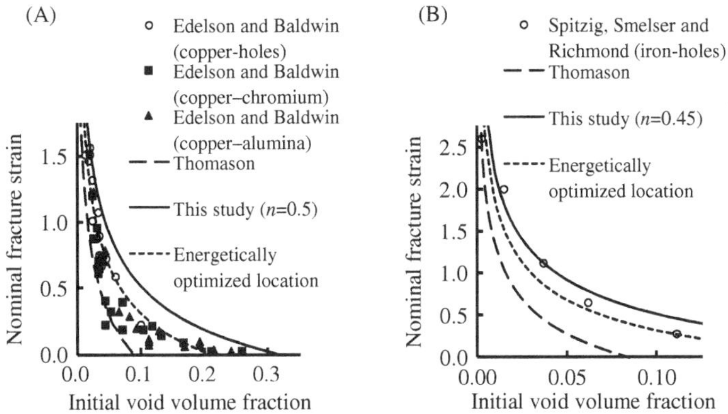

Figure 6.18 Experimental results and simulation results for uniaxial tension: (A) relationship between initial void volume fraction and nominal fracture strain for copper and (B) relationship between initial void volume fraction and nominal fracture strain for iron.

calculated using the ellipsoidal void model, in which the location of the velocity discontinuity line is determined geometrically, is compared with the following three relationships. The three relationships are the relationship calculated using the ellipsoidal void model, in which the location of the velocity discontinuity line is optimized energetically (Komori, 2017b), the relationship calculated using the Thomason model (1968) and the relationship obtained experimentally by Edelson and Baldwin (1962). Since the strain-hardening exponent of the matrix was 0.5 in the experiment, n in Eq. (6.5) is assumed to be 0.5 in the ellipsoidal void model.

The experiment is described below. The detail of experimental procedure is given in the appendix of Edelson and Baldwin (1962). Tensile specimens were prepared by powder metallurgical methods. Tensile tests were performed for two-phase copper-base alloys with various second-phase particles, including metals, nonmetals, and voids with a 20-fold range in particle size and with volume fractions ranging as high as 0.24. The logarithmic fracture strain ε_f was calculated using the formula $\varepsilon_f = \ln(S_0/S_f)$, where S_0 is the initial cross-sectional area of the specimen and S_f is the fractured cross-sectional area of the specimen. Since the strain-hardening exponent was not zero, necking deformation was expected to occur. However, the deformation was assumed to be uniform in the fractured cross section of the specimen.

When the initial void volume fraction exceeds 10%, the nominal fracture strain calculated using the Thomason model (1968) is zero.

However, even when the initial void volume fraction exceeds 20%, the nominal fracture strain obtained experimentally by Edelson and Baldwin, the nominal fracture strain calculated using the ellipsoidal void model of geometrically determined location, and the nominal fracture strain calculated using the ellipsoidal void model of energetically optimized location, are not zero. When the initial void volume fraction is specified, the nominal fracture strain calculated using the ellipsoidal void model of geometrically determined location is larger than the nominal fracture strain obtained experimentally by Edelson and Baldwin. However, when the initial void volume fraction is specified, the nominal fracture strain calculated using the ellipsoidal void model of energetically optimized location agrees with the nominal fracture strain obtained experimentally by Edelson and Baldwin. Hence, to optimize the location of the velocity discontinuity line energetically is indispensable to increase the accuracy of the simulation results in the ellipsoidal void model.

Fig. 6.18B shows the relationship between the initial void volume fraction and the nominal fracture strain for iron. The relationship calculated using the ellipsoidal void model of geometrically determined location, is compared with the relationship calculated using the ellipsoidal void model of energetically optimized location (Komori, 2017b), the relationship calculated using the Thomason model (1968) and the relationship obtained experimentally by Spitzig et al. (1988). Since the strain-hardening exponent of the matrix was 0.45 in the experiment, n in Eq. (6.5) is assumed to be 0.45 in the ellipsoidal void model.

The experiment is described below. The iron used in this study was prepared from water atomized iron powder. Compacts were prepared from the powder using varying compacting pressures to achieve different densities. After the iron bars had been molded, the iron bars were sintered in dry hydrogen. The logarithmic fracture strain ε_f was calculated using $\varepsilon_f = \ln(S_0/S_f)$, where S_0 is the initial cross-sectional area of the specimen and S_f is the fractured cross-sectional area of the specimen.

When the initial void volume fraction is specified, the nominal fracture strain calculated using the Thomason model (1968) is smaller than the nominal fracture strain obtained experimentally by Spitzig et al. However, when the initial void volume fraction is specified, the nominal fracture strain calculated using the ellipsoidal void model of geometrically determined location, and the nominal fracture strain calculated using the ellipsoidal void model of energetically optimized location agree with the nominal fracture strain obtained experimentally by Spitzig et al.

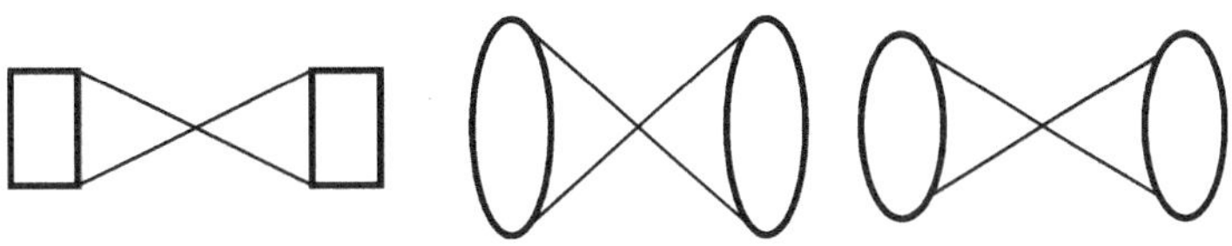

Figure 6.19 Two neighboring voids at fracture point: (A) Thomason; (B) this study ($n = 0.5$); and (C) energetically optimized location.

Fig. 6.19 shows the two neighboring voids at the fracture point when the initial void volume fraction of the material is equal to 0.05. Fig. 6.19A shows the two neighboring voids at the fracture point calculated using the Thomason model (1968), in which the void volume fraction of the material is not assumed to change and the velocity discontinuity line is assumed to pass through the vertices of the void. Fig. 6.19B shows the two neighboring voids at the fracture point calculated using the ellipsoidal void model in this study; the velocity discontinuity line is assumed to coincide with the tangent of the two neighboring voids. Fig. 6.19C shows the two neighboring voids at the fracture point calculated using the ellipsoidal void model in which the location of the velocity discontinuity line is optimized energetically. When the energy required for the internal necking deformation mode is minimized, the velocity discontinuity line does not coincide with the tangent of the two neighboring voids.

6.2 Notch tensile testing

The ductile fracture in ferrous alloys, nonferrous pure metals, and nonferrous alloys during notch tensile testing is predicted (Komori, 2016, 2017a). To obtain the simulation results precisely, the location of the velocity discontinuity line is optimized so that the energy for the internal necking deformation is minimized. Moreover, a void coalescence region is assumed. Because simulation results for various materials need to be compared, inclusions are assumed to separate from matrix and are not assumed to crack.

6.2.1 Nonferrous pure metals and alloys

Four types of nonferrous sheets and four types of nonferrous bars were used. Prior to the tensile tests, sheets were rolled to induce prestrain in the sheet specimens, whereas bars were drawn to induce prestrain in the

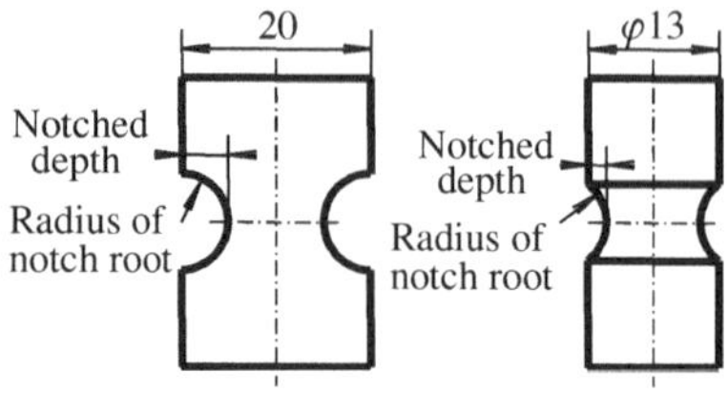

Figure 6.20 Shapes of notched sheet specimen and notched bar specimen.

bar specimens. The following nonferrous sheets were used: JIS A1050, which is a pure aluminum sheet; JIS A5052, which is an aluminum alloy sheet and is equivalent to ISO AlMg2.5; JIS C1100P, which is a tough pitch copper sheet; and JIS C2801, which is a brass sheet and is equivalent to EN CW612N. The sheets of A1050, A5052, C1100P, and C2801 were annealed and then cooled in an electric furnace prior to rolling. The following nonferrous bars were used: JIS A1070, which is a pure aluminum bar; JIS A5056, which is an aluminum alloy bar and is equivalent to ISO AlMg5Cr; JIS C1100B, which is a tough pitch copper bar; and JIS C3604, which is a free cutting brass bar and is equivalent to EN CW606L. The bars of A1070, A5056, C1100B, and C3604 were annealed and then cooled in an electric furnace prior to drawing.

The annealed sheets in thicknesses of 2 and 3 mm were rolled to obtain the prestrained sheets in thickness of 1 mm, which were used in the tensile tests. The annealed sheets in thickness of 1 mm were also used in the tensile tests. The annealed bars in diameters of 16 and 22 mm were drawn to obtain the prestrained bars in diameter of 13 mm, which were used in the tensile tests. The annealed bars in diameter of 13 mm were also used in the tensile tests.

The experiments of the tensile tests of the sheets and the bars were performed using smooth specimens and notched specimens. Fig. 6.20 shows the shapes of the notched sheet specimen and the notched bar specimen. Table 6.1 shows the dimensions of the notched sheet specimen and the notched bar specimen. Six types of notched sheet specimens with

Table 6.1 Dimensions of notched sheet specimen and notched bar specimen.

	Sheet	Bar
Notched depth (mm)	5	2
Radius of notch root (mm)	1, 2, 3, 5, 7, 9	1, 2, 3, 4, 5, 6, 7, 8, 9, 10, 13, 26, 39

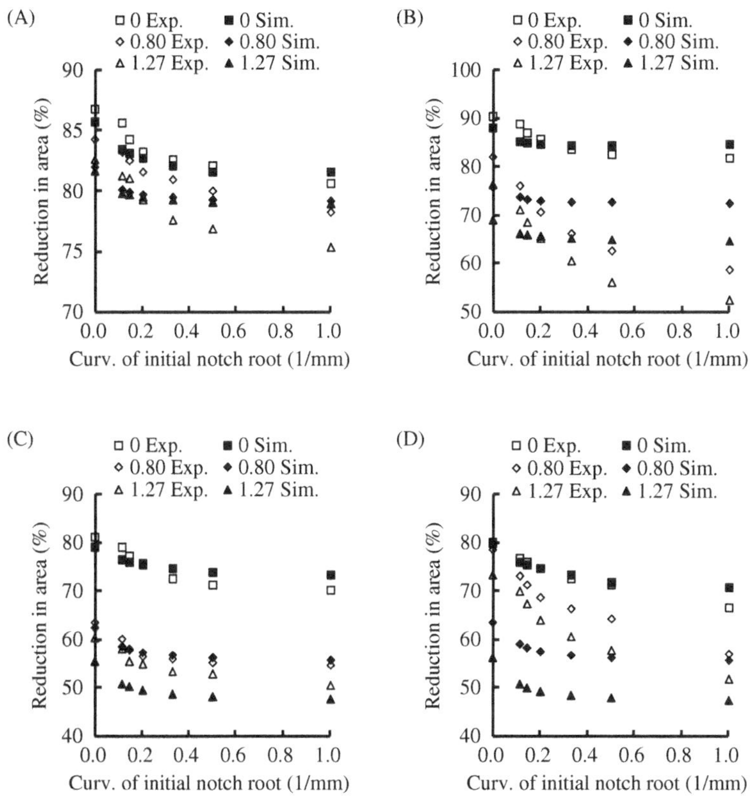

Figure 6.21 Reduction in area in the cases of sheets: (A) A1050; (B) A5052; (C) C1100P; and (D) C2801.

different notch-root radii were used, whereas thirteen types of notched bar specimens with different notch-root radii were used.

Fig. 6.21 shows the reduction in the area in the cases of sheets, and Fig. 6.22 shows the reduction in the area in the cases of bars. The curvature of the initial notch root on the horizontal axis was the reciprocal of the radius of the initial notch root. The reduction in area on the vertical axis was calculated from the area of the cross section of the fractured specimen. The experimental results are plotted using open symbols. The figures in the legend indicate the magnitude of the prestrain. With increasing the curvature of the initial notch root, the reduction in the area decreased. With increasing prestrain, the reduction in the area decreased.

The simulations of the rolling of sheets were performed using the conventional rigid-plastic finite-element method (Kobayashi et al., 1989), in which a plane-strain state was assumed. The initial void volume fraction

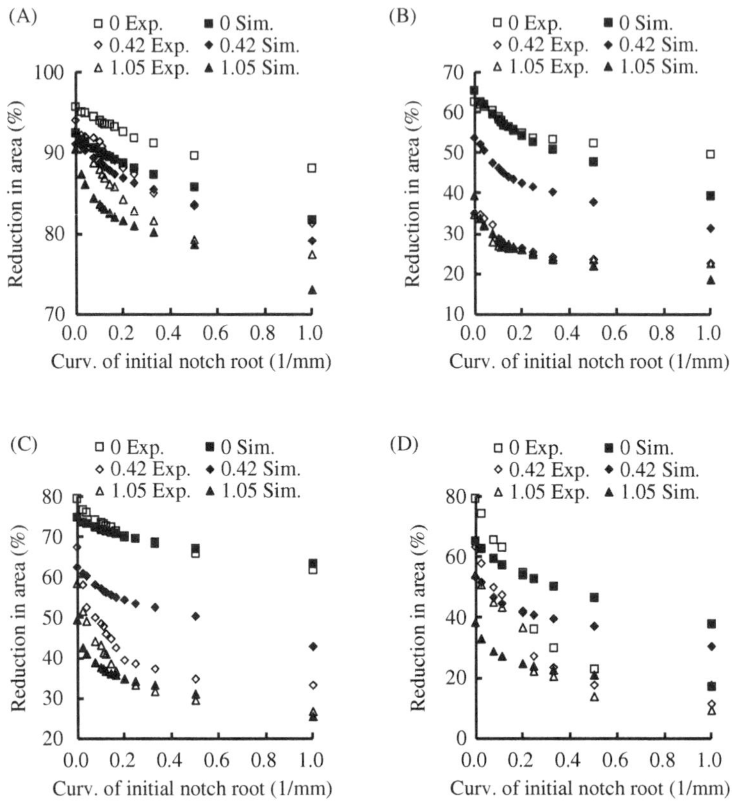

Figure 6.22 Reduction in area in the cases of bars: (A) A1070; (B) A5056; (C) C1100B; and (D) C3604.

was assumed to be 0.001 (Schmitt and Jalinier, 1982). The volume fraction of the void that existed initially was assumed to be constant during the simulation of rolling, because the volume fraction of the void that existed initially was assumed to be the minimum value. The simulations of the drawing of bars were performed using the conventional rigid-plastic finite-element method (Kobayashi et al., 1989), in which an axisymmetric state was assumed.

The simulations of the tensile tests of the sheets were performed using the conventional rigid-plastic finite-element method (Kobayashi et al., 1989), in which a plane-stress state was assumed. The simulations of the tensile tests of the bars were performed using the conventional rigid-plastic finite-element method (Kobayashi et al., 1989), in which an axisymmetric state was assumed. Although line symmetry was not observed in the experiments, line symmetry was nonetheless assumed for simplicity. The simulation was performed until the specimen ruptured (Komori, 2002).

The following evolution equations, which denote the change in the void volume fraction, were assumed.

$$\dot{f} = (1 - f)\dot{\varepsilon}_{kk} + A_0\dot{\bar{\varepsilon}} \tag{6.8}$$

$$\dot{f} = (1 - f)\dot{\varepsilon}_{kk} + A_1 R\left(\frac{\sigma_{kk}}{3\bar{\sigma}}\right)\dot{\bar{\varepsilon}} \tag{6.9}$$

Function $R(x)$ in Eq. (6.9) denotes the ramp function. The simulations were performed using either Eq. (6.8) or (6.9). The material constants A_0 and A_1 were determined such that the reduction in the area calculated from the simulation of the specimen with no prestrain and with an initial notch-root radius of 5 mm agreed with the reduction in area obtained experimentally except in the case of A1070. In the case of A1070, a minimum value A_1, at which a numerical problem did not occur, was chosen (Komori, 2016).

The selection of Eq. (6.8) or (6.9) was determined on the following concepts. The effect of prestrain on the reduction in the area calculated using Eq. (6.8) was larger than the effect of prestrain on the reduction in the area calculated using Eq. (6.9). The effect of the initial notch-root radius on the reduction in the area calculated using Eq. (6.9) was larger than the effect of the initial notch-root radius on the reduction in the area calculated using Eq. (6.8).

Table 6.2 shows the material constants used in the evolution equations. Eq. (6.8) was used in the cases of A5052, C1100P, and C1100B, whereas Eq. (6.9) was used in the cases of A1050, C2801, A1070, A5056, and C3604. Fig. 6.21 shows the reduction in the area in the cases of sheets, and Fig. 6.22 shows the reduction in the area in the cases of bars. The simulation results are plotted using solid symbols. The figures in the legend indicate the magnitude of prestrain.

With increasing the curvature of the initial notch root, the reduction in the area decreased. With increasing prestrain, the reduction in the area decreased. The reductions in the area calculated from the simulations agreed reasonably well with the reductions in the area obtained experimentally, except in the case of C2801. In the case of C2801, it is reasonable to believe that using the void configuration for a pure metal is more

Table 6.2 Material constants used in the evolution equations.

	Void configuration	A_0	A_1
A1050	Pure metal	(0.005)	0.010
A5052	Alloy	0.001	—
C1100P	Pure metal	0.02	—
	(Alloy)	(0.011)	
C2801	Alloy	—	0.026
	(Pure metal)		(0.05)
A1070	Pure metal	—	0.002
A5056	Alloy	—	0.12
C1100B	Pure metal	0.03	(0.06)
C3604	Alloy	—	0.13

appropriate than using the void configuration for an alloy (Komori, 2016).

Simulations were performed using conventional ductile fracture criteria to demonstrate the validity of the ellipsoidal void model. The Mises yield function and the Lévy–Mises constitutive equation were used. The conventional rigid-plastic finite-element method (Kobayashi et al., 1989) was also used, in which the incompressibility condition was satisfied by the penalty method. In the ellipsoidal void model, the number of material constants, A_0 or A_1, is one. Hence, a conventional ductile fracture criterion, in which the number of material constants is one, is used for comparison.

The Cockcroft and Latham fracture criterion (1968) in nondimensional form can be expressed by the following equation, in which C_2 is the material constant.

$$\int \frac{\sigma_{\max}}{\overline{\sigma}} \dot{\overline{\varepsilon}} \, \mathrm{d}t \geq C_2 \tag{6.10}$$

The Brozzo et al. fracture criterion (1972) can be expressed by the following equation, in which C_3 is the material constant.

$$\int \frac{2\sigma_{\max}}{3\left(\sigma_{\max} - \sigma_{kk}/3\right)} \dot{\overline{\varepsilon}} \, \mathrm{d}t \geq C_3 \tag{6.11}$$

Table 6.3 shows the material constants used in the conventional ductile fracture criteria.

Fig. 6.23 shows the reduction in area calculated using the conventional ductile fracture criteria in the case of C1100P. Regardless of the

Table 6.3 Material constants used in conventional ductile fracture criteria.

	C_2	C_3
C1100P	1.6	1.7
C1100B	1.4	1.4

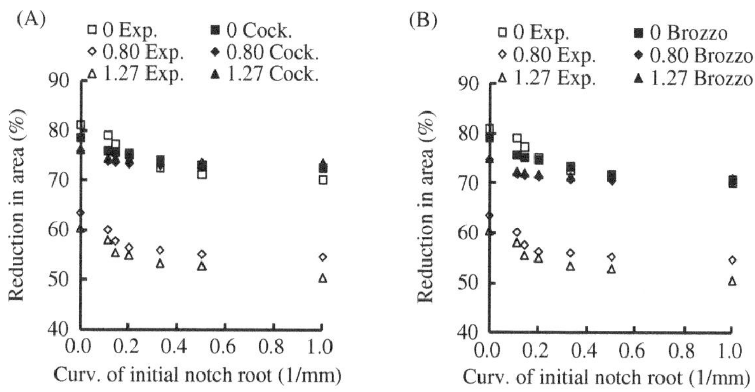

Figure 6.23 Reduction in area calculated using conventional ductile fracture criteria in the case of C1100P: (A) Cockcroft and Latham and (B) Brozzo et al.

increase in prestrain, the reductions in area calculated using the Cockcroft and Latham fracture criterion (1968) and the Brozzo et al. fracture criterion (1972) hardly changed. The reason for little change with the increase in prestrain can be explained as follows. A plane-strain rolling could be approximated for convenience as a plane-strain compression, in which the maximum principal stress $\sigma_{\max}$ is equal to zero. Hence, the left-hand sides of Eqs. (6.10) and (6.11) hardly changed during the prestraining process.

Fig. 6.24 shows the reduction in the area calculated using the conventional ductile fracture criteria in the case of C1100B. The effect of prestrain on the reductions in the area calculated using the Cockcroft and Latham fracture criterion (1968) and the Brozzo et al. fracture criterion (1972) was slightly smaller than the effect of prestrain on the reduction in the area obtained experimentally.

Hence, a comparison of Fig. 6.23 with Fig. 6.21C and a comparison of Fig. 6.24 with Fig. 6.22C revealed the validity of the ellipsoidal void model.

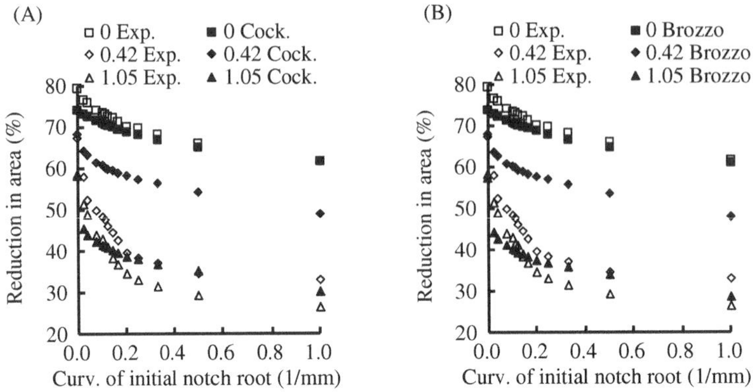

Figure 6.24 Reduction in area calculated using conventional ductile fracture criteria in the case of C1100B: (A) Cockcroft and Latham and (B) Brozzo et al.

6.2.2 Ferrous alloys

Prior to the tensile tests, sheets were rolled to induce prestrain in the sheet specimens, whereas bars were drawn to induce prestrain in the bar specimens. The following four types of as-rolled ferrous sheets were used: JIS SPCC, which is a cold-reduced carbon steel sheet and is equivalent to ISO CR1; JIS SPHC, which is a hot-rolled mild steel sheet and is equivalent to ISO HR1; JIS SUS304, which is a cold-rolled austenitic stainless steel sheet and is equivalent to ISO X5CrNi18-10; and JIS SUS430, which is a cold-rolled ferritic stainless steel sheet and is equivalent to ISO X6Cr17. The following four types of ferrous bars were used: JIS SS400, which is a rolled steel for general structures; and JIS S15C, JIS S45C, and JIS S55C, which are carbon steels for machine structural use and are equivalent to ISO C15E4, ISO C45E4, and ISO C55E4, respectively. Although the as-rolled bars of SS400 and S55C were not heat-treated prior to drawing, the cold finished bars of S15C and S45C were annealed and then cooled in a vacuum chamber prior to drawing.

The as-rolled sheets of SPCC, SUS304, and SUS430 in thicknesses of 1.5, 2, and 3 mm were rolled to obtain the prestrained sheets of SPCC, SUS304, and SUS430 in thickness of 1 mm, which were used in the tensile tests. The as-rolled sheets of SPCC, SUS304, and SUS430 in thickness of 1 mm were also used in the tensile tests. The as-rolled sheets of SPHC in thicknesses of 2.3 and 3.2 mm were rolled to obtain the prestrained sheets of SPHC in thickness of 1.6 mm, which were used in the tensile tests. The as-rolled sheets of SPHC in thickness of 1.6 mm were also used in the tensile tests. The as-rolled or annealed bars in diameters of 16 and

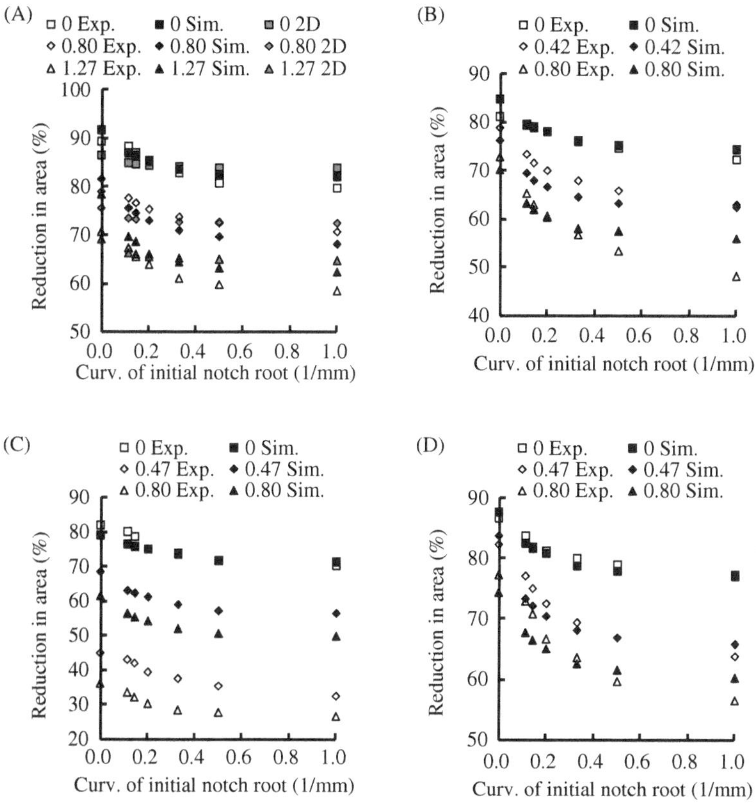

Figure 6.25 Reduction in area in the cases of sheets: (A) SPCC; (B) SPHC; (C) SUS304; and (D) SUS430.

22 mm were drawn to obtain the prestrained bars in diameter of 13 mm, which were used in the tensile tests. The as-rolled or annealed bars in diameter of 13 mm were also used in the tensile tests.

The experiments of the tensile tests of the sheets and the bars were performed using smooth specimens and notched specimens. Fig. 6.20 shows the shapes of the notched sheet specimen and the notched bar specimen. Table 6.1 shows the dimensions of the notched sheet specimen and the notched bar specimen.

Fig. 6.25 shows the reduction in the area in the cases of sheets, and Fig. 6.26 shows the reduction in the area in the cases of bars. The experimental results are plotted using open symbols. The figures in the legend indicate the magnitude of the prestrain. With increasing the curvature of the initial notch root, the reduction in the area decreased. With increasing prestrain, the reduction in the area decreased.

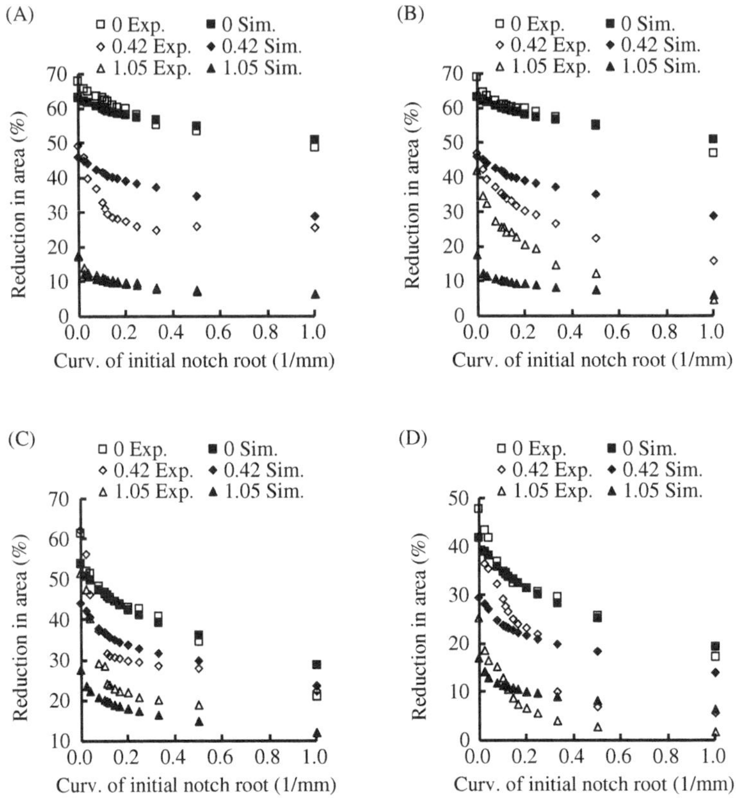

Figure 6.26 Reduction in area in the cases of bars: (A) SS400; (B) S15C; (C) S45C; and (D) S55C.

The simulations of the rolling of sheets were performed using the conventional rigid-plastic finite-element method (Kobayashi et al., 1989), in which a plane-strain state was assumed. The initial void volume fraction was assumed to be 0.001 (Schmitt and Jalinier, 1982). The volume fraction of the void that existed initially was not assumed to be constant during the simulation of rolling. The simulations of the drawing of bars were performed using the conventional rigid-plastic finite-element method (Kobayashi et al., 1989), in which an axisymmetric state was assumed.

The simulations of the tensile tests of the sheets were performed using the conventional rigid-plastic finite-element method (Kobayashi et al., 1989), in which three-dimensional deformation was assumed; the linear element of rectangular prisms was used and the number of finite elements in the thickness direction was one. The simulations of the tensile tests of the bars were performed using the conventional rigid-plastic finite-element method (Kobayashi et al., 1989), in which an axisymmetric state was assumed. Line symmetry was assumed for simplicity. The simulation was performed until the specimen ruptured (Komori, 2002).

In addition to Eqs. (6.8) and (6.9), the following evolution equation, in which Eqs. (6.8) and (6.9) are combined, was assumed.

$$\dot{f} = (1-f)\dot{\varepsilon}_{kk} + A_0\dot{\bar{\varepsilon}} + A_1 R\left(\frac{\sigma_{kk}}{3\bar{\sigma}}\right)\dot{\bar{\varepsilon}} \tag{6.12}$$

For simplicity, $A_0 = A_1$ was assumed in Eq. (6.12). Hence, the number of the material constants in Eq. (6.12) is one. The simulations were performed using either Eq. (6.8), (6.9), or (6.12). The material constants A_0 and A_1 were determined such that the reduction in the area calculated from the simulation of the specimen with no prestrain and with an initial notch-root radius of 5 mm agreed with the reduction in the area obtained experimentally.

The selection of either Eq. (6.8), (6.9), or (6.12) was determined on the following concepts. The effect of the prestrain on the reduction in the area calculated using Eq. (6.8) is the largest among the effects calculated using Eqs. (6.8), (6.9), and (6.12), whereas the effect of the prestrain on the reduction in the area calculated using Eq. (6.9) is the smallest. The effect of the initial notch-root radius on the reduction in the area calculated using Eq. (6.9) is the largest among the effects calculated using Eqs. (6.8), (6.9), and (6.12), whereas the effect of the initial notch-root radius on the reduction in the area calculated using Eq. (6.8) is the smallest.

Table 6.4 shows the material constants used in the evolution equations. Eq. (6.8) was used in the cases of SPCC, SPHC, SUS304, SUS430,

Table 6.4 Material constants used in the evolution equations.

	A_0	A_1	A_0, A_1
SPCC	0.007	–	–
SPHC	0.014	(0.027)	–
SUS304	0.019	–	–
SUS430	0.011	–	–
SS400	0.05	–	–
S15C	–	(0.08)	0.03
S45C	(0.12)	0.2	–
S55C	–	0.4	–

and SS400, Eq. (6.9) was used in the cases of S45C, and S55C, and Eq. (6.12) was used in the case of S15C.

Fig. 6.25 shows the reduction in the area in the cases of sheets, and Fig. 6.26 shows the reduction in the area in the cases of bars. The simulation results are plotted using solid symbols. The figures in the legend indicate the magnitude of prestrain. With increasing the curvature of the initial notch root, the reduction in the area decreased. With increasing prestrain, the reduction in the area decreased. The reductions in the area calculated from the simulations agreed reasonably well with the reductions in the area obtained experimentally, except in the case of SUS304. In the case of SUS304, even though Eq. (6.8) was used, the effect of prestrain on the reduction in the area calculated from the simulation was smaller than the effect of prestrain on the reduction in area obtained experimentally, because the void shape and the void configuration change during rolling and tensile tests because of the deformation-induced martensitic transformation (Angel, 1954).

In Section 6.2.1, the plane-stress state was assumed and the linear element of rectangles was used, whereas in this section, three-dimensional deformation was assumed and the linear element of rectangular prisms was used. The reduction in the area calculated using the two-dimensional finite-element method, which was used as a simulation method in Section 6.2.1, was supplemented in Fig. 6.25A. According to Fig. 6.25A, the assumption of the plane-stress state was proven to be inappropriate for the simulation of the tensile tests of the sheets.

The carbon contents of SPCC, SPHC, SS400, S15C, S45C, and S55C are 0.04%, 0.05%, 0.13%, 0.15%, 0.45%, and 0.54%, respectively. In the cases of SPCC, SPHC, and SS400, Eq. (6.8) is considered to be appropriate, whereas in the case of S15C, Eq. (6.12), that is, the combination of Eqs. (6.8) and (6.9), is considered to be more appropriate than Eq. (6.8). In contrast, in the cases of S45C and S55C, Eq. (6.9) is considered to be more appropriate than Eq. (6.12). Accordingly, Eq. (6.8) becomes inappropriate and Eq. (6.9) becomes more suitable with increasing carbon content. Eq. (6.8) indicates that void nucleation depends on equivalent plastic strain, whereas Eq. (6.9) indicates that void nucleation depends on stress triaxiality and equivalent plastic strain. Therefore with increasing carbon content of the material, void nucleation depends less on equivalent plastic strain and more on stress triaxiality. Furthermore, according to Table 6.4, the magnitudes of the material constants, A_0 and A_1, increase

Table 6.5 Material constants used in conventional ductile fracture criteria.

	C_2	C_3
SPCC	2.0	2.1
SS400	1.1	1.1

with the increase in carbon content, whereas the reduction in area decreases as the carbon content increases.

Simulations were performed using conventional ductile fracture criteria to demonstrate the validity of the ellipsoidal void model. The Mises yield function and the Lévy–Mises constitutive equation were used. The conventional rigid-plastic finite-element method (Kobayashi et al., 1989) was also used, in which the incompressibility condition was satisfied by the penalty method. In the ellipsoidal void model, the number of material constants, A_0 or A_1, is one. Hence, the Cockcroft and Latham fracture criterion (1968) and the Brozzo et al. fracture criterion (1972), in which the number of material constants is one, are used for comparison.

Table 6.5 shows the material constants used in the conventional ductile fracture criteria.

Fig. 6.27 shows the reduction in the area calculated using the conventional ductile fracture criteria in the case of SPCC. Regardless of the increase in prestrain, the reductions in area calculated using the Cockcroft and Latham fracture criterion (1968) and the Brozzo et al. fracture criterion (1972) hardly changed. The reason for little change with the increase in prestrain could be explained by the approximation that the maximum

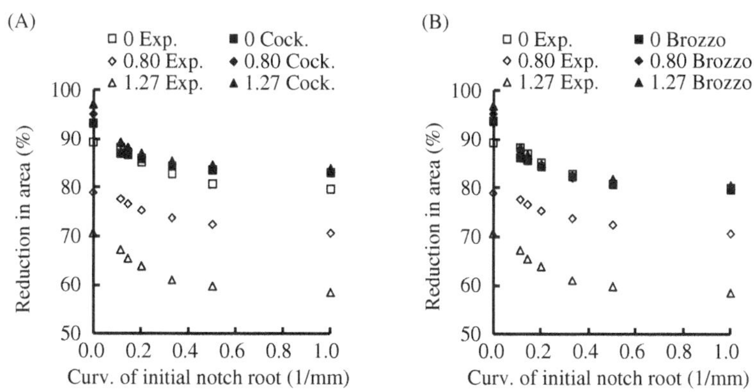

Figure 6.27 Reduction in area calculated using conventional ductile fracture criteria in the case of SPCC: (A) Cockcroft and Latham and (B) Brozzo et al.

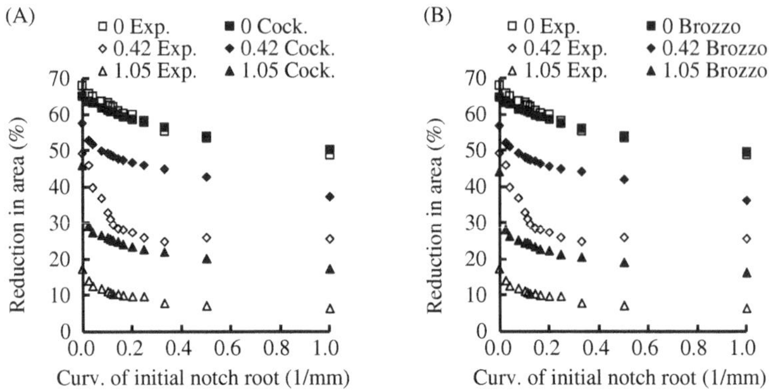

Figure 6.28 Reduction in area calculated using conventional ductile fracture criteria in the case of SS400: (A) Cockcroft and Latham and (B) Brozzo et al.

principal stress σ_{max} is equal to zero in a plane-strain rolling. Hence, the left-hand sides of Eqs. (6.10) and (6.11) hardly changed during the prestraining process.

Fig. 6.28 shows the reduction in the area calculated using the conventional ductile fracture criteria in the case of SS400. The effect of prestrain on the reductions in the area calculated using the Cockcroft and Latham fracture criterion (1968) and the Brozzo et al. fracture criterion (1972) was slightly smaller than the effect of prestrain on the reduction in area obtained experimentally.

Hence, a comparison of Fig. 6.27 with Fig. 6.25A and a comparison of Fig. 6.28 with Fig. 6.26A revealed the validity of the ellipsoidal void model.

6.3 Hole expansion test

The ductile fracture in ferrous alloys, nonferrous pure metals, and nonferrous alloys during hole expansion test is predicted (Komori, 2013b, 2014a). To obtain the simulation results immediately, the location of the velocity discontinuity line is assumed to coincide with the tangent of the two neighboring voids. Moreover, a void coalescence region is not assumed. Because simulation results for various materials need to be compared, inclusions are assumed to separate from matrix and are not assumed to crack.

6.3.1 Ferrous alloys

Prior to the hole expansion tests, sheets were rolled to induce prestrain in the sheet specimens. The following four types of as-rolled ferrous sheets were used: JIS SPCC, which is a cold-reduced carbon steel sheet and is equivalent to ISO CR1; JIS SPHC, which is a hot-rolled mild steel sheet and is equivalent to ISO HR1; JIS SUS304, which is a cold-rolled austenitic stainless steel sheet and is equivalent to ISO X5CrNi18-10; and JIS SUS430, which is a cold-rolled ferritic stainless steel sheet and is equivalent to ISO X6Cr17.

The as-rolled sheets of SPCC, SUS304, and SUS430 in thicknesses of 1.5, 2, and 3 mm were rolled to obtain the prestrained sheets of SPCC, SUS304, and SUS430 in thickness of 1 mm, which were used in the hole expansion tests. The as-rolled sheets of SPCC, SUS304, and SUS430 in thickness of 1 mm were also used in the hole expansion tests. The as-rolled sheets of SPHC in thicknesses of 2.3 and 3.2 mm were rolled to obtain the prestrained sheets of SPHC in thickness of 1.6 mm, which were used in the hole expansion tests. The as-rolled sheets of SPHC in thickness of 1.6 mm were also used in the hole expansion tests. The as-rolled sheets were rolled in the experiment in the following two directions: the direction parallel to the rolling direction of the as-rolled sheets and the direction perpendicular to the rolling direction of the as-rolled sheets.

Fig. 6.29 shows the specimen and the die of the hole expansion test. The outer diameter of the specimen was 85 mm, whereas the inner diameter of the specimen was 9 mm. The inner die was conical, with a corner angle of 45 degrees. The experiment of the hole expansion test was performed by moving the inner die upward until the surface of the hole edge of the specimen fractured. The hole expansion ratio is defined as $d/d_0 - 1$, where d is the fractured hole diameter, and d_0 is the pierced hole diameter. The normal direction of the fracture surface was parallel to the rolling direction in the experiment.

Fig. 6.30 shows the relationship between the prestrain and the hole expansion ratio. The experimental results are plotted by circles. "Parallel"

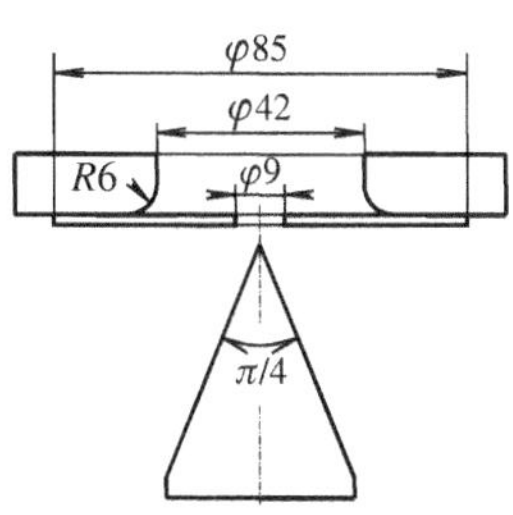

Figure 6.29 Specimen and die of hole expansion test.

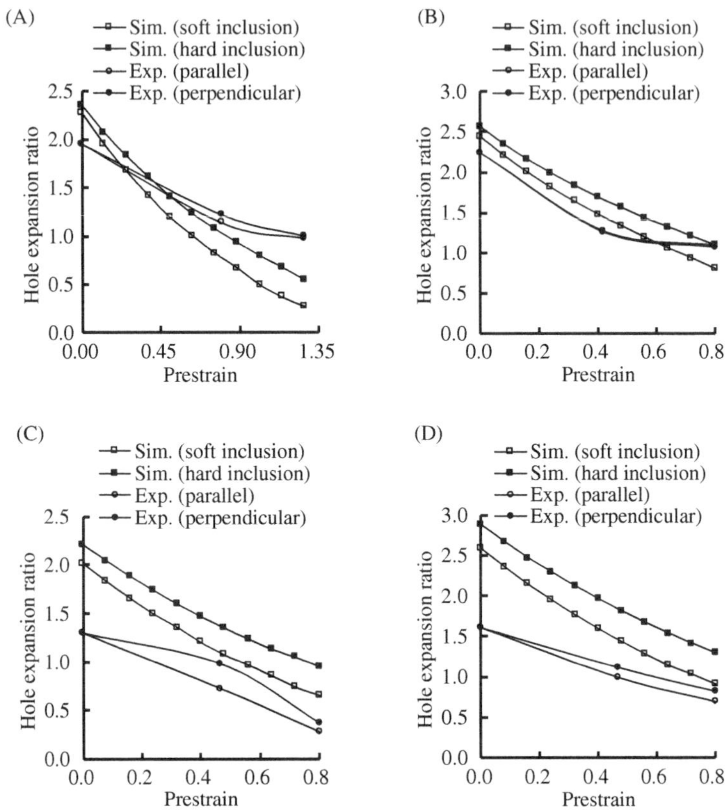

Figure 6.30 Relationship between prestrain and hole expansion ratio: (A) SPCC; (B) SPHC; (C) SUS304; and (D) SUS430.

in the legend indicates that the sheet was rolled in the direction parallel to the rolling direction of the as-rolled sheet, whereas "perpendicular" in the legend indicates that the sheet was rolled in the direction perpendicular to the rolling direction of the as-rolled sheet. With increasing prestrain, the hole expansion ratio decreased. In the case of SPCC, the hole expansion ratio of the specimen rolled in the direction parallel to the rolling direction of the as-rolled sheet was slightly smaller than the hole expansion ratio of the specimen rolled in the direction perpendicular to the rolling direction of the as-rolled sheet. This is because the cold-rolled steel sheet had a slight prestrain. In the case of SPHC, the hole expansion ratio of the specimen rolled in the direction parallel to the rolling direction of the as-rolled sheet hardly differed from the hole expansion ratio of the specimen rolled in the direction perpendicular to the rolling direction of the

Table 6.6 Material constants used in the evolution equations.

	A_1 soft inclusion	A_1 hard inclusion
SPCC	0.025	0.044
SPHC	0.020	0.037
SUS304	0.035	0.050
SUS430	0.017	0.029

as-rolled sheet. This is because the hot-rolled steel sheet hardly had a prestrain. In the cases of SUS304 and SUS430, the hole expansion ratio of the specimen rolled in the direction parallel to the rolling direction of the as-rolled sheet was smaller than the hole expansion ratio of the specimen rolled in the direction perpendicular to the rolling direction of the as-rolled sheet. This is because the cold-rolled stainless steel sheet had a prestrain.

The simulations of the tensile tests of the sheets were performed using the conventional rigid-plastic finite-element method (Kobayashi et al., 1989), in which a plane-stress state was assumed. The initial void volume fraction was assumed to be 0.001 (Schmitt and Jalinier, 1982). Although line symmetry was not observed in the experiments, line symmetry was nonetheless assumed for simplicity. The simulation was performed until the specimen ruptured (Komori, 2002). Eq. (6.9), which denotes the change in the void volume fraction, was assumed.

Table 6.6 shows the material constants used in the evolution equations. The material constant A_1 was determined such that the thickness of the material at the fracture surface calculated from the simulation of the specimen with no prestrain agreed with the thickness of the material at the fracture surface obtained in the experiment of the specimen of the as-rolled sheet.

Although prestrain was introduced by rolling in the experiment, prestrain was introduced by simple plane-strain compression in the simulation. It was assumed that the as-rolled sheet has no prestrain, and that no void nucleation and no void growth occur during prestraining. Since the normal direction of the fracture surface of the specimen is assumed to be the direction of the maximum principal strain in the plane of the sheet, the following deformation gradient of the material was assumed:

$$\frac{\partial z}{\partial Z} = \frac{t}{t_0}, \frac{\partial \theta}{\partial \Theta} = \frac{t_0}{t} \tag{6.13}$$

where t_0 denotes the thickness of the as-rolled sheet, t denotes the thickness in the hole expansion test, z-direction indicates the axial direction, and θ-direction indicates the circumferential direction. Other components of the deformation gradient of the material are assumed to be zero.

The simulations of the hole expansion tests of the sheets were performed using the conventional rigid-plastic finite-element method (Kobayashi et al., 1989), in which an axisymmetric state was assumed. The initial void volume fraction was assumed to be 0.001 (Schmitt and Jalinier, 1982). The simulation was performed until the surface of the hole edge of the specimen fractured.

Fig. 6.30 shows the relationship between the prestrain and the hole expansion ratio. The as-rolled sheet was assumed to have no prestrain. The simulation results are plotted by squares. With increasing prestrain, the hole expansion ratio decreased. The hole expansion ratio assuming a soft inclusion was smaller than the hole expansion ratio assuming a hard inclusion. In the cases of SPCC and SPHC, the hole expansion ratios calculated from the simulations agreed reasonably well with the hole expansion ratios obtained experimentally. However, in the cases of SUS304 and SUS430, the hole expansion ratios calculated from the simulations fairly agreed with the hole expansion ratios obtained experimentally. This is because the as-rolled sheet was assumed to have no prestrain, although the cold-rolled stainless steel sheet had a prestrain. Hence, the as-rolled sheet was assumed to have a prestrain in the following (Komori, 2014a).

Fig. 6.31 shows the relationship between the prestrain and the hole expansion ratio for various prestrains in the as-rolled sheet. Soft inclusions were assumed. t_1 denotes the thickness of the sheet with no prestrain. Attention is paid to the difference between the hole expansion ratio of the specimen rolled in the direction parallel to the rolling direction of the as-rolled sheet and the hole expansion ratio of the specimen rolled in the direction perpendicular to the rolling direction of the as-rolled sheet.

In the case of SUS304, in the experiment, the difference between the two hole expansion ratios at the prestrain of 0.47 was larger than the difference between the two hole expansion ratios at the prestrain of 0.80. When t_1/t_0 is equal to 1.5, the difference between the two hole expansion ratios at the prestrain of 0.47 was larger than the difference between the two hole expansion ratios at the prestrain of 0.80. When t_1/t_0 is equal to 2, the difference between the two hole expansion ratios at the prestrain of 0.47 was smaller than the difference between the two hole expansion ratios at the prestrain of 0.80. Hence, the prestrain of the as-rolled sheet

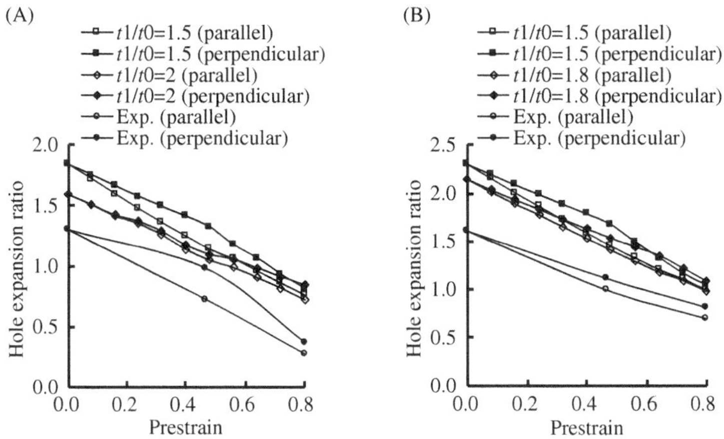

Figure 6.31 Relationship between prestrain and hole expansion ratio for various prestrains in as-rolled sheet: (A) SUS304 and (B) SUS430.

of SUS304 is conjectured to be approximately 0.5 because $(2/\sqrt{3})\ \ln\ (t_1/t_0)$ is equal to 0.47 when t_1/t_0 is equal to 1.5.

In the case of SUS430, in the experiment, the difference between the two hole expansion ratios at the prestrain of 0.47 was almost the same as the difference between the two hole expansion ratios at the prestrain of 0.80. When t_1/t_0 is equal to 1.5, the difference between the two hole expansion ratios at the prestrain of 0.47 was larger than the difference between the two hole expansion ratios at the prestrain of 0.80. When t_1/t_0 is equal to 1.8, the difference between the two hole expansion ratios at the prestrain of 0.47 was almost the same as the difference between the two hole expansion ratios at the prestrain of 0.80. Hence, the prestrain of the as-rolled sheet of SUS430 is conjectured to be approximately 0.7 because $(2/\sqrt{3})\ \ln\ (t_1/t_0)$ is equal to 0.68 when t_1/t_0 is equal to 1.8.

According to the comparison of Fig. 6.30C with Fig. 6.31A and the comparison of Fig. 6.30D with Fig. 6.31B, by introducing the assumption that the as-rolled sheet has a prestrain, the relationship between the prestrain and the hole expansion ratio calculated from the simulation approached to the relationship between the prestrain and the hole expansion ratio obtained in the experiment.

Simulations were performed using conventional ductile fracture criteria to demonstrate the validity of the ellipsoidal void model. The Mises yield function and the Lévy–Mises constitutive equation were used. The conventional rigid-plastic finite-element method (Kobayashi et al., 1989) was also used, in which the incompressibility condition was satisfied by the penalty method. In the ellipsoidal void model, the number of material constants, A_1, is one. Hence, the fracture criteria, in which the number of material constants is one, were used for comparison.

The Freudenthal fracture criterion (1950) in nondimensional form can be expressed by the following equation, in which C_1 is the material constant.

$$\int \dot{\bar{\varepsilon}}\,\mathrm{d}t \geq C_1 \tag{6.14}$$

The Oyane fracture criterion (1972) can be expressed by the following equation, in which B_4 and C_4 are the material constants.

$$\int \left(\frac{\sigma_{kk}}{3\bar{\sigma}} + B_4 \right) \dot{\bar{\varepsilon}}\,\mathrm{d}t \geq C_4 \tag{6.15}$$

Because the number of material constants in Eq. (6.15) is two, the value of the material constant B_4 is determined in the following to make the number of material constants in Eq. (6.15) one. During prestraining of plane-strain compression, stress triaxiality $\sigma_{kk}/3\bar{\sigma}$ is equal to $-1/\sqrt{3}$. Since B_4 is generally smaller than $1/\sqrt{3}$ (Oyane et al., 1980), this causes the left-hand side of Eq. (6.15) to become negative during prestraining. Hence, this causes prestraining to decrease the occurrence of fracture, whereas in the experiment, prestraining increases the occurrence of fracture. Therefore for simplicity, B_4 is assumed to be $1/\sqrt{3}$ so that the left-hand side of Eq. (6.15) becomes zero during prestraining. Hence, this causes the occurrence of fracture not to depend on prestraining.

The Cockcroft and Latham fracture criterion (1968) and the Brozzo et al. fracture criterion (1972) were also used for comparison.

Table 6.7 shows the material constants used in the conventional ductile fracture criteria.

Fig. 6.32 shows the relationship between the prestrain and the hole expansion ratio for different conventional ductile fracture criteria. When the prestrain is equal to zero, the hole expansion ratios calculated using

Table 6.7 Material constants used in conventional ductile fracture criteria.

	C_1	C_2	C_3	C_4
SPCC	1.04	1.11	1.16	1.04
SPHC	1.06	1.15	1.21	1.08

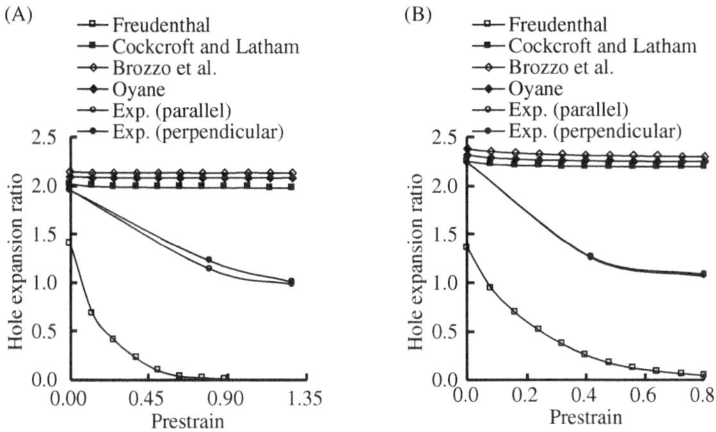

Figure 6.32 Relationship between prestrain and hole expansion ratio for different conventional ductile fracture criteria: (A) SPCC and (B) SPHC.

the Cockcroft and Latham fracture criterion (1968), the Brozzo et al. fracture criterion (1972), and the Oyane fracture criterion (1972) agreed with the hole expansion ratio obtained in the experiment. However, with increasing prestrain, the hole expansion ratio obtained in the experiment decreases, whereas the hole expansion ratios calculated using the Cockcroft and Latham fracture criterion, the Brozzo et al. fracture criterion, and the Oyane fracture criterion remained almost constant. The reason for remaining constant can be explained as follows. During prestraining of plane-strain compression, $\sigma_{\max}$ is equal to zero and stress triaxiality $\sigma_{kk}/3\overline{\sigma}$ is equal to $-1/\sqrt{3}$. Hence, the left-hand sides of Eqs. (6.10), (6.11), and (6.15) are zero during prestraining.

With increasing prestrain, both the hole expansion ratio obtained in the experiment and the hole expansion ratio calculated using the Freudenthal fracture criterion (1950) decreased. However, the hole expansion ratio calculated using the Freudenthal fracture criterion differed from the hole expansion ratio obtained in the experiment.

Hence, the comparison of Fig. 6.32A with Fig. 6.30A and the comparison of Fig. 6.32B with Fig. 6.30B revealed the validity of the ellipsoidal void model.

6.3.2 Nonferrous pure metals and alloys

Prior to the hole expansion tests, sheets were rolled to induce prestrain in the sheet specimens. The following four types of nonferrous pure metals and alloys were used: JIS A1050, which is a pure aluminum sheet; JIS A5052, which is an aluminum–magnesium alloy sheet; JIS C1100, which is a pure copper sheet; and JIS C2801, which is a copper–zinc alloy (brass) sheet. The sheets of A1050, A5052, C1100, and C2801 were annealed and then cooled in an electric furnace prior to hole expansion test.

The annealed sheets of A1050, A5052, C1100, and C2801 in thicknesses of 2, and 3 mm were rolled to obtain the prestrained sheets of A1050, A5052, C1100, and C2801 in thickness of 1 mm, which were used in the hole expansion tests. The annealed sheets of A1050, A5052, C1100, and C2801 in thickness of 1 mm were also used in the hole expansion tests.

Fig. 6.29 shows the specimen and the die of the hole expansion test. The experiment of the hole expansion test was performed by moving the inner die upward until the surface of the hole edge of the specimen fractured. The hole expansion ratio is defined as $d/d_0 - 1$, where d is the fractured hole diameter, and d_0 is the pierced hole diameter. The normal direction of the fracture surface was parallel to the rolling direction in the experiment.

Fig. 6.33 shows the relationship between the prestrain and the hole expansion ratio. The experimental results are plotted by circles. Because all the sheets were annealed, the terms "parallel" and "perpendicular" defined in Section 6.3.1 were not required. With increasing prestrain, the hole expansion ratio decreased slightly in the cases of A1050 and C1100, whereas the hole expansion ratio decreased considerably in the cases of A5052 and C2801.

The simulations of the tensile tests of the sheets were performed using the conventional rigid-plastic finite-element method (Kobayashi et al., 1989), in which a plane-stress state was assumed. The initial void volume fraction was assumed to be 0.001 (Schmitt and Jalinier, 1982). Line symmetry was assumed for simplicity. The simulation was performed until the specimen ruptured (Komori, 2002). Eq. (6.9), which denotes the change in the void volume fraction, was assumed.

Table 6.8 shows the material constants used in the evolution equations. The material constant A_1 was determined such that the thickness of the material at the fracture surface calculated from the simulation of the specimen with no prestrain agreed with the thickness of the material at the fracture surface obtained in the experiment of the specimen of the annealed sheet. Although A1050 and C1100 are not alloys but pure

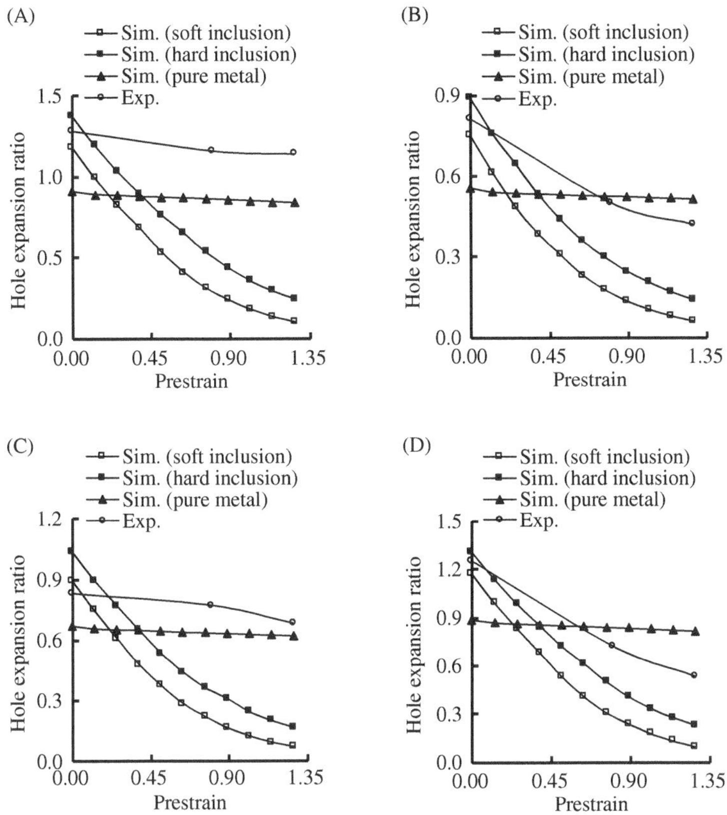

Figure 6.33 Relationship between prestrain and hole expansion ratio: (A) A1050; (B) A5052; (C) C1100; and (D) C2801.

Table 6.8 Material constants used in the evolution equations.

	A_1 soft inclusion	A_1 hard inclusion	A_1 pure metal
A1050	0.12	0.12	0.30
A5052	0.27	0.24	0.55
C1100	0.20	0.19	0.44
C2801	0.12	0.13	0.31

metals, A1050 and C1100 were assumed to be pure metals and also alloys for comparison. Although A5052 and C2801 are not pure metals but alloys, A5052 and C2801 were assumed to be not only alloys but also pure metals for comparison.

Prestrain was introduced by simple plane-strain compression in the simulation for simplicity. It was assumed that the annealed sheet has no prestrain, and that no void nucleation and no void growth occur during prestraining. Since the normal direction of the fracture surface of the specimen is assumed to be the direction of the maximum principal strain in the plane of the sheet, Eq. (6.13) was assumed, where t_0 denotes the thickness of the annealed sheet.

The simulations of the hole expansion tests of the sheets were performed using the conventional rigid-plastic finite-element method (Kobayashi et al., 1989), in which an axisymmetric state was assumed. The initial void volume fraction was assumed to be 0.001 (Schmitt and Jalinier, 1982). The simulation was performed until the surface of the hole edge of the specimen fractured.

Fig. 6.33 shows the relationship between the prestrain and the hole expansion ratio. The annealed sheet was assumed to have no prestrain. The simulation results are plotted by squares for alloys and by triangles for pure metals. With increasing prestrain, the hole expansion ratio assuming a pure metal decreased slightly, whereas the hole expansion ratio assuming an alloy decreased considerably. The hole expansion ratio assuming a soft inclusion was smaller than the hole expansion ratio assuming a hard inclusion.

In the cases of A1050 and C1100, the hole expansion ratios assuming pure metals calculated from the simulations agreed reasonably well with the hole expansion ratios obtained experimentally, whereas the hole expansion ratios assuming alloys calculated from the simulations did not agree with the hole expansion ratios obtained experimentally. In the cases of A5052 and C2801, the hole expansion ratios assuming alloys calculated from the simulations fairly agreed with the hole expansion ratios obtained experimentally, whereas the hole expansion ratios assuming pure metals calculated from the simulations did not agree with the hole expansion ratios obtained experimentally.

6.4 Central burst in wire drawing

The central burst in the multipass drawing of ferrite–pearlite steels is predicted (Komori, 2014b). In ferrite–pearlite steels, the nucleation of voids are caused by the fracture of pearlite nodules, and the growth of voids and the coalescence of voids are caused by the propagation of cracks (Miller and Smith, 1970; Rosenfield et al., 1972; Inoue and Kinoshita, 1977a,b).

Hence, the void shape and the void configuration due to the cracking of the inclusions are principally used, whereas the void shape and the void configuration due to the separation of the inclusions from the matrix are used for comparison. To obtain the simulation results precisely, the location of the velocity discontinuity line is optimized so that the energy for the internal necking deformation is minimized. Moreover, a void coalescence region is assumed.

The following four types of carbon steels for machine structural use were used: JIS S15C, JIS S35C, JIS S45C, and JIS S55C, which are equivalent to ISO C15E4, ISO C35E4, ISO C45E4, and ISO C55E4, respectively. The cold finished bars of S15C, S35C, and S45C were annealed and then cooled in a vacuum chamber prior to drawing, whereas the as-rolled bar of S55C was not heat-treated prior to drawing.

The experiment on the multipass drawing of a bar was performed prior to the simulation. The initial diameter of a bar was 13 mm. Fig. 6.34 shows the coordinates and the notations used in drawing. r_b denotes the radius of the specimen before drawing, r_f denotes the radius of the specimen after drawing, and α denotes the die semiangle. The experiment was performed for a specified reduction in the area in each die $(r_b^2 - r_f^2)/r_b^2$ and for a specified die angle 2α. The specified reductions in area in each die $(r_b^2 - r_f^2)/r_b^2$ were 5%, 10%, 15%, 20%, and 25%, whereas the specified die angle 2α were 15 degrees, 22.5 degrees, and 30 degrees. The specified die angle of 22.5 degrees implies the alternate use of the die whose angle is 15 degrees and the die whose angle is 30 degrees.

The experiment was continued until the material fractured. However, when the diameter of the specimen was less than 5 mm, the experiment was dangerous. Hence, the experiment, in which the diameter of the specimen was less than 5 mm, was not performed.

Table 6.9 shows the inner diameter of the die at which the material fractured. In each material, when the die angle is specified, the inner

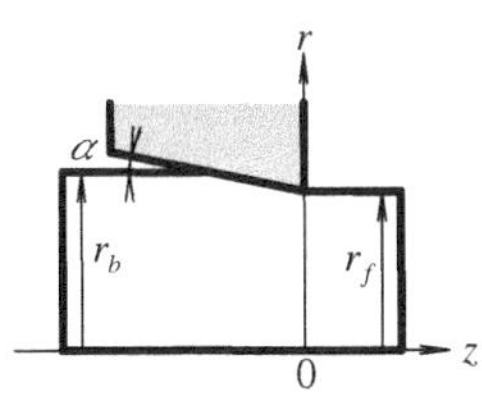

Figure 6.34 Coordinates and notations used in drawing.

Table 6.9 Inner diameter of die at which material fractured.

Exp.	$(r_b^2 - r_f^2)/r_b^2$					Sim.	$(r_b^2 - r_f^2)/r_b^2$				
(a) S15C											
2α (degrees)	5%	10%	15%	20%	25%	2α(degrees)	5%	10%	15%	20%	25%
15	$\varphi 8.1$	$\varphi 6.6$	$\leq \varphi 5$	$\leq \varphi 5$		15	$\varphi 9.0$	$\varphi 8.1$	$\varphi 7.5$	$\varphi 6.9$	
22.5		$\varphi 6.9$	$\varphi 6.9$	$\varphi 5.6$		22.5		$\varphi 8.5$	$\varphi 8.1$	$\varphi 7.7$	
30		$\varphi 8.5$	$\varphi 7.5$	$\varphi 6.9$	$\varphi 6.7$	30		$\varphi 9.0$	$\varphi 8.1$	$\varphi 7.7$	$\varphi 7.7$
(b) S35C											
2α (degrees)	5%	10%	15%	20%	25%	2α(degrees)	5%	10%	15%	20%	25%
15	$\varphi 8.8$	$\varphi 8.1$	$\varphi 6.9$	$\varphi 6.2$		15	$\varphi 11.4$	$\varphi 10.5$	$\varphi 10.3$	$\varphi 8.5$	
22.5		$\varphi 9.0$	$\varphi 8.1$	$\varphi 8.5$		22.5		$\varphi 11.1$	$\varphi 10.3$	$\varphi 10.5$	
30		$\varphi 9.5$	$\varphi 9.5$	$\varphi 8.5$	$\varphi 7.7$	30		$\varphi 11.1$	$\varphi 11.1$	$\varphi 10.5$	$\varphi 8.8$
(c) S45C											
2α (degrees)	5%	10%	15%	20%	25%	2α(degrees)	5%	10%	15%	20%	25%
15	$\varphi 9.5$	$\varphi 8.5$	$\varphi 8.1$	$\varphi 7.7$		15	$\varphi 11.4$	$\varphi 11.1$	$\varphi 10.3$	$\varphi 9.5$	
22.5		$\varphi 9.5$	$\varphi 9.5$	$\varphi 8.5$		22.5		$\varphi 11.1$	$\varphi 10.3$	$\varphi 10.5$	
30		$\varphi 10.0$	$\varphi 10.3$	$\varphi 9.5$	$\varphi 8.8$	30		$\varphi 11.1$	$\varphi 11.1$	$\varphi 10.5$	$\varphi 8.8$
(d) S55C											
2α (degrees)	5%	10%	15%	20%	25%	2α(degrees)	5%	10%	15%	20%	25%
15	$\varphi 9.7$	$\varphi 8.5$	$\varphi 6.9$	$\leq \varphi 5$		15	$\varphi 11.4$	$\varphi 11.1$	$\varphi 10.3$	$\varphi 9.5$	
22.5		$\varphi 9.5$	$\varphi 9.5$	$\varphi 8.5$		22.5		$\varphi 11.1$	$\varphi 10.3$	$\varphi 10.5$	
30		$\varphi 11.1$	$\varphi 10.3$	$\varphi 9.5$	$\varphi 8.8$	30		$\varphi 11.1$	$\varphi 11.1$	$\varphi 10.5$	$\varphi 10.0$

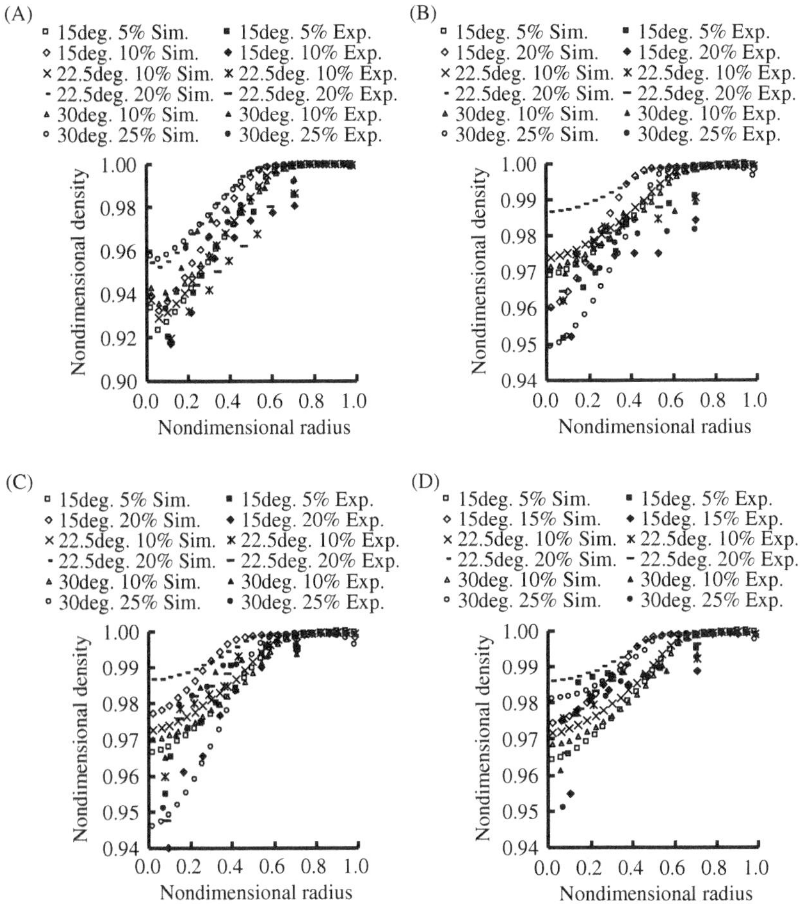

Figure 6.35 Material density distribution in radial direction after drawing through die preceding die at which material fractured: (A) S15C; (B) S35C; (C) S45C; and (D) S55C.

diameter of the die decreased with increasing the reduction in the area in each die, whereas when the reduction in area in each die is specified, the inner diameter of the die increased with increasing the die angle. Under each drawing condition, the inner diameters of the dies in the cases of S15C, S35C, S45C, and S55C were the smallest, the second smallest, the second largest, and the largest of the materials, respectively.

Fig. 6.35 shows the material density distribution in the radial direction after drawing through the die preceding the die at which the material fractured. In each die angle, the material density distribution in the minimum reduction in the area in each die, and the material density

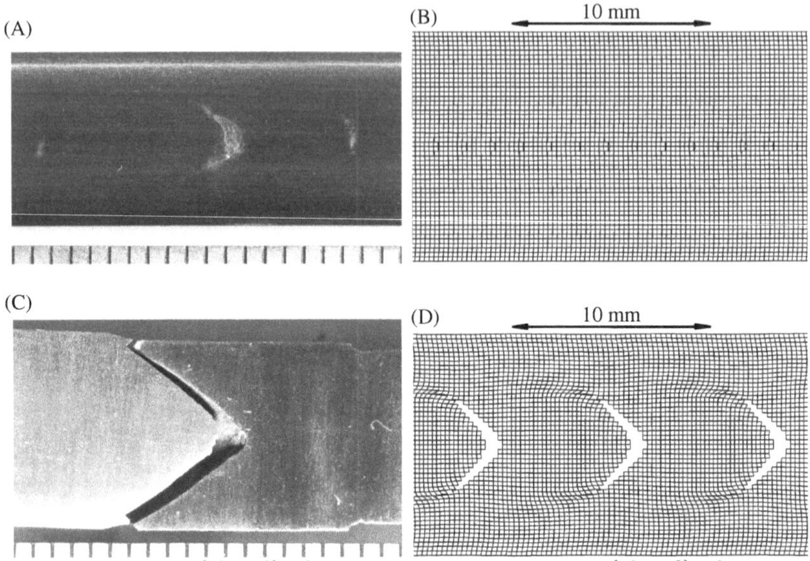

Figure 6.36 Material shape in longitudinal section after drawing through die at which material fractured in the case of S35C. (A) Exp. ($2\alpha = 15$ degrees, $(r_b^2 - r_f^2)/r_b^2 = 5\%$); (B) Sim. ($2\alpha = 15$ degrees, $(r_b^2 - r_f^2)/r_b^2 = 5\%$); (C) Exp. ($2\alpha = 30$ degrees, $(r_b^2 - r_f^2)/r_b^2 = 10\%$); and (D) Sim. ($2\alpha = 30$ degrees, $(r_b^2 - r_f^2)/r_b^2 = 10\%$).

distribution in the maximum reduction in the area in each die were shown. The nondimensional radius in the horizontal axis is the quotient of the coordinate in the radial direction divided by the radius of the specimen, whereas the nondimensional density in the vertical axis is the quotient of the material density divided by the initial material density. In each material and under each drawing condition, the material density near the center of the specimen in the radial direction was lower than the material density near the surface of the specimen in the radial direction.

Fig. 6.36 shows the material shape in the longitudinal section after drawing through the die at which the material fractured in the case of S35C. Fig. 6.36A shows the material shape when the die angle was equal to 15 degrees and the reduction in area in each die was equal to 5%. The central burst appeared periodically in the drawing direction. The period of the central burst was approximately 8 mm. Fig. 6.36C shows the material shape when the die angle was equal to 30 degrees and the reduction in area in each die was equal to 10%. The specimen was broken into two pieces.

The simulations of the tensile tests of the bars were performed using the conventional rigid-plastic finite-element method (Kobayashi et al., 1989), in which an axisymmetric state was assumed. Line symmetry was assumed for simplicity. The initial void volume fraction was assumed to be 0.001 (Schmitt and Jalinier, 1982). The simulation was performed until the specimen ruptured (Komori, 2002). The material constant A_1 in Eq. (6.9) was assumed to have a value such that the diameter of the specimen at rupture calculated from the simulation coincided with the diameter of the specimen at rupture obtained experimentally.

The flattening of the void f^* was assumed to be as follows for simplicity, with reference to the experimental observation (Inoue and Kinoshita, 1977a,b).

$$f^* = 0.9 \tag{6.16}$$

Table 6.10 shows the material constants used in the evolution equations. The material constants for the void shape and the void configuration due to the cracking of the inclusions, and also the material constants for the void shape and the void configuration due to the separation of the inclusions from the matrix, were shown.

The simulations of the multipass drawing of the bars were performed using the conventional rigid-plastic finite-element method (Kobayashi et al., 1989), in which an axisymmetric state was assumed. The initial void volume fraction was assumed to be 0.001 (Schmitt and Jalinier, 1982). The simulation was performed until the center of the specimen in the radial direction fractured.

Table 6.9 shows the inner diameter of the die at which the material fractured. In each material, when the die angle is specified, the inner diameter of the die decreased with increasing the reduction in the area in each die, whereas when the reduction in the area in each die is specified, the inner diameter of the die increased with increasing the die angle. Under each drawing condition, the inner diameters of the dies in the cases

Table 6.10 Material constants used in the evolution equations.

	A_1 inclusion cracking	A_1 soft inclusion	A_1 hard inclusion
S15C	0.04	0.06	0.08
S35C	0.10	0.21	0.19
S45C	0.11	0.28	0.26
S55C	0.12	0.30	0.27

of S15C, S35C, S45C, and S55C were the smallest, the second smallest, the second largest, and the largest of the materials, respectively. In each material and under each drawing condition, the inner diameter of the die calculated from the simulation fairly agreed with the inner diameter of the die obtained experimentally.

Fig. 6.35 shows the material density distribution in the radial direction after drawing through the die preceding the die at which the material fractured. In each material and under each drawing condition, the material density near the center of the specimen in the radial direction was lower than the material density near the surface of the specimen in the radial direction. In each material and under each drawing condition, the material density distribution in the radial direction calculated from the simulation agreed with the material density distribution in the radial direction obtained experimentally.

Fig. 6.36B shows the material shape when the die angle was equal to 15 degrees and the reduction in the area in each die was equal to 5%. Fig. 6.36D shows the material shape when the die angle was equal to 30 degrees and the reduction in area in each die was equal to 10%. The central burst appeared periodically in the drawing direction. The periods of the central burst were approximately 1.4 mm in Fig. 6.36B, and 7.5 mm in Fig. 6.36D. The magnitude of the central burst calculated from the simulation was slightly smaller than the magnitude of the central burst obtained experimentally.

The simulations of the multipass drawing of the bars were performed using the void shape and the void configuration due to the separation of the inclusions from the matrix.

Fig. 6.37 shows the material density distribution in the radial direction after drawing through the die preceding the die at which the material fractured calculated using the void shape and the void configuration due to the separation of the inclusions from the matrix. In the case of S15C, under each drawing condition, the material density distribution in the radial direction calculated from the simulation agreed with the material density distribution in the radial direction obtained experimentally. However, in the case of S55C, under each drawing condition, the material density distribution in the radial direction calculated from the simulation did not agree with the material density distribution in the radial direction obtained experimentally.

Fig. 6.38 shows the void configuration and the void shape at fracture in the case of S15C. The die angle was equal to 22.5 degrees, and the

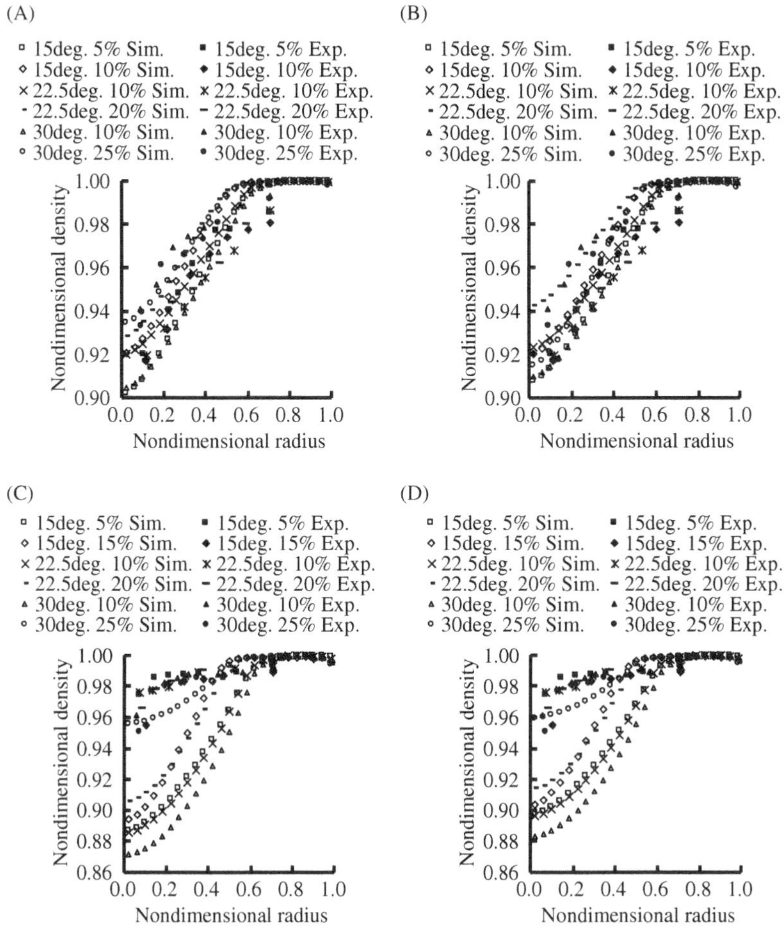

Figure 6.37 Material density distribution in radial direction after drawing through die preceding die at which material fractured calculated using void shape and void configuration due to separation of inclusions from matrix: (A) S15C (soft inclusion); (B) S15C (hard inclusion); (C) S55C (soft inclusion); and (D) S55C (hard inclusion).

reduction in area in each die was equal to 15%. Fig. 6.38A shows the void configuration and the void shape due to the cracking of the inclusions. The pearlite nodule was broken into two pieces. Fig. 6.38B and C shows the void configuration and the void shape due to the separation of the inclusions from the matrix. The flattening of the void at fracture due to the cracking of the inclusions was larger than the flattening of the void at fracture due to the separation of the inclusions from the matrix. However, the void volume fraction at fracture due to the cracking of the

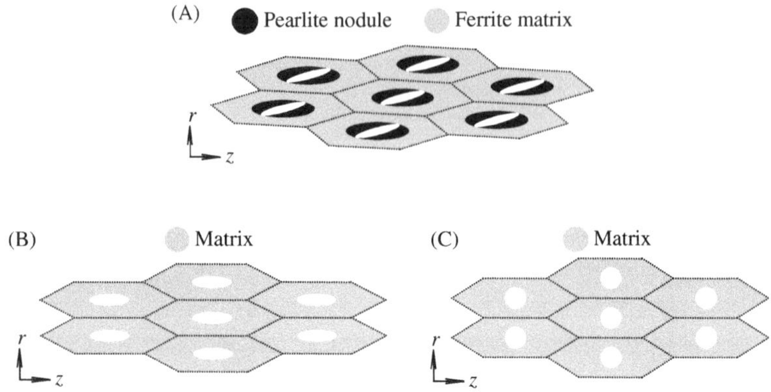

Figure 6.38 Void configuration and void shape at fracture in the case of S15C ($2\alpha = 22.5$ degrees, $(r_b^2 - r_f^2)/r_b^2 = 15\%$): (A) inclusion cracking; (B) soft inclusion; and (C) hard inclusion.

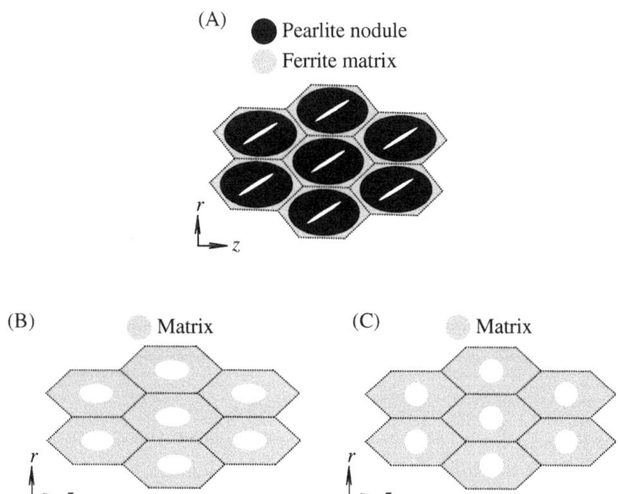

Figure 6.39 Void configuration and void shape at fracture in the case of S55C ($2\alpha = 22.5$ degrees, $(r_b^2 - r_f^2)/r_b^2 = 15\%$): (A) inclusion cracking; (B) soft inclusion; and (C) hard inclusion.

inclusions was almost the same as the void volume fraction at fracture due to the separation of the inclusions from the matrix.

Fig. 6.39 shows the void configuration and the void shape at fracture in the case of S55C. The die angle was 22.5 degrees, and the reduction in area in each die was 15%. Fig. 6.39A shows the void configuration and the void shape due to the cracking of the inclusions. The pearlite nodule was not broken into two pieces. Fig. 6.39B and C shows the void configuration and the void shape due to the separation of the inclusions from the matrix. The flattening of the void at fracture due to the cracking of the inclusions was much larger than the flattening of the void at fracture due to the separation of the inclusions from the matrix. However, the void volume fraction at fracture due to the cracking of the inclusions was much smaller than the void volume fraction at fracture due to the separation of the inclusions from the matrix. The reason for the difference in the void volume fraction at fracture is that, in the ellipsoidal void model, the energy required for the internal necking deformation mode depends on the length of the major axis of the void, as shown in Fig. 5.7B.

6.5 Surface cracking in cylinder upsetting and future work

The surface cracking in the cylinder upsetting of a ferrite–pearlite steel is predicted. Although cylinders that were not prestrained were upset in previous studies (Kudo and Aoi, 1967; Thomason, 1969; Kobayashi, 1970; Kuhn and Lee, 1971), cylinders that are prestrained in the multipass drawing of the ferrite–pearlite steel are upset, and the effect of prestrain on the strain to fracture is predicted. In ferrite–pearlite steels, because the nucleation of voids is caused by the fracture of pearlite nodules, the void shape and the void configuration due to the cracking of the inclusions are used. To obtain the simulation results precisely, the location of the velocity discontinuity line is optimized so that the energy for the internal necking deformation is minimized. Moreover, a void coalescence region is assumed.

The following one type of carbon steel for machine structural use was used: JIS S55C, which is equivalent to ISO C55E4. The as-rolled bar of S55C was not heat-treated prior to drawing. The experiment on the multipass drawing of a bar was performed prior to the experiment on the upsetting of a cylinder. The initial diameter of a bar was 13 mm. The coordinates and the notations used in drawing are shown in Fig. 6.34. The experiment on the multipass drawing of a bar was performed for a specified reduction in the area in each die $(r_b^2 - r_f^2)/r_b^2$ of 15% and for a specified die angle 2α of 15 degrees. The experiment on the upsetting of a cylinder was performed using a bar which was prestrained due to the multipass drawing of a bar and the diameters of which were 13.0, 11.1, 9.5, 8.1, 6.9, 5.9, and 5.0 mm. The initial height/diameter ratios of a cylinder were 1.0, 1.25, 1.5, 2.0, and 2.5.

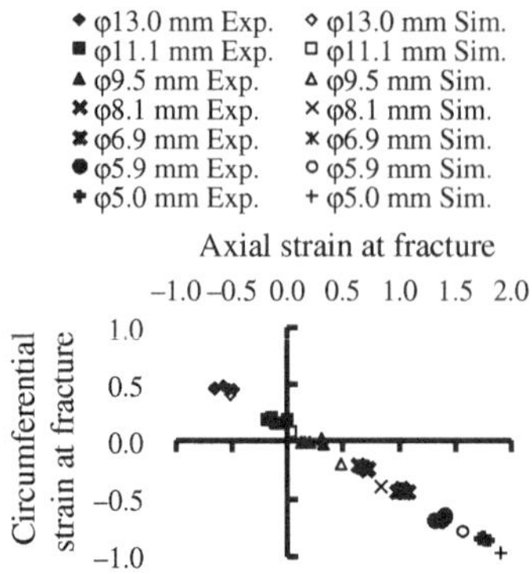

Figure 6.40 Relationship between axial strain at fracture and circumferential strain at fracture.

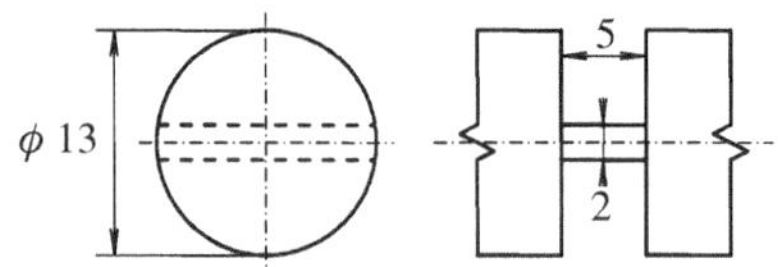

Figure 6.41 Specimen for plane-strain tensile test of bar.

Fig. 6.40 shows the relationship between the axial strain at fracture and the circumferential strain at fracture. The axial strain and the circumferential strain of the prestrained material were calculated using the deformation gradient of the material. With increasing the magnitude of the prestrain, in other words, with decreasing the diameter of the material in cylinder upsetting, the axial strain at fracture increased, whereas the circumferential strain at fracture decreased.

The simulation of the uniaxial tensile test of a bar was performed using the conventional rigid-plastic finite-element method (Kobayashi et al., 1989), in which an axisymmetric state was assumed. However, the strains at fracture calculated from the simulation of the upsetting of a cylinder using the material constant identified in the simulation of the uniaxial tensile test of a bar disagreed with the strains at fracture obtained by the experiment on the upsetting of a cylinder. Hence, the simulation of the plane-strain tensile test of a bar was performed using the conventional rigid-plastic finite-element method, in which a plane-strain state was assumed. Fig. 6.41 shows the specimen for the plane-strain tensile test of a bar. Line symmetry was assumed for simplicity. The initial void volume fraction was assumed to be 0.001 (Schmitt and Jalinier, 1982).

The simulation was performed until the specimen ruptured (Komori, 2002). The material constants A_0 and A_1 in Eq. (6.12) were assumed to have a value such that the thickness of the specimen at rupture calculated from the simulation coincided with the thickness of the specimen at rupture obtained by the experiment, and were identified to be 0.06. The flattening of the void f^* was assumed to be 0.9 (Inoue and Kinoshita, 1977a,b).

The simulation of the upsetting of a cylinder was performed using the conventional rigid-plastic finite-element method (Kobayashi et al., 1989), in which an axisymmetric state was assumed. The initial void volume fraction was assumed to be 0.001 (Schmitt and Jalinier, 1982). The initial height/diameter ratio of a cylinder was assumed to be 1.0. The simulation was performed until the surface of the specimen fractured.

Fig. 6.40 shows the relationship between the axial strain at fracture and the circumferential strain at fracture. With increasing the magnitude of the prestrain, in other words, with decreasing the diameter of the material in cylinder upsetting, the axial strain at fracture increased, whereas the circumferential strain at fracture decreased. The axial and circumferential strains at fracture calculated from the simulation agreed with the axial and circumferential strains at fracture obtained by the experiment. Because the comparison of the simulation result and the experimental result was performed in only one type of ferrite–pearlite steel, the comparison of the simulation result and the experimental result should be performed in other types of ferrite–pearlite steels in future work.

The Lode parameter is equal to negative one in the uniaxial tensile test of a bar, whereas the Lode parameter is equal to zero in the plane-strain tensile test of a bar. Plane-strain deformation, in which the Lode parameter is equal to zero, occurred immediately before fracture in cylinder upsetting (Kuhn and Lee, 1971). The strains at fracture calculated from the simulation of the upsetting of a cylinder using the material constant identified in the simulation of the uniaxial tensile test of a bar disagreed with the strains at fracture obtained by the experiment on the upsetting of a cylinder. The strains at fracture calculated from the simulation of the upsetting of a cylinder using the material constant identified in the simulation of the plane-strain tensile test of a bar agreed with the strains at fracture obtained by the experiment on the upsetting of a cylinder. These disagreement and agreements are conjectured from the fact that the Lode parameter in the uniaxial tensile test of a bar is not identical to the Lode parameter in the upsetting of a cylinder and that the Lode parameter in the plane-strain tensile test of a bar is identical to the Lode parameter in the upsetting of a cylinder. Since ductile fracture depends on the Lode parameter, the effect of the Lode parameter on ductile fracture in the simulation using the ellipsoidal void model should be clarified in future work.

The simulations of the central burst in wire drawing of high-carbon steels (Komori, 2008) and the rupture in shearing (Komori, 2006b) in which the void shape was assumed to be a parallelogram were performed using the conventional rigid-plastic finite-element method (Kobayashi et al., 1989). However, the simulations of the central burst in wire drawing of high-carbon steels and the rupture in shearing in which the void shape is assumed to be an ellipsoid should be performed in future work to improve the reliability of the simulation results.

In the ellipsoidal void model, the deformation gradient of the matrix is assumed to be identical to the deformation gradient of the void. However, this assumption is not necessarily appropriate. Although the difference between the deformation gradient of the matrix and the deformation gradient of the void is examined by the comparison of the simulation result calculated using the ellipsoidal void model and the simulation result calculated using the representative volume element (Komori, 2014c, 2017a), further researches on the difference should be performed in the future.

Bibliography

Al-Mousawi, M.M., Daragheh, A.M., Ghosh, S.K., Harrison, D.K., 1992. Some physical defects in metal forming processes and creation of a data base. J. Mater. Process. Technol. 32 (1–2), 461–470.

Anand, L., Spitzig, W.A., 1980. Initiation of localized shear bands in plane strain. J. Mech. Phys. Solids 28 (2), 113–128.

Angel, T., 1954. Formation of martensite in austenitic stainless steel. J. Iron Steel Inst. 177 (1), 165–174.

Aravas, N., 1986. The analysis of void growth that leads to central bursts during extrusion. J. Mech. Phys. Solids 34 (1), 55–79.

Arcan, M., Hashin, Z., Voloshin, A., 1978. A method to produce uniform plane-stress states with applications of fiber-reinforced materials. Exp. Mech. 18 (4), 141–146.

Argon, A.S., Im, J., 1975. Separation of second phase particles in spheroidized 1045 steel, Cu–0.6pct Cr alloy, and maraging steel in plastic straining. Metall. Trans. A 6 (4), 839–851.

Argon, A.S., Im, J., Needleman, A., 1975a. Distribution of plastic strain and negative pressure in necked steel and copper bars. Metall. Trans. A 6 (4), 815–824.

Argon, A.S., Im, J., Safoglu, R., 1975b. Cavity formation from inclusions in ductile fracture. Metall. Trans. A 6 (4), 825–837.

Arrieux, R., Bedrin, C., Boivin, M., 1982. Determination of an intrinsic forming limit stress diagram for isotropic metal sheets. In: Proceedings of the Twelfth Biennial Congress of the International Deep Drawing Research Group.

Ashby, M.F., 1966. Work hardening of dispersion-hardened crystals. Philos. Mag. 14 (132), 1157–1178.

Atkins, A.G., 1981. Possible explanation for unexpected departures in hydrostatic tension–fracture strain relations. Met. Sci. 15 (2), 81–83.

Atkinson, J.D., Brown, L.M., Stobbs, W.M., 1974. The work-hardening of copper–silica IV. The Bauschinger effect and plastic relaxation. Philos. Mag. 30 (6), 1247–1280.

Avitzur, B., 1965. Hydrostatic extrusion. Trans. ASME J. Eng. Ind. 87 (4), 487–494.

Avitzur, B., 1968a. Metal Forming: Processes and Analysis. McGraw-Hill, New York.

Avitzur, B., 1968b. Analysis of central bursting defects in extrusion and wire drawing. Trans. ASME J. Eng. Ind. 90 (1), 79–91.

Avitzur, B., 1980. Metal Forming: The Application of Limit Analysis. Marcel Dekker, New York.

Avitzur, B., Choi, J.C., 1986. Analysis of central bursting defects in plane strain drawing and extrusion. Trans. ASME J. Eng. Ind. 108 (4), 317–321.

Avitzur, B., Van Tyne, C.J., Turczyn, S., 1988. The prevention of central bursts during rolling. Trans. ASME J. Eng. Ind. 110 (2), 173–178.

Azrin, M., Backofen, W.A., 1970. The deformation and failure of a biaxially stretched sheet. Metall. Trans. 1 (10), 2857–2865.

Babout, L., Ludwig, W., Maire, E., Buffière, J.Y., 2003. Damage assessment in metallic structural materials using high resolution synchrotron X-ray tomography. Nucl. Inst. Methods Phys. Res. B 200, 303–307.

Babout, L., Maire, E., Fougères, R., 2004a. Damage initiation in model metallic materials: X-ray tomography and modelling. Acta Mater. 52 (8), 2475–2487.

Babout, L., Brechet, Y., Maire, E., Fougères, R., 2004b. On the competition between particle fracture and particle decohesion in metal matrix composites. Acta Mater. 52 (15), 4517–4525.

Bai, Y., Dodd, B., 1992. Adiabatic Shear Localization. Pergamon Press, Oxford.

Bai, Y., Wierzbicki, T., 2010. Application of extended Mohr–Coulomb criterion to ductile fracture. Int. J. Fract. 161 (1), 1–20.

Bandstra, J.P., Goto, D.M., Koss, D.A., 1998. Ductile failure as a result of a void-sheet instability: experiment and computational modeling. Mater. Sci. Eng. A 249 (1–2), 46–54.

Bao, Y., Wierzbicki, T., 2004. On fracture locus in the equivalent strain and stress triaxiality space. Int. J. Mech. Sci. 46 (1), 81–98.

Barenblatt, G.I., 1962. The mathematical theory of equilibrium cracks in brittle fracture. Adv. Appl. Mech. 7, 55–129.

Barsoum, I., Faleskog, J., 2007a. Rupture mechanisms in combined tension and shear—experiments. Int. J. Solids Struct. 44 (6), 1768–1786.

Barsoum, I., Faleskog, J., 2007b. Rupture mechanisms in combined tension and shear—micromechanics. Int. J. Solids Struct. 44 (17), 5481–5498.

Bathe, K.J., 1996. Finite Element Procedures. Prentice-Hall, Upper saddle river, NJ.

Beachem, C.D., 1963. An electron fractographic study of the influence of plastic strain conditions upon ductile rupture processes in metals. Trans. ASM 56, 318–326.

Beaver, P.W., 1983. Localized thinning, fracture and formability of aluminium sheet alloys in biaxial tension. J. Mech. Work. Technol. 7 (3), 215–231.

Becker, R., Needleman, A., Richmond, O., Tvergaard, V., 1988. Void growth and failure in notched bars. J. Mech. Phys. Solids 36 (3), 317–351.

Beevers, C.J., Honeycombe, R.W.K., 1962. The initiation of ductile fracture in pure metals. Philos. Mag. 7 (77), 763–773.

Benzerga, A.A., 2002. Micromechanics of coalescence in ductile fracture. J. Mech. Phys. Solids 50 (6), 1331–1362.

Benzerga, A.A., Leblond, J.B., 2010. Ductile fracture by void growth to coalescence. Adv. Appl. Mech. 44, 169–305.

Benzerga, A.A., Besson, J., Pineau, A., 1999. Coalescence-controlled anisotropic ductile fracture. Trans. ASME J. Eng. Mater. Technol. 121 (2), 221–229.

Benzerga, A.A., Besson, J., Batisse, R., Pineau, A., 2002. Synergistic effects of plastic anisotropy and void coalescence on fracture mode in plane strain. Modell. Simul. Mater. Sci. Eng. 10 (1), 73–102.

Benzerga, A.A., Besson, J., Pineau, A., 2004a. Anisotropic ductile fracture. Part I. Experiments. Acta Mater. 52 (15), 4623–4638.

Benzerga, A.A., Besson, J., Pineau, A., 2004b. Anisotropic ductile fracture. Part II. Theory. Acta Mater. 52 (15), 4639–4650.

Benzerga, A.A., Leblond, J.B., Needleman, A., Tvergaard, V., 2016. Ductile failure modeling. Int. J. Fract. 201 (1), 29–80.

Beremin, F.M., 1980. Elastoplastic calculation of circumferentially notched specimens using the finite element method. J. Mech. Appl. 4 (3), 307–325 (in French).

Beremin, F.M., 1981. Cavity formation from inclusions in ductile fracture of A508 steel. Metall. Trans. A 12 (5), 723–731.

Berg, C.A., 1962. The motion of cracks in plane viscous deformation. In: Rosenberg, R. M. (Ed.), Proceedings of the Fourth U.S. National Congress on Applied Mechanics. ASME, New York, pp. 885–892.

Berveiller, M., Zaoui, A., 1978. An extension of the self-consistent scheme to plastically-flowing polycrystals. J. Mech. Phys. Solids 26 (5–6), 325–344.

Besson, J., 2010. Continuum models of ductile fracture: a review. Int. J. Damage Mech. 19 (1), 3–52.

Besson, J., Steglich, D., Brocks, W., 2001. Modeling of crack growth in round bars and plane strain specimens. Int. J. Solids Struct. 38 (46–47), 8259–8284.

Besson, J., Steglich, D., Brocks, W., 2003. Modeling of plane strain ductile rupture. Int. J. Plast. 19 (10), 1517–1541.

Bishop, J.F.W., Hill, R., 1951. A theory of the plastic distortion of a polycrystalline aggregate under combined stresses. Philos. Mag. 42 (327), 414–427.

Bontcheva, N., Iankov, R., 1991. Numerical investigation of the damage process in metal forming. Eng. Fract. Mech. 40 (2), 387–393.

Bourcier, R.J., Koss, D.A., Smelser, R.E., Richmond, O., 1986. The influence of porosity on the deformation and fracture of alloys. Acta Metall. 34 (12), 2443–2453.

Bressan, J.D., Williams, J.A., 1983. The use of a shear instability criterion to predict local necking in sheet metal deformation. Int. J. Mech. Sci. 25 (3), 155–168.

Bridgman, P.W., 1952. Studies in Large Plastic Flow and Fracture With Special Emphasis on the Effects of Hydrostatic Pressure. McGraw-Hill, New York.

Broek, D., 1973. The role of inclusions in ductile fracture and fracture toughness. Eng. Fract. Mech. 5 (1), 55–66.

Brokken, D., Brekelmans, W.A.M., Baaijens, F.P.T., 1998. Numerical modelling of the metal blanking process. J. Mater. Process. Technol. 83 (1–3), 192–199.

Brokken, D., Brekelmans, W.A.M., Baaijens, F.P.T., 2000. Predicting the shape of blanked products: a finite element approach. J. Mater. Process. Technol. 103 (1), 51–56.

Brown, L.M., Clarke, D.R., 1975. Work hardening due to internal stresses in composite materials. Acta Metall. 23 (7), 821–830.

Brown, L.M., Embury, J.D., 1973. The initiation and growth of voids at second phase particles. In: Proceedings of the Third International Conference on the Strength of Metals and Alloys. pp. 164–169.

Brown, L.M., Stobbs, W.M., 1971. The work-hardening of copper–silica II. The role of plastic relaxation. Philos. Mag. 23 (185), 1201–1233.

Brown, L.M., Stobbs, W.M., 1976. The work-hardening of copper–silica V. Equilibrium plastic relaxation by secondary dislocations. Philos. Mag. 34 (3), 351–372.

Brownrigg, A., Spitzig, W.A., Richmond, O., Teirlinck, D., Embury, J.D., 1983. The influence of hydrostatic pressure on the flow stress and ductility of a spheroidized 1045 steel. Acta Metall. 31 (8), 1141–1150.

Brozzo, P., De Luca, B., Rendina, R., 1972. A new method for the prediction of the formability limits of metal sheets. In: Proceedings of the Seventh Biennial Congress of the International Deep Drawing Research Group.

Budiansky, B., Hutchinson, J.W., Slutsky, S., 1982. Void growth and collapse in viscous solids. In: Hopkins, H.G., Sewell, M.J. (Eds.), Mechanics of Solids The Rodney Hill 60th Anniversary Volume. Pergamon Press, Oxford, pp. 13–45.

Chae, D., Koss, D.A., 2004. Damage accumulation and failure of HSLA-100 steel. Mater. Sci. Eng. A 366 (2), 299–309.

Chae, D., Bandstra, J.P., Koss, D.A., 2000. The effect of pre-strain and strain-path changes on ductile fracture: experiment and computational modeling. Mater. Sci. Eng. A 285 (1–2), 165–171.

Chang, T.M., 1951. Shearing of metal blanks. J. Inst. Met. 78, 393–414.

Chang, T.M., Swift, H.W., 1950. Shearing of metal bars. J. Inst. Met. 78, 119–146.

Chaouadi, R., De Meester, P., Vandermeulen, W., 1994. Damage work as ductile fracture criterion. Int. J. Fract. 66 (2), 155–164.

Chen, C.C., Oh, S.I., Kobayashi, S., 1979. Ductile fracture in axisymmetric extrusion and drawing. Part 1. Deformation mechanics of extrusion and drawing. Trans. ASME J. Eng. Ind. 101 (1), 23–35.

Chitkara, N.R., Johnson, W., 1974. Hole flanging and piercing of circular plates. Sheet Met. Ind. 51 (10), 635–640.

Chow, C.L., Yu, L.G., Demeri, M.Y., 1997. A unified damage approach for predicting forming limit diagrams. Trans. ASME J. Eng. Mater. Technol. 119 (4), 346–353.

Chu, C.C., Needleman, A., 1980. Void nucleation effects in biaxially stretched sheets. Trans. ASME J. Eng. Mater. Technol. 102 (3), 249–256.

Clausing, D.P., 1967. The development of fibrous fracture in a mild steel. Trans. ASM 60, 504–515.

Clausing, D.P., 1970. Effect of plastic strain state on ductility and toughness. Int. J. Fract. Mech. 6 (1), 71–85.

Clift, S.E., Hartley, P., Sturgess, C.E.N., Rowe, G.W., 1990. Fracture prediction in plastic deformation processes. Int. J. Mech. Sci. 32 (1), 1–17.

Cockcroft, M.G., Latham, D.J., 1968. Ductility and the workability of metals. J. Inst. Met. 96, 33–39.

Coffin, L.F., Rogers, H.C., 1967. Influence of pressure on the structural damage in metal forming processes. Trans. ASM 60, 672–686.

Cox, T.B., Low, J.R., 1974. An investigation of the plastic fracture of AISI 4340 and 18 Nickel-200 grade maraging steels. Metall. Trans. 5 (6), 1457–1470.

Cusminsky, G., Ellis, F., 1967. An investigation into the influence of edge shape on cracking during rolling. J. Inst. Met. 95, 33–37.

Dadras, P., 1981. A semi-empirical solution to upset forging. Trans. ASME J. Eng. Ind. 103 (4), 478–483.

Dieter, G.E., 1988. Mechanical Metallurgy, SI Metric Edition. McGraw-Hill, London.

Dodd, B., Bai, Y., 1987. Ductile Fracture and Ductility. Academic Press, London.

Dodd, B., Boddington, P., 1980. The causes of edge cracking in cold rolling. J. Mech. Work. Technol. 3 (3–4), 239–252.

Donea, J., 1983. Arbitrary Lagrangian–Eulerian finite element methods. In: Belytschko, T., Hughes, T.J.R. (Eds.), Computational Methods for Transient Analysis. Elsevier, Amsterdam, pp. 473–516.

Dubensky, E.M., Koss, D.A., 1987. Void/pore distributions and ductile fracture. Metall. Trans. A 18 (11), 1887–1895.

Edelson, B.I., Baldwin, W.M., 1962. The effect of second phases on the mechanical properties of alloys. Trans. ASM 55, 230–250.

Embury, J.D., Duncan, J.L., 1981. Formability maps. Ann. Rev. Mater. Sci. 11, 505–521.

Embury, J.D., LeRoy, G.H., 1978. Failure maps applied to metal deformation processes. In: Taplin, D.M.R. (Ed.), Advances in Research on the Strength and Fracture of Materials. Pergamon Press, New York, U.S.A., pp. 15–42.

Eshelby, J.D., 1957. The determination of the elastic field of an ellipsoidal inclusion, and related problems. Proc. Roy. Soc. A 241 (1226), 376–396.

Fischer, F.D., Kolednik, O., Shan, G.X., Rammerstorfer, F.G., 1995. A note on calibration of ductile failure damage indicators. Int. J. Fract. 73 (4), 345–357.

Fisher, J.R., Gurland, J., 1981a. Void nucleation in spheroidized carbon steels. Part 1. Experimental. Met. Sci. 15 (5), 185–192.

Fisher, J.R., Gurland, J., 1981b. Void nucleation in spheroidized carbon steels. Part 2. Model. Met. Sci. 15 (5), 193–202.

French, I.E., Weinrich, P.F., 1973. The effect of hydrostatic pressure on the tensile fracture of α-brass. Acta Metall. 21 (11), 1533–1537.

French, I.E., Weinrich, P.F., 1975a. The influence of hydrostatic pressure on the tensile deformation and fracture of copper. Metall. Trans. A 6 (4), 785–790.

French, I.E., Weinrich, P.F., 1975b. The effects of hydrostatic pressure on the mechanism of tensile fracture of aluminum. Metall. Trans. A 6 (6), 1165–1169.

French, I.E., Weinrich, P.F., 1976. The shear mode of ductile fracture in materials with few inclusions. Metall. Trans. A 7 (12), 1841–1845.

French, I.E., Weinrich, P.F., 1979. The shear mode of ductile fracture in a spheroidized steel. Metall. Trans. A 10 (3), 297–304.

French, I.E., Weinrich, P.F., Weaver, C.W., 1973. Tensile fracture of free machining brass as a function of hydrostatic pressure. Acta Metall. 21 (8), 1045–1049.

Freudenthal, A.M., 1950. The Inelastic Behavior of Engineering Materials and Structures. John Wiley & Sons, New York.

Gänser, H.P., Atkins, A.G., Kolednik, O., Fischer, F.D., Richard, O., 2001. Upsetting of cylinders: a comparison of two different damage indicators. Trans. ASME J. Eng. Mater. Technol. 123 (1), 94–99.

Gao, X., Kim, J., 2006. Modeling of ductile fracture: significance of void coalescence. Int. J. Solids Struct. 43 (20), 6277–6293.

Garrison, W.M., Moody, N.R., 1987. Ductile fracture. J. Phys. Chem. Solids 48 (11), 1035–1074.

Geltmacher, A.B., Koss, D.A., Matic, P., Stout, M.G., 1996. A modeling study of the effect of stress state on void linking during ductile fracture. Acta Mater. 44 (6), 2201–2210.

Ghosh, A.K., 1976. A criterion for ductile fracture in sheets under biaxial loading. Metall. Trans. A 7 (4), 523–533.

Ghosh, S.K., 1981. Principally on sheet metal forming defects as described in the eleventh biennial congress of the International Deep Drawing Research Group (IDDRG). Int. J. Mech. Sci. 23 (4), 195–211.

Ghosh, S., Moorthy, S., 1995. Elastic–plastic analysis of arbitrary heterogeneous materials with the Voronoi Cell finite element method. Comput. Methods Appl. Mech. Eng. 121 (1–4), 373–409.

Ghosh, S., Moorthy, S., 1998. Particle fracture simulation in non-uniform microstructures of metal–matrix composites. Acta Mater. 46 (3), 965–982.

Glover, G., Duncan, J.L., Embury, J.D., 1977. Failure maps for sheet metal. Met. Technol. 4 (1), 153–159.

Goijaerts, A.M., Govaert, L.E., Baaijens, F.P.T., 2000. Prediction of ductile fracture in metal blanking. Trans. ASME J. Manuf. Sci. Eng. 122 (3), 476–483.

Goijaerts, A.M., Govaert, L.E., Baaijens, F.P.T., 2001. Evaluation of ductile fracture models for different metals in blanking. J. Mater. Process. Technol. 110 (3), 312–323.

Gologanu, M., Leblond, J.B., Devaux, J., 1993. Approximate models for ductile metals containing non-spherical voids—case of axisymmetric prolate ellipsoidal cavities. J. Mech. Phys. Solids 41 (11), 1723–1754.

Gologanu, M., Leblond, J.B., Devaux, J., 1994. Approximate models for ductile metals containing nonspherical voids—case of axisymmetric oblate ellipsoidal cavities. Trans. ASME J. Eng. Mater. Technol. 116 (3), 290–297.

Gologanu, M., Leblond, J.B., Perrin, G., Devaux, J., 1997. Recent extensions of Gurson's model for porous ductile metals. In: Suquet, P. (Ed.), Continuum Micromechanics. Springer, Vienna, Austria, pp. 61–130.

Gologanu, M., Leblond, J.B., Perrin, G., Devaux, J., 2001a. Theoretical models for void coalescence in porous ductile solids. I. Coalescence in layers. Int. J. Solids Struct. 38 (32–33), 5581–5594.

Gologanu, M., Leblond, J.B., Devaux, J., 2001b. Theoretical models for void coalescence in porous ductile solids. II. Coalescence in columns. Int. J. Solids Struct. 38 (32–33), 5595–5604.

Goods, S.H., Brown, L.M., 1979. Overview No. 1. The nucleation of cavities by plastic deformation. Acta Metall. 27 (1), 1–15.

Goodwin, G.M., 1968. Application of strain analysis to sheet metal forming problems in the press shop. La Metallurgia Italiana 60 (8), 767–774.

Goto, D.M., Koss, D.A., Jablokov, V., 1999. The influence of tensile stress states on the failure of HY-100 steel. Metall. Mater. Trans. A 30 (11), 2835–2842.

Gouveia, B.P.P.A., Rodrigues, J.M.C., Martins, P.A.F., 1996. Fracture predicting in bulk metal forming. Int. J. Mech. Sci. 38 (4), 361–372.

Grumbach, M., Sanz, G., 1972. Influences des grilles de mesure et des trajectoires de deformation sur les courbes limites d'emboutissage. In: Proceedings of the Seventh Biennial Congress of the International Deep Drawing Research Group (in French).

Gunawardena, S.R., Jansson, S., Leckie, F.A., 1993. Modeling of anisotropic behavior of weakly bonded fiber reinforced MMC's. Acta Metall. Mater. 41 (11), 3147–3156.

Gurland, J., 1972. Observations on the fracture of cementite particles in a spheroidized 1.05% C steel deformed at room temperature. Acta Metall. 20 (5), 735–741.

Gurland, J., Plateau, J., 1963. The mechanism of ductile rupture of metals containing inclusions. Trans. ASM 56, 442–454.

Gurson, A.L., 1977. Continuum theory of ductile rupture by void nucleation and growth. Part I. Yield criteria and flow rules for porous ductile media. Trans. ASME J. Eng. Mater. Technol. 99 (1), 2–15.

Hambli, R., 2001a. Finite element model fracture prediction during sheet-metal blanking processes. Eng. Fract. Mech. 68 (3), 365–378.

Hambli, R., 2001b. Comparison between Lemaitre and Gurson damage models in crack growth simulation during blanking process. Int. J. Mech. Sci. 43 (12), 2769–2790.

Hambli, R., Badie-Levet, D., 2000. Damage and fracture simulation during the extrusion processes. Comput. Methods Appl. Mech. Eng. 186 (1), 109–120.

Hambli, R., Potiron, A., 2000. Finite element modeling of sheet-metal blanking operations with experimental verification. J. Mater. Process. Technol. 102 (1–3), 257–265.

Han, H.N., Kim, K.H., 2003. A ductile fracture criterion in sheet metal forming process. J. Mater. Process. Technol. 142 (1), 231–238.

Hancock, J.W., Brown, D.K., 1983. On the role of strain and stress state in ductile failure. J. Mech. Phys. Solids 31 (1), 1–24.

Hancock, J.W., Mackenzie, A.C., 1976. On the mechanisms of ductile failure in high-strength steels subjected to multi-axial stress-states. J. Mech. Phys. Solids 24 (2–3), 147–169.

Hao, S., Liu, W.K., Chang, C.T., 2000. Computer implementation of damage models by finite element and meshfree methods. Comput. Methods Appl. Mech. Eng. 187 (3–4), 401–440.

Hellan, K., 1975. An approximate study of void expansion by ductility or creep. Int. J. Mech. Sci. 17 (5), 369–374.

Hill, R., 1950. The Mathematical Theory of Plasticity. Clarendon Press, Oxford.

Hill, R., 1952. On discontinuous plastic states, with special reference to localized necking in thin sheets. J. Mech. Phys. Solids 1 (1), 19–30.

Hill, R., 1958. A general theory of uniqueness and stability in elastic–plastic solids. J. Mech. Phys. Solids 6 (3), 236–249.

Hill, R., 1979. Theoretical plasticity of textured aggregates. Math. Proc. Camb. Philos. Soc. 85 (1), 179–191.

Hill, R., Hutchinson, J.W., 1975. Bifurcation phenomena in the plane tension test. J. Mech. Phys. Solids 23 (4–5), 239–264.

Hill, R., Tupper, S.J., 1948. A new theory of the plastic deformation in wire-drawing. J. Iron Steel Inst. 159 (4), 353–359.

Hirth, J.P., Lothe, J., 1982. Theory of Dislocations, second ed. John Wiley & Sons, New York.

Hoffmanner, A.L., 1971. The use of workability test results to predict processing limits. In: Hoffmanner, A.L. (Ed.), Metal Forming: Interrelation Between Theory and Practice. Plenum Press, New York, pp. 349–391.

Hom, C.L., McMeeking, R.M., 1989. Void growth in elastic–plastic materials. Trans. ASME J. Appl. Mech. 56 (2), 309–317.

Huang, Y., 1991. Accurate dilatation rates for spherical voids in triaxial stress fields. Trans. ASME J. Appl. Mech. 58 (4), 1084–1086.

Hutchinson, J.W., Tvergaard, V., 1980. Surface instabilities on statically strained plastic solids. Int. J. Mech. Sci. 22 (6), 339–354.

Inoue, T., Kinoshita, S., 1977a. Strain partitioning and void formation in ferrite–pearlite steels deformed in tension. Trans. ISIJ 17 (5), 245–251.

Inoue, T., Kinoshita, S., 1977b. Three stages of ductile fracture process and criteria of void inititation in spheroidized and ferrite/pearlite steels. Trans. ISIJ 17 (9), 523–531.

Iosipescu, N., 1967. New accurate procedure for single shear testing of metals. J. Mater. 2 (3), 537–566.

Irwin, G.R., 1957. Analysis of stresses and strains near the end of a crack traversing a plate. Trans. ASME J. Appl. Mech. 24 (3), 361–364.

Jain, M., Allin, J., Lloyd, D.J., 1999. Fracture limit prediction using ductile fracture criteria for forming of an automotive aluminum sheet. Int. J. Mech. Sci. 41 (10), 1273–1288.

Jalinier, J.M., 1983. Calculation of the forming limit curve at fracture. J. Mater. Sci. 18 (6), 1794–1802.

Jenner, A., Dodd, B., 1981. Cold upsetting and free surface ductility. J. Mech. Work. Technol. 5 (1–2), 31–43.

Jennison, H.C., 1930. Certain types of defects in copper wire caused by improper dies and drawing practice. Trans. Am. Inst. Mining Metall. Eng. 89, 121–139.

Johnson, G.R., Cook, W.H., 1985. Fracture characteristics of three metals subjected to various strains, strain rates, temperatures and pressures. Eng. Fract. Mech. 21 (1), 31–48.

Johnson, W., Kudo, H., 1962. The Mechanics of Metal Extrusion. Manchester University Press, Manchester (Chapter 10).

Johnson, W., Mamalis, A.G., 1977. A survey of some physical defects arising in metal working processes. In: Tobias, S.A. (Ed.), Proceedings of the Seventeenth International Machin Tool Design and Research Conference. Macmillan Press, London, pp. 607–621.

Johnson, W., Mamalis, A.G., 1985. Common defects in the processing of metals and composite materials. In: Sawczuk, A., Bianchi, G. (Eds.), Plasticity Today: Modelling, Methods and Applications. Elsevier, London, pp. 745–766.

Johnson, W., Slater, R.A.C., 1967. A survey of the slow and fast blanking of metals at ambient and high temperatures. In: Colwell, L.V. (Ed.), Proceedings of the International Conference on Manufacturing Technology. American Society of Tool and Manufacturing Engineers, Southfield, MI, pp. 825–852.

Johnson, W., Chitkara, N.R., Ibrahim, A.H., Dasgupta, A.K., 1973. Hole flanging and punching of circular plates with conically headed cylindrical punches. J. Strain Anal. 8 (3), 228–241.

Johnson, W., Ghosh, S.K., Reid, S.R., 1980. Piercing and hole-flanging of sheet metals: a survey. Mémoires Scientifiques de la Revue de Métallurgie 77 (3–4), 585–606.

Johnson, W., Sowerby, R., Venter, R.D., 1982. Plane-Strain Slip-Line Fields for Metal-Deformation Processes: A Source Book and Bibliography. Pergamon Press, Oxford.

Kao, A.S., Kuhn, H.A., Richmond, O., Spitzig, W.A., 1990. Tensile fracture and fractographic analysis of 1045 spheroidized steel under hydrostatic pressure. J. Mater. Res. 5 (1), 83–91.

Kasz, F., Varley, P.C., 1949. The hot rolling of aluminium and its alloys. J. Inst. Met. 76, 407–428.

Keeler, S.P., 1965. Determination of forming limits in automotive stampings. Sheet Met. Ind. 42 (459), 683–691.

Keeler, S.P., Backofen, W.A., 1963. Plastic instability and fracture in sheets stretched over rigid punches. Trans. ASM 56, 25–48.

Kim, H.S., Im, Y.T., Geiger, M., 1999. Prediction of ductile fracture in cold forging of aluminum alloy. Trans. ASME J. Manuf. Sci. Eng. 121 (3), 336–344.

Ko, D.C., Kim, B.M., 2000. The prediction of central burst defects in extrusion and wire drawing. J. Mater. Process. Technol. 102 (1–3), 19–24.

Ko, Y.K., Lee, J.S., Huh, H., Kim, H.K., Park, S.H., 2007. Prediction of fracture in hub-hole expanding process using a new ductile fracture criterion. J. Mater. Process. Technol. 187–188, 358–362.

Kobayashi, S., 1970. Deformation characteristics and ductile fracture of 1040 steel in simple upsetting of solid cylinders and rings. Trans. ASME J. Eng. Ind. 92 (2), 391–399.

Kobayashi, S., Lee, C.H., 1973. Deformation mechanics and workability in upsetting solid circular cylinders. In: Proceedings of the First North American Metalworking Research Conference. pp. 185–204.

Kobayashi, S., Lee, C.H., Shah, S.N., 1973. Analysis of rigid-plastic deformation problems by the matrix method. J. Jpn. Soc. Technol. Plast. 14 (153), 770–778 (in Japanese).

Kobayashi, S., Oh, S.I., Altan, T., 1989. Metal Forming and the Finite-Element Method. Oxford University Press, New York.

Komori, K., 1999a. Simulation of chevron crack formation and evolution in drawing. Int. J. Mech. Sci. 41 (12), 1499–1513.

Komori, K., 1999b. Proposal and use of a void model for the simulation of ductile fracture behavior. Acta Mater. 47 (10), 3069–3077.

Komori, K., 2001. Simulation of shearing by node separation method. Comput. Struct. 79 (2), 197–207.

Komori, K., 2002. Simulation of tensile test by node separation method. J. Mater. Process. Technol. 125–126, 608–612.

Komori, K., 2003. Effect of ductile fracture criteria on chevron crack formation and evolution in drawing. Int. J. Mech. Sci. 45 (1), 141–160.

Komori, K., 2005. Ductile fracture criteria for simulating shear by node separation method. Theor. Appl. Fract. Mech. 43 (1), 101–114.

Komori, K., 2006a. Proposal and use of a void model for the simulation of inner fracture defect in drawing. Acta Mater. 54 (16), 4351–4364.

Komori, K., 2006b. Proposal and use of a void model for the simulation of shearing. Mater. Sci. Eng. A 421 (1–2), 226–237.

Komori, K., 2008. Evolution of void volume fraction for simulating inner fracture defects related to in drawing of high-carbon steel wire rods. Theor. Appl. Fract. Mech. 50 (2), 157–166.

Komori, K., 2011. Improved void model for simulating inner fracture defects in drawing. In: Steel Research International Special Issue on the 10th International Conference on Technology of Plasticity. pp. 913–918.

Komori, K., 2013a. An ellipsoidal void model for simulating ductile fracture behavior. Mech. Mater. 60, 36–54.

Komori, K., 2013b. An ellipsoidal void model for the evaluation of ductile fracture in sheet metal forming. Comput. Mater. Sci. 67, 408–416.

Komori, K., 2014a. Evaluation of ductile fracture in sheet metal forming using the ellipsoidal void model. Mech. Mater. 77, 67–79.

Komori, K., 2014b. Proposal and use of a void model for inclusion cracking for simulating inner fracture defects in drawing of ferrite–pearlite steels. Mech. Mater. 77, 98–109.

Komori, K., 2014c. Improvement of an ellipsoidal void model for simulating ductile fracture behavior. Key Eng. Mater. 577–578, 93–96.

Komori, K., 2016. Predicting ductile fracture in pure metals and alloys using notched tensile specimens by an ellipsoidal void model. Eng. Fract. Mech. 151, 51–69.

Komori, K., 2017a. Predicting ductile fracture in ferrous materials during tensile tests using an ellipsoidal void model. Mech. Mater. 113, 24–43.

Komori, K., 2017b. Improvement and validation of the ellipsoidal void model for predicting ductile fracture. Proc. Eng. 207, 2036–2041.

Koplik, J., Needleman, A., 1988. Void growth and coalescence in porous plastic solids. Int. J. Solids Struct. 24 (8), 835–853.

Kudo, H., 1960a. An upper-bound approach to plane-strain forging and extrusion-I. Int. J. Mech. Sci. 1 (1), 57–83.

Kudo, H., 1960b. Some analytical and experimental studies of axi-symmetric cold forging and extrusion-I. Int. J. Mech. Sci. 2 (1–2), 102–127.

Kudo, H., Aoi, K., 1967. Effect of compression test condition upon fracturing of a medium carbon steel. J. Jpn. Soc. Technol. Plast. 8 (72), 17–27 (in Japanese).

Kuhn, H.A., Lee, P.W., 1971. Strain instability and fracture at the surface of upset cylinders. Metall. Trans. 2 (11), 3197–3202.

Kuhn, H.A., Lee, P.W., Erturk, T., 1973. A fracture criterion for cold forming. Trans. ASME J. Eng. Mater. Technol. 95 (4), 213–218.

Kuna, M., Sun, D.Z., 1996. Three-dimensional cell model analyses of void growth in ductile materials. Int. J. Fract. 81 (3), 235–258.

Leblond, J.B., Mottet, J., 2008. A theoretical approach of strain localization within thin planar bands in porous ductile materials. Comptes Rendus Mécanique 336 (1–2), 176–189.

Leblond, J.B., Perrin, G., Devaux, J., 1995. An improved Gurson-type model for hardenable ductile metals. Eur. J. Mech. A/Solids 14 (4), 499–527.

Lee, C.H., Kobayashi, S., 1973. New solutions to rigid-plastic deformation problems using a matrix method. Trans. ASME J. Eng. Ind. 95 (3), 865–873.

Lee, P.W., Kuhn, H.A., 1973. Fracture in cold upset forging—a criterion and model. Metall. Trans. 4 (4), 969–974.

Lee, B.J., Mear, M.E., 1992. Axisymmetric deformation of power-law solids containing a dilute concentration of aligned spheroidal voids. J. Mech. Phys. Solids 40 (8), 1805–1836.

Lee, Y.W., Woertz, J.C., Wierzbicki, T., 2004. Fracture prediction of thin plates under hemi-spherical punch with calibration and experimental verification. Int. J. Mech. Sci. 46 (5), 751–781.

Lei, L.P., Kim, J., Kang, B.S., 2002. Bursting failure prediction in tube hydroforming processes by using rigid-plastic FEM combined with ductile fracture criterion. Int. J. Mech. Sci. 44 (7), 1411–1428.

Lemaitre, J., 1985. A continuous damage mechanics model for ductile fracture. Trans. ASME J. Eng. Mater. Technol. 107 (1), 83–89.

Lemaitre, J., Dufailly, J., 1987. Damage measurements. Eng. Fract. Mech. 28 (5/6), 643–661.

Le Roy, G., Embury, J.D., Edwards, G., Ashby, M.F., 1981. A model of ductile fracture based on the nucleation and growth of voids. Acta Metall. 29 (8), 1509–1522.

Liu, C.T., Gurland, J., 1968. The fracture behavior of spheroidized carbon steels. Trans. ASM 61, 156–167.

Mackenzie, A.C., Hancock, J.W., Brown, D.K., 1977. On the influence of state of stress on ductile failure initiation in high strength steels. Eng. Fract. Mech. 9 (1), 167–188.

Magnusen, P.E., Dubensky, E.M., Koss, D.A., 1988. The effect of void arrays on void linking during ductile fracture. Acta Metall. 36 (6), 1503–1509.

Magnusen, P.E., Srolovitz, D.J., Koss, D.A., 1990. A simulation of void linking during ductile microvoid fracture. Acta Metall. Mater. 38 (6), 1013–1022.

Maire, E., Buffière, J.Y., Salvo, L., Blandin, J.J., Ludwig, W., Létang, J.M., 2001. On the application of X-ray microtomography in the field of materials science. Adv. Eng. Mater. 3 (8), 539–546.

Maire, E., Bouaziz, O., Di Michiel, M., Verdu, C., 2008. Initiation and growth of damage in a dual-phase steel observed by X-ray microtomography. Acta Mater. 56 (18), 4954–4964.

Mamalis, A.G., Johnson, W., 1987. Defects in the processing of metals and composites. In: Predeleanu, M. (Ed.), Computational Methods for Predicting Material Processing Defects. Elsevier, Amsterdam, pp. 231–250.

Marciniak, Z., Kuczyński, K., 1967. Limit strains in the processes of stretch-forming sheet metal. Int. J. Mech. Sci. 9 (9), 609–620.

Mariage, J.F., Saanouni, K., Lestriez, P., Cherouat, A., 2002. Numerical simulation of ductile damage in metal forming processes: a simple predictive model. 1. Theoretical and numerical aspects. Int. J. Form. Process. 5 (2-3-4), 363–376.

Marini, B., Mudry, F., Pineau, A., 1985a. Ductile rupture of A508 steel under nonradial loading. Eng. Fract. Mech. 22 (3), 375–386.

Marini, B., Mudry, F., Pineau, A., 1985b. Experimental study of cavity growth in ductile rupture. Eng. Fract. Mech. 22 (6), 989–996.

Mase, G.E., 1970. Continuum Mechanics. McGraw-Hill, New York (Chapter 3).

McAllen, P., Phelan, P., 2005. A method for the prediction of ductile fracture by central bursts in axisymmetric extrusion. Proc. Inst. Mech. Eng. C: J. Mech. Eng. Sci. 219 (3), 237–250.

McAllen, P.J., Phelan, P., 2007. Numerical analysis of axisymmetric wire drawing by means of a coupled damage model. J. Mater. Process. Technol. 183 (2–3), 210–218.

McClintock, F.A., 1968a. A criterion for ductile fracture by the growth of holes. Trans. ASME J. Appl. Mech. 35 (2), 363–371.

McClintock, F.A., 1968b. On the mechanics of fracture from inclusions. Ductility. American Society for Metals, Novelty, OH, pp. 255–277.

McClintock, F.A., Kaplan, S.M., Berg, C.A., 1966. Ductile fracture by hole growth in shear bands. Int. J. Fract. Mech. 2 (4), 614–627.

McVeigh, C., Liu, W.K., 2006. Prediction of central bursting during axisymmetric cold extrusion of a metal alloy containing particles. Int. J. Solids Struct. 43 (10), 3087–3105.

Melander, A., 1979. Computer simulation of ductile fracture in a random distribution of voids. Mater. Sci. Eng. 39 (1), 57–63.

Melander, A., Ståhlberg, U., 1980. The effect of void size and distribution on ductile fracture. Int. J. Fract. 16 (5), 431–439.

Miller, L.E., Smith, G.C., 1970. Tensile fracture in carbon steels. J. Iron Steel Inst. 208 (11), 998–1005.

Miyauchi, K., 1984. A proposal of a planar simple shear test in sheet metals. Sci. Pap. Inst. Phys. Chem. Res. 78 (3), 27–40.

Mohr, D., Henn, S., 2007. Calibration of stress-triaxiality dependent crack formation criteria: a new hybrid experimental–numerical method. Exp. Mech. 47 (6), 805–820.

Moorthy, S., Ghosh, S., 1996. A model for analysis of arbitrary composite and porous microstructures with Voronoi Cell finite elements. Int. J. Numer. Methods Eng. 39 (14), 2363–2398.

Moritoki, H., 1991. The criterion for central bursting and its occurrence in drawing and extrusion under plane strain. Int. J. Plast. 7 (7), 713–731.

Moritoki, H., 1993. Free surface ductility in upsetting. Int. J. Plast. 9 (4), 507–523.

Nagumo, M., 1973. A criterion of ductile fracture on the tensile testing of a perforated mild steel sheet. Acta Metall. 21 (12), 1661–1667.

Needleman, A., 1972. Void growth in an elastic–plastic medium. Trans. ASME J. Appl. Mech. 39 (4), 964–970.

Needleman, A., 1987. A continuum model for void nucleation by inclusion debonding. Trans. ASME J. Appl. Mech. 54 (3), 525–531.

Needleman, A., Tvergaard, V., 1984. An analysis of ductile rupture in notched bars. J. Mech. Phys. Solids 32 (6), 461–490.

Norris, D.M., Reaugh, J.E., Moran, B., Quiñones, D.F., 1978. A plastic-strain mean-stress criterion for ductile fracture. Trans. ASME J. Eng. Mater. Technol. 100 (3), 279–286.

Oh, S.I., Kobayashi, S., 1975. An approximate method for a three-dimensional analysis of rolling. Int. J. Mech. Sci. 17 (4), 293–305.

Oh, S.I., Kobayashi, S., 1976. Workability of aluminum alloy 7075-T6 in upsetting and rolling. Trans. ASME J. Eng. Ind. 98 (3), 800–806.

Oh, S.I., Chen, C.C., Kobayashi, S., 1979. Ductile fracture in axisymmetric extrusion and drawing. Part 2. Workability in extrusion and drawing. Trans. ASME J. Eng. Ind. 101 (1), 36–44.

Orbegozo, J.I., 1968. Fracture in wire drawing. Ann. CIRP 16, 319–330.

Orr, J., Brown, D.K., 1974. Elasto-plastic solution for a cylindrical inclusion in plane strain. Eng. Fract. Mech. 6 (2), 261–274.

Osakada, K., Watadani, A., Sekiguchi, H., 1977. Ductile fracture of cabin steel under cold metal forming conditions: 1st report. Tension and torsion tests under pressure. Bull. JSME 20 (150), 1557–1565.

Osakada, K., Watadani, A., Sekiguchi, H., 1978. Ductile fracture of cabin steel under cold metal forming conditions: 2nd report. Effect of metallurgical structure. Bull. JSME 21 (158), 1236–1243.

Oyane, M., 1972. Criteria of ductile fracture strain. Bull. JSME 15 (90), 1507–1513.

Oyane, M., Sato, T., Okimoto, K., Shima, S., 1980. Criteria for ductile fracture and their applications. J. Mech. Work. Technol. 4 (1), 65–81.

Palmer, I.G., Smith, G.C., 1968. Fracture of internally oxidized copper alloys. In: Ansell, G.S., Cooper, T.D., Lenel, F.V. (Eds.), Oxide Dispersion Strengthening. Gordon and Breach, New York, pp. 253–290.

Pardoen, T., 2006. Numerical simulation of low stress triaxiality ductile fracture. Comput. Struct. 84 (26–27), 1641–1650.

Pardoen, T., Delannay, F., 1998a. Assessment of void growth models from porosity measurements in cold-drawn copper bars. Metall. Mater. Trans. A 29 (7), 1895–1909.

Pardoen, T., Delannay, F., 1998b. The coalescence of voids in prestrained notched round copper bars. Fatig. Fract. Eng. Mater. Struct. 21 (12), 1459–1472.

Pardoen, T., Hutchinson, J.W., 2000. An extended model for void growth and coalescence. J. Mech. Phys. Solids 48 (12), 2467–2512.

Pardoen, T., Doghri, I., Delannay, F., 1998. Experimental and numerical comparison of void growth models and void coalescence criteria for the prediction of ductile fracture in copper bars. Acta Mater. 46 (2), 541–552.

Pepe, J.J., 1976. Central burst formation during hydrostatic extrusion. Met. Eng. Quart. 16 (1), 46–58.

Pickering, F.B., 1978. Physical Metallurgy and the Design of Steels. Applied Science Publishers, London (Chapter 3).

Pineau, A., Benzerga, A.A., Pardoen, T., 2016. Failure of metals I. Brittle and ductile fracture. Acta Mater. 107, 424–483.

Prager, W., Hodge, P.G., 1951. Theory of Perfectly Plastic Solids. John Wiley & Sons, New York.

Puttick, K.E., 1959. Ductile fracture in metals. Philos. Mag. 4 (44), 964–969.

Ragab, A.R., 2000. Prediction of ductile fracture in axisymmetric tension by void coalescence. Int. J. Fract. 105 (4), 391–409.

Ragab, A.R., 2002. Fracture limit curve in upset forging of cylinders. Mater. Sci. Eng. A 334 (1–2), 114–119.

Ragab, A.R., 2004a. Application of an extended void growth model with strain hardening and void shape evolution to ductile fracture under axisymmetric tension. Eng. Fract. Mech. 71 (11), 1515–1534.

Ragab, A.R., 2004b. A model for ductile fracture based on internal necking of spheroidal voids. Acta Mater. 52 (13), 3997–4009.

Ragab, A.R., 2008. Prediction of fracture limit curves in sheet metals using a void growth and coalescence model. J. Mater. Process. Technol. 199 (1–3), 206–213.

Ragab, A.R., Saleh, Ch.A., 2005. Evaluation of bendability of sheet metals using void coalescence models. Mater. Sci. Eng. A 395 (1–2), 102–109.

Reddy, N.V., Dixit, P.M., Lal, G.K., 1996. Central bursting and optimal die profile for axisymmetric extrusion. Trans. ASME J. Manuf. Sci. Eng. 118 (4), 579–584.

Reddy, N.V., Dixit, P.M., Lal, G.K., 2000. Ductile fracture criteria and its prediction in axisymmetric drawing. Int. J. Mach. Tools Manuf. 40 (1), 95–111.

Remmers, W.E., 1930. Causes of cuppy wire. Trans. Am. Inst. Mining Metall. Eng. 89, 107–120.

Rice, J.R., 1977. The localization of plastic deformation. In: Koiter, W.T. (Ed.), Theoretical and Applied Mechanics. North-Holland Publishing Company, Amsterdam, pp. 207–220.

Rice, J.R., Rudnicki, J.W., 1980. A note on some features of the theory of localization of deformation. Int. J. Solids Struct. 16 (7), 597–605.

Rice, J.R., Tracey, D.M., 1969. On the ductile enlargement of voids in triaxial stress fields. J. Mech. Phys. Solids 17 (3), 201–217.

Richards, W., 1955a. The production of components with extruded necks. Machinery 86, 298–300.

Richards, W., 1955b. The production of components with extruded necks. Machinery 86, 586–589.

Richards, W., 1955c. The production of components with extruded necks. Machinery 86, 1303–1305.

Richards, W., 1955d. The production of components with extruded necks. Machinery 87, 679–682.

Richards, W., 1955e. The production of components with extruded necks. Machinery 87, 1432–1437.

Roberts, W., Lehtinen, B., Easterling, K.E., 1976. An in situ SEM study of void development around inclusions in steel during plastic deformation. Acta Metall. 24 (8), 745–758.

Rogers, H.C., 1960. The tensile fracture of ductile metals. Trans. Metall. Soc. AIME 218, 498–506.

Rosenfield, A.R., 1968. Criteria for ductile fracture of two-phase alloys. Metall. Rev. 13 (1), 29–40.

Rosenfield, A.R., Hahn, G.T., 1966. Numerical descriptions of the ambient low-temperature, and high-strain rate flow and fracture behavior of plane carbon steel. Trans. ASM 59, 962–980.

Rosenfield, A.R., Hahn, G.T., Embury, J.D., 1972. Fracture of steels containing pearlite. Metall. Trans. 3 (11), 2797–2804.

Rousselier, G., 1981. Finite deformation constitutive relations including ductile fracture damage. In: Nemat-Nasser, S. (Ed.), Three-Dimensional Constitutive Relations and Ductile Fracture. North-Holland, Amsterdam, pp. 331–355.

Rousselier, G., 1987. Ductile fracture models and their potential in local approach of fracture. Nucl. Eng. Des. 105 (1), 97–111.

Rudnicki, J.W., Rice, J.R., 1975. Conditions for the localization of deformation in pressure-sensitive dilatant materials. J. Mech. Phys. Solids 23 (6), 371–394.

Saanouni, K., 2012. Damage Mechanics in Metal Forming: Advanced Modeling and Numerical Simulation. John Wiley & Sons, Hoboken, NJ.

Saanouni, K., Mariage, J.F., Cherouat, A., Lestriez, P., 2004. Numerical prediction of discontinuous central bursting in axisymmetric forward extrusion by continuum damage mechanics. Comput. Struct. 82 (27), 2309–2332.

Saje, M., Pan, J., Needleman, A., 1982. Void nucleation effects on shear localization in porous plastic solids. Int. J. Fract. 19 (3), 163–182.

Scheider, I., Brocks, W., 2003. Simulation of cup–cone fracture using the cohesive model. Eng. Fract. Mech. 70 (14), 1943–1961.

Schey, J.A., 1966. Prevention of edge cracking in rolling by means of edge restraint. J. Inst. Met. 94, 193–200.

Schey, J.A., 1980. Fracture in rolling processes. J. Appl. Met. Work. 1 (2), 48–59.

Schmitt, J.H., Jalinier, J.M., 1982. Damage in sheet metal forming-I. Physical behavior. Acta Metall. 30 (9), 1789–1798.

Shabrov, M.N., Sylven, E., Kim, S., Sherman, D.H., Chuzhoy, L., Briant, C.L., et al., 2004. Void nucleation by inclusion cracking. Metall. Mater. Trans. A 35 (6), 1745–1755.

Smith, E., 1968. The piling up of screw dislocations against a circular cylindrical rigid inclusion. J. Mech. Phys. Solids 16 (3), 163–167.

Sowerby, R., Chandrasekaran, N., 1984. The cold upsetting and free surface ductility of some commercial steels. J. Appl. Met. Work. 3 (3), 257–263.

Sowerby, R., O'Reilly, I., Chandrasekaran, N., Dung, N.L., 1984. Materials testing for cold forging. Trans. ASME J. Eng. Mater. Technol. 106 (1), 101–106.

Sowerby, R., Chandrasekaran, N., Dung, N.L., Mahrenholtz, O., 1985. The prediction of damage accumulation during upsetting tests based on McClintock's model. Forschung im Ingenieurwesen 51 (5), 147–150.

Spitzig, W.A., Smelser, R.E., Richmond, O., 1988. The evolution of damage and fracture in iron compacts with various initial porosities. Acta Metall. 36 (5), 1201–1211.

Stören, S., Rice, J.R., 1975. Localized necking in thin sheets. J. Mech. Phys. Solids 23 (6), 421–441.

Sun, J., 1991. Effect of stress triaxiality on micro-mechanisms of void coalescence and micro-fracture ductility of materials. Eng. Fract. Mech. 39 (5), 799–805.

Swift, H.W., 1952. Plastic instability under plane stress. J. Mech. Phys. Solids 1 (1), 1–18.

Takuda, H., Mori, K., Fujimoto, H., Hatta, N., 1999. Prediction of forming limit in bore-expanding of sheet metals using ductile fracture criterion. J. Mater. Process. Technol. 92–93, 433–438.

Takuda, H., Mori, K., Takakura, N., Yamaguchi, K., 2000. Finite element analysis of limit strains in biaxial stretching of sheet metals allowing for ductile fracture. Int. J. Mech. Sci. 42 (4), 785–798.

Tanaka, H., 1952. On the causes of cuppy defects in a drawn material. J. Jpn. Inst. Met. Mater. 16 (10), 567–571.

Tanaka, K., Mori, T., Nakamura, T., 1970. Cavity formation at the interface of a spherical inclusion in a plastically deformed matrix. Philos. Mag. 21 (170), 267–279.

Taupin, E., Breitling, J., Wu, W.T., Altan, T., 1996. Material fracture and burr formation in blanking results of FEM simulations and comparison with experiments. J. Mater. Process. Technol. 59 (1–2), 68–78.

Thomason, P.F., 1968. A theory for ductile fracture by internal necking of cavities. J. Inst. Met. 96, 360–365.

Thomason, P.F., 1969. The free surface ductility of uniaxial compression specimens with longitudinal surface defects. Int. J. Mech. Sci. 11 (1), 65–73.

Thomason, P.F., 1978. Riemann-integral solutions for the plastic slip-line fields around elliptical holes. Trans. ASME J. Appl. Mech. 45 (3), 678–679.

Thomason, P.F., 1985a. Three-dimensional models for the plastic limit-loads at incipient failure of the intervoid matrix in ductile porous solids. Acta Metall. 33 (6), 1079–1085.

Thomason, P.F., 1985b. A three-dimensional model for ductile fracture by the growth and coalescence of microvoids. Acta Metall. 33 (6), 1087–1095.

Thomason, P.F., 1993. Ductile fracture by the growth and coalescence of microvoids of non-uniform size and spacing. Acta Metall. Mater. 41 (7), 2127–2134.

Thomason, P.F., 1998. A view on ductile-fracture modelling. Fatig. Fract. Eng. Mater. Struct. 21 (9), 1105–1122.

Thompson, A.W., 1987. Modeling of local strains in ductile fracture. Metall. Mater. Trans. A 18 (11), 1877–1886.

Thomson, P.F., Burman, N.M., 1980. Edge cracking in hot-rolled Al–Mg alloys. Mater. Sci. Eng. 45 (1), 95–107.

Thomson, R.D., Hancock, J.W., 1984. Ductile failure by void nucleation, growth and coalescence. Int. J. Fract. 26 (2), 99–112.

Thomson, P.F., Nayak, P.U., 1980. An experimental investigation into the development of thickness non-uniformities leading to failure in sheet steel. J. Mech. Work. Technol. 4 (3), 223–232.

Tipper, C.F., 1949. The fracture of metals. Metallurgia 39 (1), 133–137.

Tomita, Y., 1990. Numerical Elasticity and Plasticity. Yokendo, Tokyo (Chapter 3) (in Japanese).

Tracey, D.M., 1971. Strain-hardening and interaction effects on the growth of voids in ductile fracture. Eng. Fract. Mech. 3 (3), 301–315.

Tvergaard, V., 1981. Influence of voids on shear band instabilities under plane strain conditions. Int. J. Fract. 17 (4), 389–407.

Tvergaard, V., 1982a. On localization in ductile materials containing spherical voids. Int. J. Fract. 18 (4), 237–252.

Tvergaard, V., 1982b. Ductile fracture by cavity nucleation between larger voids. J. Mech. Phys. Solids 30 (4), 265–286.

Tvergaard, V., 1990. Material failure by void growth to coalescence. Adv. Appl. Mech. 27, 83–151.

Tvergaard, V., Needleman, A., 1984. Analysis of the cup–cone fracture in a round tensile bar. Acta Metall. 32 (1), 157–169.

Van Stone, R.H., Cox, T.B., Low, J.R., Psioda, J.A., 1985. Microstructural aspects of fracture by dimpled rupture. Int. Met. Rev. 30 (4), 157–179.

Weck, A., Wilkinson, D.S., 2008. Experimental investigation of void coalescence in metallic sheets containing laser drilled holes. Acta Mater. 56 (8), 1774–1784.

Weck, A., Wilkinson, D.S., Maire, E., 2008a. Observation of void nucleation, growth and coalescence in a model metal matrix composite using X-ray tomography. Mater. Sci. Eng. A 488 (1–2), 435–445.

Weck, A., Wilkinson, D.S., Maire, E., Toda, H., 2008b. Visualization by X-ray tomography of void growth and coalescence leading to fracture in model materials. Acta Mater. 56 (12), 2919–2928.

Weinrich, P.F., French, I.E., 1976. The influence of hydrostatic pressure on the fracture mechanisms of sheet tensile specimens of copper and brass. Acta Metall. 24 (4), 317–322.

Wierzbicki, T., Bao, Y., Lee, Y.W., Bai, Y., 2005. Calibration and evaluation of seven fracture models. Int. J. Mech. Sci. 47 (4–5), 719–743.

Wilcox, R.J., Whitton, P.W., 1959. Further experiments on the cold extrusion of metals using lubrication at slow speed. J. Inst. Met. 88, 145–149.

Wilsdorf, H.G.F., 1983. The ductile fracture of metals: a microstructural viewpoint. Mater. Sci. Eng. 59 (1), 1–39.

Worswick, M.J., Pick, R.J., 1990. Void growth and constitutive softening in a periodically voided solid. J. Mech. Phys. Solids 38 (5), 601–625.

Yoshida, K., Abe, K., Hosono, K., Takezoe, A., 1968. An experimental study of ultimate ductility and average ductility in steel sheet forming. Rep. Inst. Phys. Chem. Res. 44 (3), 128–139 (in Japanese).

Zener, C., Hollomon, J.H., 1944. Effect of strain rate upon plastic flow of steel. J. Appl. Phys. 15 (1), 22–32.

Zhang, Z.L., Niemi, E., 1994a. Studies on the ductility predictions by different local failure criteria. Eng. Fract. Mech. 48 (4), 529–540.

Zhang, Z.L., Niemi, E., 1994b. Analyzing ductile fracture using dual dilational constitutive equations. Fatig. Fract. Eng. Mater. Struct. 17 (6), 695–707.

Zhang, Z.L., Niemi, E., 1995. A new failure criterion for the Gurson–Tvergaard dilational constitutive model. Int. J. Fract. 70 (4), 321–334.

Zhang, Z.L., Thaulow, C., Ødegård, J., 2000. A complete Gurson model approach for ductile fracture. Eng. Fract. Mech. 67 (2), 155–168.

Zhang, K.S., Bai, J.B., François, D., 2001. Numerical analysis of the influence of the Lode parameter on void growth. Int. J. Solids Struct. 38 (32–33), 5847–5856.

Zhu, Y.D., Avitzur, B., 1988. Criteria for the prevention of split ends. Trans. ASME J. Eng. Ind. 110 (2), 162–172.

Zimerman, Z., Avitzur, B., 1970. Analysis of the effect of strain hardening on central bursting defects in drawing and extrusion. Trans. ASME J. Eng. Ind. 92 (1), 135–145.

Zimerman, Z., Darlington, H., Kottcamp, E.H., 1971. Selection of operating parameters to prevent central bursting defects during cold extrusion. In: Hoffmanner, A.L. (Ed.), Metal Forming: Interrelation Between Theory and Practice. Plenum Press, New York, pp. 47–62.

Author Index

Note: Page numbers followed by "*f*" and "*t*" refer to figures and tables, respectively.

Subject Index

Note: Page numbers followed by "*f*" and "*t*" refer to figures and tables, respectively.

Kolibri 360 GmbH
Ferdinand-Jühlke-Straße 7
99095 Erfurt, Deutschland
produktsicherheit@kolibri360.de